ROUTLEDGE HANDBOOK ON LABOUR IN CONSTRUCTION AND HUMAN SETTLEMENTS

Routledge Handbook on Labour in Construction and Human Settlements presents a detailed and comprehensive examination of the relationship between labour and the built environment, and synergises these critical focus areas in innovative ways. This unrivalled edited collection of chapters analyses problems and presents possible solutions related to the employment and conditions of workers in the construction industry. It provides comprehensive coverage of the relationship between the global workforce and the built environment and is divided into four topical areas: how labour and the built environment relate to development; employment generation in the built environment; quality of employment in the built environment; and the impact of the built environment on labour in other sectors.

Underpinning the entire book is the premise that the way the built environment is produced, and its main products – buildings, cities and towns – have an impact on large numbers of workers. At the same time, the quality of the built environment requires construction workers who are well trained and with good working conditions. While cities and towns are the engines of economic growth, they will not be able to fulfil their economic potential if poverty in the workforce is not addressed. Those who are unemployed, underemployed or work in unfavourable conditions cannot fully contribute to production, and at the same time are limited in their ability to purchase goods and services – therefore limiting economic growth and restricting improvements in their living standards. In addition, investments in infrastructure, housing and inner-city redevelopment cannot be sustainable if labour issues – i.e., poverty – are not addressed. This book aims at analysing this complex set of issues comprehensively and will be essential reading to a wide range of researchers across the interdisciplinary intersections of construction, business and management, economic development, urban studies, sociology, political science and project management.

Edmundo Werna is a specialist in the built environment, including the construction industry and urbanism. He is widely recognised from his work on labour in construction and in human settlements during his 16 years at the ILO (International Labour Office, United Nations). He joined the London South Bank University in July 2020, after 22 years at the United Nations (United Nations Development Programme and ILO). He is currently also a visiting researcher at Westminster and Oxford Brookes universities, and a member of the editorial boards of *Habitat International* and *International Journal of Urban Sustainable Development*.

George Ofori has been working on research in construction industry development since 1980. He has written four books on the construction industry and its development, and a more recent book on leadership in construction. He has experience in curating and editing collections of chapters in world-leading books on subjects in construction. Ofori is a Professor in the School of the Built Environment and Architecture at London South Bank University, UK. He was, until recently, the Deputy Chair of the Board of CoST, the Infrastructure Transparency Initiative and a Member of the Board of Trustees of Engineers Against Poverty.

ROUTLEDGE HANDBOOK ON LABOUR IN CONSTRUCTION AND HUMAN SETTLEMENTS

The Built Environment at Work

Edited by Edmundo Werna and George Ofori

LONDON AND NEW YORK

Designed cover image: © Getty Images

First published 2024
by Routledge
4 Park Square, Milton Park, Abingdon, Oxon OX14 4RN

and by Routledge
605 Third Avenue, New York, NY 10158

Routledge is an imprint of the Taylor & Francis Group, an informa business

British Library Cataloguing-in-Publication Data
A catalogue record for this book is available from the British Library

Library of Congress Cataloging-in-Publication Data
Names: Werna, Edmundo, editor. | Ofori, George, editor.
Title: Routledge handbook on labour in construction and human settlements : the built environment at work / edited by Edmundo Werna and George Ofori.
Description: New York, NY : Routledge, 2024. | Includes bibliographical references and index.
Identifiers: LCCN 2023030298 (print) | LCCN 2023030299 (ebook) | ISBN 9781032201863 (hardback) | ISBN 9781032202082 (paperback) | ISBN 9781003262671 (ebook)
Subjects: LCSH: Construction workers--Employment. | Construction industry--Employees--Health and hygiene. | Work environment. | Labor policy. | Human settlements--Social aspects.
Classification: LCC HD9715.A2 R6838 2024 (print) | LCC HD9715.A2 (ebook) | DDC 331.2/046900973--dc23/eng/20230731
LC record available at https://lccn.loc.gov/2023030298
LC ebook record available at https://lccn.loc.gov/2023030299

ISBN: 978-1-032-20186-3 (hbk)
ISBN: 978-1-032-20208-2 (pbk)
ISBN: 978-1-003-26267-1 (ebk)

DOI: 10.1201/9781003262671

Typeset in Times New Roman
by MPS Limited, Dehradun

(from Edmundo Werna)

This is for Aquilea and Augusto. For doing all you could in the family. It worked!

(from George Ofori)

This is for Jonah and Cassius.

You have been an inspiration and bright sparks in our family.

CONTENTS

List of tables — *x*
List of figures — *xi*
List of boxes — *xiii*
List of abbreviations — *xiv*
List of contributors — *xix*
Preface — *xxvi*

1 Introducing the *Routledge Handbook on Labour in Construction and Human Settlements: The Built Environment at Work* — 1
George Ofori and Edmundo Werna

2 The built environment, construction, development and their implications for labour — 12
George Ofori

3 Characteristics of the construction industry and construction process, and implications for labour policies and practices — 39
George Ofori

4 Employment generation in the construction sector: Volatility and prospects — 63
Arup Mitra

5 Differing approaches to embedding low energy construction and climate literacy into vocational education and training — 76
Linda Clarke, Melahat Sahin-Dikmen, Christopher Winch, Vivian Price, John Calvert, Pier-Luc Bilodeau, and Evelyn Dionne

6 Green jobs and climate justice in the built environment: Lessons from American cities 97
Edmundo Werna, Mônica A. Haddad, Erin Ritter, and Anne Wurtenberger

7 Labour contracting, migration and wage theft in the construction industry in Qatar, China, India, US and the EU 114
Jill Wells

8 The precariat of the built environment: Decent work and the myth of Sisyphus 137
Andrés Mella

9 Human resource management and development in construction: Strategic considerations 158
Yaw A. Debrah, Aziz Christian Jabaru, Richard B. Nyuur, Florence Ellis, and Juliet Banoeng-Yakubo

10 Culture in construction: A driver as well as a barrier for the improvement of labour situations within construction industry 177
Wilco Tijhuis

11 Gender, construction work, and organisation 192
Maria Johansson and Kristina Johansson

12 Mistreatment of migrant construction workers: Trajectory from the past to the present and into the future 207
Abdul-Rashid Abdul-Aziz, AbdulLateef Olanrewaju, and Poline Bala

13 The smart city as the factory of the twenty-first century? How urban platforms reshape the nexus between the built environment, livelihoods, and labour 224
Jeroen Klink and Ângela Cristina Tepassê

14 Housing markets and labour markets: Towards a new research agenda for the Global South 242
Edmundo Werna, Ramin Keivani, and Youngha Cho

15 Social innovation of workplaces in the built environment: How public spaces have become central workplaces – Lessons from Kampala City, Uganda 263
Andrew Gilbert Were, Stephen Mukiibi, Michael Majale, and Barnabas Nawangwe

16 Building resilient workplaces: Prioritizing safety and disaster risk reduction for the global workforce 275
Jane Katz, Emma Harwood, and Olivia Nielsen

17 Impact of community on construction projects: Lessons from South Africa 297
C.O. Aigbavboa and I.O. Akinradewo

18 Labour, the built environment and human settlements: Lessons from the book 313
George Ofori and Edmundo Werna

Index *323*

TABLES

2.1 Linkages between infrastructure and broader development goals 18
4.1 Workforce structure in India 65
4.2 Percentage share of agriculture and construction sectors in total value added (2011–2012 prices) 67
4.3 Average daily wages of casual worker employed across the industries in Urban India (in INR) 68
8.1 Categories of workplace accidents according to selected occupational prevalence and causes 142
8.2 Types of contract work 142
8.3 Women's participation in the construction sector 146
8.4 Child labour: Common tasks, hazards, and injuries 148
11.1 Craft and related trades workers: building and manufacturing workers in Sweden (by sex and average age 2018) 194
12.1 Immigrants in selected construction-related occupations in the United States 210
12.2 Number of foreign construction trainees under the EPS, March 2015 211

FIGURES

2.1	Range of competences in construction, importance of construction in job creation and parallel competences which construction relies on	23
10.1	Representation of the causal relations between the three areas 'contact, contract, and conflict', being prominent fields of attention, and their major importance within the business processes and the communication-structure therein, also known as the 3C-Model™	181
10.2	The NEDU-Matrix™ with conclusions and recommendations given as keywords, placed in the matrix fields	182
10.3	Impression of the One Za'abeel building and its surroundings in Dubai, United Arab Emirates	186
10.4	Schematic representation of individuals A and B standing beside each other; they might behave different or the same or could even be connected, but will still have different mindsets	186
10.5	Impression of the typical shape of the One Za'abeel building and its world-record cantilever-beam design structure in Dubai, United Arab Emirates	187
10.6	Schematic representation of individuals A and B standing beside each other but explicitly interconnected by their 'culture'. However, this interconnection also creates a kind of distance, which symbolises the multi-faceted aspects of 'culture' by also distinguishing individuals, too	188
10.7	Impression of the typical facade and its reflection of sunlight of one of the towers of the One Za'abeel building in Dubai, United Arab Emirates	188
10.8	Schematic representation of an individual A or B, being individuals with their specific qualities, i.e. talents, knowledge, and skills	188

10.9 Model, representing individuals A and B, standing beside each other within a labour environment, being influenced by different aspects of 'culture', i.e.: Culture is 'helpful' or culture 'hurts', leading to happy or angry people 189
11.1 Female participation in the construction industry in a selection of countries 193
17.1 Local labour and businesses management framework 308

BOXES

2.1 Profile of the construction workforce in the United States 25
3.1 Social value in Dominvs group, UK 43
8.1 Health and safety aspects during the COVID-19 pandemic 141
8.2 Unlawful work in the construction industry 142
8.3 Women in informal construction work in India 147
8.4 Precarious working conditions of children in Ulaanbaatar, Mongolia 149
8.5 The Spanish experience in reducing workplace accidents 151
8.6 Children working in brick kilns 152
11.1 Women becoming skilled trade workers, an example from India 200
11.2 Stop macho culture, a Swedish example 202
11.3 Women's awards 202

ABBREVIATIONS

AC	Air Conditioning
ACFTU	All-China Federation of Trade Unions
ADVANCE	USAID Advancing Private Sector Engagement in Disaster Preparedness and Response in Indonesia (Indonesia)
AGC	Associated General Contractors of America (US)
AI	Artificial Intelligence
AIRC	Australian Industrial Relations Commission
AMCHAM	American Indonesian Chamber of Commerce (Indonesia)
ARRA	American Recovery and Reinvestment Act (US)
BCA	Building and Construction Authority (Singapore)
BCAWU	Building, Construction and Allied Workers Union (South Africa)
BEIS	Department for Business, Energy and Industrial Strategy
BHA	Bali Hotels Association (Indonesia)
BHRRC	Business and Human Rights Resource Centre
BIM	Building Information Modelling
BtR	Built to Rent
BUS	Build-up Skills (BUS)
BWI	Building and Wood Workers International
CBD	Central Business District
CBTU	Canadian Building Trades Unions
CBMWU	Construction and Building Materials Workers' Union (Ghana)
CECE	Committee for European Construction Equipment
CIB	International Council for Research and Innovation in Building and Construction
CIB W112	CIB Working Commission on Culture in Construction
CIOB	Chartered Institute of Building
COVID-19	Coronavirus Disease
CITB	Construction Industry Training Board (UK)
CME	Coordinated Market Economy

CQ	Certificate of Qualification (Canada)
CSIR	Centre for Scientific and Industrial Research
CSN	Construction Skills Network (UK)
DFID	Department for International Development (UK)
DHA	Department of Human Settlements (South Africa)
DLSA	Division of Labor Standards Enforcement (California, US)
DOL	Department of Labor (US)
DPW	Department of Public Works
DRM	Disaster Risk Management
DRR	Disaster Risk Reduction
EAP	Engineers Against Poverty
EC	European Commission
ECA	Electrical Contractors Association
EERI	Earthquake Engineering Research Institute
EFBWW	European Federation of Building and Woodworkers
EFQM	European Foundation for Quality Management
EIIP	Employment Intensive Investment Programme
EPA	Employment Placement Agencies (EU)
EPA	Environment Protection Agency
EPBD	European Performance of Buildings Directive
EPS	Republic of Korea Employment Permit System
ESG	Environmental, Social and Governance
EU	European Union
EWPOSS	Enterprise Wage Payment Online Supervision System (China)
FCL	Fundación Laboral de la Construcción (Spain)
FE	Further Education (FE) College
FES	Foundation Energy Skills Programme
FIC	Factory Investigating Commission (New York State, US)
FIEC	European Construction Industry Federation
GATS	General Agreement on Trade in Services
GCC	Gulf Cooperation Council
GDFCF	Gross Domestic Capital Formation
GDP	Gross Domestic Product
GIPC	Ghana Investment Promotion Centre
GHG	Greenhouse Gas
GPRO	Green Professional Programme (Canada)
GUD	Global Urban Development
GVA	Gross Value Added
HBE	Home-based Enterprises
HR	Human Resource
HRM	Human Resource Management
HSE	Health and Safety Executive (UK)
HUD	Department of Housing and Urban Development (US)
HVAC	Heating, Ventilation and Air-Conditioning
ICED	Infrastructure and Cities for Economic Development
ICT	Information and Communications Technology

IFC	International Finance Corporation
ILGWU	International Ladies' Garment Workers' Union
ILO	International Labour Organisation
IOM	International Organisation for Migration
IoT	Internet of Things
IPA	Infrastructure and Projects Authority
4IR	Fourth Industrial Revolution (or Industry 4.0)
IT	Information Technology
JIB	Electrical Joint Industry Board
KADIN	Indonesian Chamber of Commerce and Industry
KCCA	Kampala Capital City Authority
KPI	Key Performance Indicator
LEC	Low Energy Construction
LEED	Leadership in Energy and Environmental Design
LID	Labour Inspection Department (Qatar)
LID	Low-Impact Development
LME	Liberal Market Economies
MADLSA	Ministry of Administrative Development, Labour and Social Affairs (Qatar)
MCEDC	Montgomery County Economic Development Corporation
MGNREGS	Mahatma Gandhi National Rural Employment Guarantee Scheme (India)
MIERA	More Income and Employment in Rural Areas (Malawi)
MLIT	Ministry of Land, Infrastructure, Transport and Tourism (Japan)
MMC	Modern Methods of Construction
MOHURD	Ministry of Housing and Urban-Rural Development (China)
MOSPI	Ministry of Statistics and Programme Implementation (India)
NAA	National Apprenticeship Act (US)
NABTU	North American Building Trades Unions
NEPA	National Environmental Policy Act (US)
NGO	Non-Government Organisation
NIST	National Institute of Standards and Technology (US)
NREGA	National Rural Employment Guarantee Act 2005 (India)
NZEB	Nearly Zero Energy Building
NCWM&SISystem	National Construction Workers Management and Service Information System (China)
NOC	No-objection Certificate
NUA	New Urban Agenda
NWA	National Water Agency (Singapore)
NZIC	New Zealand Infrastructure Commission
ODA	Official Development Assistance
OECD	Organisation for Economic Cooperation and Development
ONS	Office for National Statistics (UK)
OSH	Occupational Safety and Health
OSHA	Occupational Safety and Health Administration (US)
PHP	People's Housing Process (South Africa)

PLFS	Periodic Labour Force Survey (India)
PMAY	Pradhan Mantra Awas Yojana (India)
PPE	Personal Protective Equipment
PPP	Purchasing Power Parity
PWD	Posting of Workers Directive (EU)
QCB	Qatar Central Bank
REIF	Real Estate Investment Fund
REIT	Real Estate Investment Trust
RIBA	Royal Institute of British Architects
RICS	Royal Institution of Chartered Surveyors
SAFCE	South African Forum of Civil Engineering Contractors (SAFCEC)
SAP	Structural Adjustment Programme
SBA	Small Business Administration (US)
SDG	Sustainable Development Goal
SEWA	Self-Employed Women's Association
SHARP	Safety and Health Achievement Recognition Program (US)
SIF	Salary Information File (Qatar)
SME	Small and Medium-Sized Enterprise
SOAS	School of Oriental and African Studies, University of London
SOE	State-Owned Enterprise
SFH	Single-Family Housing Unit
SFR	Single-Family Rental
STR	Short-Term Rental
TUWIC	Tackling Undeclared Work in the Construction industry (EU project)
TVET	Technical Vocational Education and Training
TWA	Temporary Work Agencies (EU)
UA Canada	United Association of Journeymen and Apprentices of the Plumbing and Pipe Fitting Industry of the United States and Canada
UAE	United Arab Emirates
UBI	Universal Basic Income
UN-Habitat	United Nations Human Settlements Programme
UNCTAD	United Nations Centre for Tariffs and Trade
UNDP	United Nations Development Programme
UNDRR	United Nations Office for Disaster Risk Reduction
UNEP	United Nations Environment Programme
UNFCCC	United Nations Framework Convention on Climate Change
UNICEF	United Nations Children's Emergency Fund
UNICRI	UN Interregional Crime and Justice Research Institute
UNOCHA	United Nations Office for the Coordination of Humanitarian Affairs
USAID	United States Agency for International Development
USMID	Uganda Support to Municipal Infrastructure Development Programme
UTIP	Union Training and Innovation Program
VET	Vocational Education and Training
VET4LEC	Inclusive Vocational Education and Training for Low Energy Construction

VoC	Varieties of Capitalism
VPP	Voluntary Protection Program (US)
WHO	World Health Organisation
WIEGO	Women in Informal Employment: Globalizing and Organizing
WPS	Wage Protection System (Qatar)
WSIZ	Wheaton Sustainable Innovation Zone
WTO	World Trade Organisation

CONTRIBUTORS

Abdul-Rashid Abdul-Aziz is a professor at Wawasan Open University, Malaysia. He was previously with Universiti Sains Malaysia for more than 30 years. He was commissioned twice by the International Labour Organisation to research into foreign workers in the Malaysian construction industry. Apart from construction labour, his research interests include international construction and low-cost housing.

C.O. Aigbavboa is the Director of the Research Chair in Sustainable Construction Management and Leadership in the Built Environment at the University of Johannesburg, South Africa. He specialises in sustainable construction industry development, focusing on areas such as sustainable housing regeneration, construction management, and the application of artificial intelligence in the built environment sector. He has authored 19 research books and mentored numerous postgraduate and postdoctoral researchers. He is the chief editor of the *Journal of Construction Project Management and Innovation*, and a visiting professor at Shandong University, China.

I.O. Akinradewo is a PhD candidate in Construction Management at the University of Johannesburg, South Africa, where he obtained his Master's in Quantity Surveying with distinction in 2019. He is a certified blockchain architect with the Blockchain Council, US. As a testament to his research activities, he won the "Academic Rising Star" award at The BIM Africa Innovation Awards 2021.

Poline Bala is the Director of the Institute of Borneo Studies and Professor of Anthropology at the Faculty of Social Sciences and Humanities at Universiti Malaysia Sarawak. Her focus area of research has been quite broad. Foremost is the anthropology and history of Bornean societies. This includes population movement, and interaction and labour relations between communities at the frontiers of international boundaries. Poline has also been examining the interplay between technologies (i.e., information and communication technologies), government policies and strategies for development intervention in Malaysia.

Juliet Banoeng-Yakubo is a Lecturer in Business Management at De Montfort University, UK. Her PhD, undertaken at the Warwick Business School, UK, examined shared leadership in intra-organisational work teams in the global south. Her research interests include leadership, creativity, innovation and work teams. Her research has appeared in high-ranking peer-reviewed journals, including the *Leadership Quarterly*.

Pier-Luc Bilodeau is Professor in the Department of Industrial Relations at Université Laval (Quebec City, Canada), where he earned his PhD. His research focuses on collective bargaining and labour policy in the construction industry. He is a Construction Industry Research Team member of the *Building it Green* project for the Canadian Building Trade Unions, a co-researcher at the Interuniversity Research Centre on Globalization and Work, and a research scholar at the Institute for Construction Employment Research.

John Calvert was an Associate Professor at Simon Fraser University in Vancouver, Canada prior to his retirement in 2021. He was a co-director of a major Social Science and Humanities research project, "Adapting Canadian Work and Workplaces to Climate Change," which examined the role of labour in contributing to the efforts to find solutions to the climate crisis. His current research involves promoting climate literacy in Canada's building trades. He has a PhD from the London School of Economics in the United Kingdom.

Youngha Cho is Reader in the School of the Built Environment and joint lead of the Healthy Ageing and Care Research, Innovation and Knowledge Exchange Network at Oxford Brookes University, United Kingdom. Youngha is an established researcher with an international reputation in housing policy, and private and public housing market analysis. She gained her PhD from London School of Economics and Political Science, United Kingdom. She collaborates with leading academics and institutions in Hong Kong, Japan, South Korea, Singapore and the United Kingdom, and some African countries. Her research interests include integration of subjects in housing, health and social care, and development of age-friendly communities.

Linda Clarke, Professor in the Centre for the Study of the Production of the Built Environment (ProBE), University of Westminster, London, is an expert to the European Construction Social Dialogue Working Group, Vocational Education and Training (VET) and was awarded the European Commission 2022 VET Researcher Excellence Award. She has long experience of comparative research across Europe on VET, employment, wage relations, equality, labour, and climate change, focusing on the construction industry, including *Inclusive VET for Low Energy Construction* (2019).

Yaw A. Debrah was a Professor of International Human Resource Management at Swansea University, United Kingdom. He holds a PhD from Warwick University, United Kingdom. His scholarly work has appeared in leading journals. Professor Debrah has consulted or conducted training for many organisations including the Commonwealth Secretariat, ILO, Singapore Telecom, Nigerian National Petroleum Corporation, Central Bank of Nigeria and private sector firms in Africa. His publications have been used by the International Monetary Fund and the World Bank in their dealings with African governments.

Evelyn Dionne is a PhD candidate in Industrial Relations at Université Laval, Canada, interested in work and unionism, notably in the construction industry. Her research focuses on environmental labour studies as well as diversity and inclusion. She is currently completing a thesis on unions as actors in the fight against climate change.

Florence Ellis is a Senior Lecturer in the Department of Human Resources and Organisational Development at the KNUST School of Business, Ghana. She holds a PhD in Management Studies from Swansea University, United Kingdom. Her research interests include Organisational Studies, Organisational Behaviour, Organisational Leadership, Conflict and Conflict Management in Teams and Organisations, Diversity in Team and Organisations, and Human Resource Management. Her research has appeared in a number of internationally recognised peer-reviewed journals.

Andrew Gilbert Were is a Lecturer at the Department of Architecture and Physical Planning, Makerere University. He holds a PhD in Urban Planning, Design and Management. His research includes informality and the urban micro-economy, slum upgrading, urban governance and management, and resettlement planning of displaced persons. He is a member of the Uganda Inter-Agency Committee on Slum Improvement Research Project and a UN-Habitat slum upgrading expert. He is the co-author of works on a moralist theory and a theory of spontaneous location.

Mônica A. Haddad is a Professor of Community and Regional Planning at Iowa State University. She specializes in geographic information systems (GIS) and spatial analysis applications for planning issues. Climate justice and social justice are the driving forces of her scholarship. Her work has been published in journals such as *Applied Geography*, *Frontiers in Sustainable Cities*, *The Professional Geographer*, and *Economic Development Quarterly*. She received her PhD in regional planning from the University of Illinois at Urbana-Champaign, United States.

Emma Harwood is a Program Coordinator at Miyamoto International, a global multi-hazard engineering and disaster risk management firm that sustains industries and communities worldwide. As a recent graduate, Emma brings a fresh perspective to her role at Miyamoto, where she supports her team's diverse project portfolio to realize development objectives in affordable housing and climate adaptation. Emma is instrumental in expanding the global footprint of the firm's non-profit arm, Miyamoto Relief, to positively impact communities affected by disaster and climate change.

Aziz Christian Jabaru, MPhil, is a Research Assistant at the University of Ghana Business School, Ghana. His research interest is in Human Resource Management and organizational behaviour. He holds an MPhil Degree in Human Resource Management from the University of Ghana Business School. He has presented his work at a conference.

Kristina Johansson is an Assistant Professor in Human Work Science at the Department of Social Science, Technology, and Arts at Luleå University of Technology in Sweden. Her research interest concerns gender and equality at work, focusing on both service and industry organization. Her work has appeared in journals such as *Gender, Work and Organization, Gender, Place & Culture,* and *Men and Masculinities.*

Maria Johansson is an associate senior lecturer in Human Work Sciences at the Department of Social Science, Technology, and Arts at Luleå University of Technology in Sweden. Her research interest concerns occupational health and safety as well as gender and equality in the construction industry and other male-dominated industries.

Jane Katz is Board Vice Chair of Global Urban Development and an advisor to the UN and other global organizations on housing, land, and urban issues. She serves on the UNDRR Stakeholder Engagement Advisory Group, and the Steering Committee for UN Habitat's World Urban Campaign, among others. She was Director of International Affairs and Programs at Habitat for Humanity for close to twenty years. Previously, Ms. Katz worked in the public and private sectors at Fannie Mae and in US government agencies. She received her Master of Arts degree in Government and Politics from the University of Maryland, and Bachelor of Arts in Foreign Affairs from the University of Cincinnati.

Ramin Keivani is Head of School of the Built Environment and Professor of International Land Policy and Urban Development at Oxford Brookes University. He is an urbanist with a particular research interest on low income and affordable housing policy, urban social sustainability, and impact of land and property markets on urban development and urban equity in both developed and transition economies. He has a strong track record of international comparative urban research and has published widely on various aspects of housing, land markets and urban development in different parts of the world including Brazil, China, India, countries in the Asia Pacific, Central/Eastern Europe, the Middle East and Southern Africa, and the United Kingdom.

Jeroen Klink (BSc and MSc in economics from Tilburg University, the Netherlands; PhD in urban planning from University of São Paulo, Brazil) has worked as a staff member in the Institute for Housing and Urban Development Studies (Erasmus University, the Netherlands) and as a secretary for regional planning in the city of Santo André (Metropolitan São Paulo). Nowadays, he is an Associate Professor at the Federal University of the ABC region (UFABC, located in São Paulo) and board member of the Urban Studies Foundation (Glasgow, UK).

Michael Majale is an independent consultant/researcher and holds a Bachelors Honours degree in Architecture and a Master's degree in Urban and Regional Planning from the University of Nairobi, Kenya and a PhD from Newcastle University, United Kingdom He has over 30 years of international experience in the field of housing and urban development. He has expertise in research, policy and practice related to housing and urban development, and in particular housing for the urban poor and slum upgrading.

Andrés Mella is a development and policy specialist with a background in Economics, Law, and Development Economics (double MSc in Economic Development and Growth from Carlos III University, Spain and Lund University, Sweden). He manages green skills and multi-stakeholder projects for the Fundación Laboral de la Construcción and has implemented social dialogue and private-sector development programmes for ILO-HQ and GIZ Malawi. He has also published research and policy notes addressing diverse aspects of working conditions in the built environment for ILO, African Development Bank; UK Foreign, Commonwealth and Development Office; International Trades Union Confederation; Building and Wood Workers International; and European Federation of Building and Woodworkers.

Arup Mitra is Professor of Economics at South Asian University in India. His research interest includes development economics, issues in urban development, labour and welfare, corruption, industrial growth and productivity, and gender inequality. He was awarded the Mahalanobis Memorial Gold Medal by the Indian Econometric Society for his outstanding contribution to quantitative economics. His book, *Insights into Inclusive Growth, Employment and Wellbeing in India*, published by Springer, received the Professor S. R. Sen Best Book Award, 2019. His book, *Urban Headway and Upward Mobility in India* (2020) is published by Cambridge University Press.

Stephen Mukiibi is an academician and a professional architect with thirty-one years' working experience, in the academia and architectural practice. He is an Associate Professor and former Head of Department of Architecture and Physical Planning at Makerere University. He has taught both at the undergraduate and graduate levels, supervising Masters and PhD students. He has carried out extensive research in areas of housing, housing policy, slum upgrading, human settlements and urban development in the East African region.

Barnabas Nawangwe has been the Vice Chancellor of Makerere University, one of the oldest and premier universities in Africa, since August 2017 and he is currently serving his second term of five years. Prior to that, he was Deputy Vice Chancellor, Dean and Principal of the College of Engineering, and Head of the Department of Architecture, all at Makerere University. His research interests are in the areas of vernacular architecture, urbanism and sustainable human settlements.

Olivia Nielsen is a Principal at Miyamoto International where she focuses on resilient housing solutions. From post-disaster Haiti to Papua New Guinea, she has developed and worked on critical housing programs in over 45 countries. Olivia has over a decade of experience in housing finance, housing public-private partnerships, post-disaster reconstruction and green construction. Through her work, she hopes to make safe and affordable housing available to all.

Richard B. Nyuur is a Professor in International Business at the School of Management, University of Bradford, United Kingdom. His research interests lie at the intersection of strategy and international business in the areas of international business strategy, international human resource management, and corporate social responsibility. His works have been published in journals such as *British Journal of Management*, *International Journal of HRM*, *Journal of International Management*, *International Marketing Review*, and *Journal of Business Research*, *Journal of Small Business Management*, and *Multinational Business Review*.

AbdulLateef Olanrewaju is affiliated with Universiti Tunku Abdul Rahman in Malaysia. He holds a Master of Science degree in the Built Environment from the International Islamic University Malaysia and a Doctorate from Universiti Teknologi PETRONAS Malaysia. He received his quantity surveying education at Kaduna Polytechnic in Nigeria. His academic pursuits focus on lecturing and research in the fields of quantity surveying, construction health and safety management, sustainable construction, value management, building maintenance, research methodology, and construction management.

Vivian Price, PhD, Professor at California State University, Dominguez Hills, former union electrician, is a researcher and filmmaker for local and international projects on labour and climate justice. Recent joint publications include "Defence sector worker views on the political-economic barriers to a Just Transition" and "A Green New Deal for all: The centrality of a worker and community-led just transition in the US." Her film, "Talking Union, Talking Climate," centres views of oil workers in the United States, Norway and Nigeria.

Erin Ritter is a Program Assistant for the Conservation Law Foundation. Based in the Rhode Island office, she works in environmental justice advocacy and research. She holds a BA in Religious Studies from Grinnell College, United States.

Melahat Sahin-Dikmen is a lecturer at the School of Organisations, Economy and Society, University of Westminster. She is an inter-disciplinary scholar whose expertise lies in the sociology of work and employment. Her recent research is concerned with the implications of climate change for the world of work with particular reference to the built environment occupations, and representations of gender. Her recent publications include 'Climate change, inequality and work in the construction industry' in Healy et al (2023).

Ângela Cristina Tepassê is a Brazilian economist with a master's degree in political economics from the PUC University in São Paulo, Brazil. Since 2014, she has been working as a senior analyst at the inter-syndical Department for Statistics and Socioeconomic Studies (DIEESE). She has had works published on issues related to: public policies, work, employment, and income; labor market dynamics; and the entanglements between trade, growth and development, among others. She is currently doing PhD research on real estate platforms and housing affordability in Brazilian cities at the Federal University of the ABC region.

Wilco Tijhuis (PhD, MSc, BSc) is a part-time Associate Professor at the University of Twente in The Netherlands. His current research is on Construction Management and Procurement, Internationalisation and Strategies, Business Development and Business Cultures in the International Construction Industry. He is also joint coordinator of the Working Commission W112 "Culture in Construction" of the international research- and industry-platform, International Council for Research and Innovation on Building and Construction (CIB). Parallel to his academic position, Wilco has management positions in international industry and is an advisor to businesses.

Christopher Winch, Professor of Educational Philosophy and Policy at King's College London, has written extensively on the philosophy of education and vocational education, with a particular focus on epistemological issues related to professional knowledge and judgement and learning as well as taking part in projects on comparative European vocational education and training. Books he has authored include: *Dimensions of Expertise* (2010), '*Teachers' Know-how* (2017), *Educational Explanations* (2022) and *Knowledge, Skills and Competence in the European Labour Market* (2010) with Linda Clarke and others.

Anne Wurtenberger is an Iowa State University Community and Regional Planning Alumna. While currently working for the U.S. Army Corps of Engineers on civil works and military planning projects, she also is in the final year of her graduate program in Homeland Security and Intelligence at American Military University.

PREFACE

We first met each other at the United Nations Headquarters in New York where we had both been invited to attend the Expert Group Meeting and Follow-Up Workshop on "Changing Consumption Patterns in Human Settlements", organised by the then United Nations Centre for Human Settlements in New York in April 1997, but had known of each other's work, and had been in communication with each other before then. Edmundo was working at the University of Sao Paulo in Brazil, and George was then with the National University of Singapore. A previous collaboration between us was in 2003, when George was the lead consultant on a project on green jobs in construction in Malaysia for the International Labour Office (ILO). At the time, Edmundo worked at the ILO and was the focal point for this project. In 2017, George joined the London South Bank University. In 2020, Edmundo joined the university.

The book arose out of a discussion between us in 2021. We felt that it would make sense to consider: how construction workers benefit from the work they do in putting in place the countries' physical basis for development, and building some of the most luxurious buildings, some of which become symbols of their nations, and give the citizens pride. We also wanted to explore how workers in other sectors of the economy benefit from the items provided for them by their counterparts in the construction industry – i.e. the products of the built environment. George was in the final stages of completing a book of which he was the Editor, the *Research Companion to Construction Economics*. He had initially promised himself that that would be the last of such books he would do. To his dismay, he had allowed himself to be persuaded to do another book, *Building a Body of Knowledge for Project Management in Developing countries*. This would be another stretch. Edmundo made a convincing case about how the present book would fill a huge vacuum. George was persuaded.

We would bring together Edmundo's background in Development Economics and the Construction Industry, his wide experience, and in particular, his knowledge and experience of labour in construction from his time as the Construction Policy Adviser at the ILO in Geneva; and George's research and practical experience in Construction Industry Development. We had several meetings to discuss and debate the subject. We reached consensus on the possible topics such a book would cover. The book covers the three themes we had in mind. What we are seeking to achieve is an inter- and multi-disciplinary work which brings the three themes together.

After the proposal was approved, we set out to invite potential authors who were leading experts on the particular themes. We were gratified that most of the people we invited agreed to write chapters. We were off to a good start. However, there were gaps in the list of topics for which it took some time to find authors. In the end, there were two topics that we could not find authors for. As new authors joined the endeavour and two found it no longer possible to contribute, the process was 'organic', as George described it to Edmundo.

We had good support from the publishers, as we had to keep moving the deadline for submitting the agreed manuscript. The main enabler to the process of coordinating the contributions was the willingness of some of the other authors to review the draft chapters. In particular, our mutual long-time friend, Jill Wells, was most helpful. She was very thorough in her reviews. We write more about her later in this note.

The problems and challenges were what one would expect in setting out to edit a collection of chapters. For a book with specifically selected authors, who would be expected to be busy, we had to be patient, and expect delays. We had to revise the dates each time a new author or group of authors agreed to join the process.

Would we have done anything differently? We believe the path we took was the right one.

We believe that we have made a new thrust into an area of knowledge built on three interrelated themes which should be studied together in order to find ways to realise the possible synergies among them. We will monitor developments in the field by following up on citations of the chapters in the book to see how the subject develops. We will write a short synopsis for the governments, especially of the developing countries, multilateral agencies such as the ILO and the World Bank, the regional development banks, and bilateral development partners. We hope the contributors of chapters, after they have seen the whole book, will be enthused to do more on the subject. We also hope they will continue to work towards the further development of the area.

We now end this short note by expressing our thanks. We start by thanking our families for putting up with long periods when we denied them chunks of our time. We continue with Jill Wells. George first met Jill in the 1980s, when both of them had completed their doctoral studies. They both have a strong interest in the construction industries in developing countries. Edmundo has also known Jill for many decades. An important point to reveal is that Edmundo was Jill's successor as the Construction Sector Adviser at the ILO. We are really grateful for the huge support Jill gave us in our work on this book. We thank our co-contributors. We think, together with our co-contributors, we have produced a book which will open up discussions on, and further development of, the triad of themes: *how labour and the built environment relate to development; workers of the built environment; and the impact of the built environment on labour in other sectors.*

We are also grateful to Ed Needle, Publisher, Engineering at Taylor and Francis and his colleagues including Martha Luke, for the support they have given us in our work on this book.

Finally, we thank you, our reader, for picking up this book to read. We hope you find it useful for your purposes.

Edmundo Werna and George Ofori
London, October 2023

1

INTRODUCING THE *ROUTLEDGE HANDBOOK ON LABOUR IN CONSTRUCTION AND HUMAN SETTLEMENTS*

The built environment at work

George Ofori and Edmundo Werna

London South Bank University

Status of and challenges related to labour in the built environment

Cities and towns are engines of economic development. However, they will not be able to fulfil their economic potential if poverty among their residents and other citizens of the countries whose economic and social progress the cities and towns are to foster is not addressed. People who are unemployed or underemployed and those who work under unfavourable conditions cannot fully contribute to production in the economy. They also face difficulties in purchasing goods and services. This, in turn, can limit economic development in the cities and towns in which they live and prevent these groups of people, and their dependants, from having a decent standard of living and, therefore, an improvement in the quality of their lives. In addition, investments in infrastructure, housing, inner-city renewal, and the like will not be sustainable if labour issues are not addressed.

Since the 1970s, there has been mounting evidence of the pivotal importance of construction activities in socio-economic development (Turin, 1973). The relationship between the two is increasingly being better understood (Ofori, 2022). The construction industry is the largest component of the built environment sector which forms the cities, towns, and other human settlements. Some authors use 'construction' and 'the built environment' as synonyms. However, they are different. A crucial distinction is that the built environment includes construction – the segment of which produces the built environment – but also includes the use of the products of the construction industry (for example, how people use the buildings where they live and work, including issues such as heating and cooling). The built environment also includes real estate development. Chapter 2 elaborates on the definitions of construction and the built environment.

One conceptual challenge is that a comprehensive approach to the built environment should include everything that human beings change and produce. In this respect, even agriculture is part of the built environment; a car is part of the built environment, and so on.

DOI: 10.1201/9781003262671-1

However, for the purposes of this book, the built environment is approached only with reference to the issues mentioned in the previous paragraph – the physical buildings and other items of infrastructure and their use for productive and other purposes.

Workers are central in the discussion on the contribution of construction in particular, and the built environment in general, to national socio-economic development, considering that they are the agents of production and, therefore, need proper training, effective deployment, continuous development, and appropriate working conditions, employment relations, and overall welfare. Also, considering the large number of workers the built environment employs, policies related to the creation of employment (or having implications for maintaining current levels of employment) and improvements in the quality of employment have a significant impact on development. It is estimated that there are some 330 million workers (formal and informal) engaged in the production of the built environment around the world (Mella and Savage, 2017). This figure increases if those working in the production and supply of construction materials and equipment, others undertaking other activities in the value-chain, and those providing services to construction companies such as accountants and lawyers, are taken into account.

Construction has been changing rapidly, especially during this decade, due to the pace of technological development brought about by the Fourth Industrial Revolution (World Economic Forum, 2019) and the critical and heightened need to protect the environment which construction activity adversely affects in many respects (UNEP, 2019). At the same time, the traditional view of 'developed' and 'developing' countries is also changing. Some countries in the Global South (conventionally seen as 'developing') are now leading in some areas of development. For example, some South-based multinational construction companies are major players in the international market. In parallel, issues which were regarded as occurring only in developing countries (such as the informal sector of the economy, slums in cities and towns, and significant levels of homelessness) are now evident in countries in the Global North. It is clear that it is timely to revisit the construction-development nexus and pay special attention to labour. The role of labour now faces major challenges which could lead to massive unemployment and related consequences such as increase in poverty, anomie, and insecurity. It is necessary to analyse these issues as well as explore opportunities for improvements. For example, it is pertinent to consider a socially conscious use of technological advancements coupled with redistribution, especially with regard to investments in social security. Considering another issue, while environmental policies are necessary, they may also lead to technological substitution and eventually to economic slowdown, with uncertain consequences for levels of employment, especially considering the prospect of population growth in some countries. It is imperative to understand the present status of, and the future prospects for, employment in the built environment and to make recommendations to address the current and likely future challenges.

A large number of people (the workers and their families) depend on jobs related to construction. However, such jobs are inherently unstable. Some of the many reasons for this are as follows: construction activity is undertaken in the form of one-off projects; the construction industry typically goes through periods of boom and bust; and broad changes in employment relations in many countries have weakened the bargaining position of workers (Wells, 2006). The advent of the Fourth Industrial Revolution is likely to lead to a 'migration' of jobs from construction to other sectors, such as manufacturing and information and communications technology, and also among countries, and the increase in capital-intensive methods (including automation) may lead to the laying off of many workers. It is imperative to understand the present status of, and the future prospects for, employment in construction and to make appropriate recommendations for necessary interventions.

In addition to the spectre of unemployment, many who are now working in the construction industry face deficits in the quality of their jobs related to a broad range of issues such as poor safety and health, low wages, long and often unsocial working hours, lack of freedom of association, discrimination on various grounds, especially in countries where construction has a poor social image, and abuse of rights (Wells, 2013). There is even a new concept, 'precariat' ('precarious proletariat') which refers to an emerging class of people facing insecurity, moving in and out of precarious work that gives little meaning to their lives (Standing, 2011).

Previous studies have indicated the connection between these deficits and the structure of the construction industry (ILO, 2001; Werna and Klink, 2022). For example, several layers of subcontracting are commonly encountered on projects. This makes it more difficult to establish responsibility in a site when something goes wrong. Employers seem to be pushing not only for zero-hour contracts, but also that every worker becomes a one person enterprise (Werna and Klink, 2022). This means that in a construction site, at any one time, there would be an intricate web of enterprises hiring, and subcontracting work to, others. In the situation, there would be only contracts between 'enterprises'. There would be no employment relations. The 'employer', and its responsibilities as such, would cease to exist in legal terms.

Innovations brought about by the Fourth Industrial Revolution are exacerbating the impact of these trends. They are intensifying the casualization of labour, making social dialogue more difficult, and further weakening employment relations. It is clear that technological innovations will bring opportunities for labour, if the relationships between technology and labour are well explored. It is necessary to analyse both these challenges and opportunities in order to provide a basis for appropriate informed action.

Finally, the built environment also affects workers in many other sectors of the economy in many ways. The workers need proper workplace facilities; appropriate housing; transportation systems; education, health, and leisure facilities; and other forms of infrastructure. Examples of deficits in the built environment are easily seen in many cities, towns, and settlements throughout the world: buildings, including offices and factories, in a poor state of repair, affecting the health and productivity of workers and sometimes leading to fatal accidents, including collapse of buildings; large numbers of workers from several trades working on the pavements or in the open as they cannot afford a proper workplace; and homeless workers who cannot afford a roof over their heads, including some construction workers who might have built some of the luxury homes in the city or human settlement concerned. Key subjects covered in this book include territorial development and livelihoods; housing and labour markets – how, together, they can influence the development of a city or human settlement; creation and maintenance of housing as a place of work (home-based enterprises); workers who have to ply their trade in public spaces which they adopt as workplaces; and challenges faced by those who work in unhealthy or unsafe workplaces. These are examples of issues that the chapters in the book analyse, together with the set of different policies and implications of each of them. Crucially, the chapters seek to explore the potential synergies among the phenomena.

The book

The way the built environment is produced and its main products – cities, towns, and other human settlements – have an impact on large numbers of workers in all sectors of the economy in every country. The construction workers are better- or worse-off depending on the characteristics of the process of production of buildings, the other types of infrastructure, and the items within and connecting them that make them useful for operations. At the same time, workers in other sectors of the economy need buildings, infrastructure, and other elements of the built environment. The quality of such facilities influences their work and all other aspects of their lives.

The book has two parts, preceded by Chapter 2 by Ofori, which deals with the relationships between the built environment, construction, development, and labour. This discussion is extended to the consideration of opportunities for improvements – for example, a socially conscious use of technological advancements coupled with redistribution.

The two parts of the book correspond to what can be considered the 'two sides' of the relationship between labour and the built environment: (i) the workers who produce the built environment – the problems and possible solutions related to employment and working conditions of construction workers; and (ii) how the built environment, particularly the products of construction (buildings and other items of infrastructure), impacts workers in all sectors of the economy. As the built environment is the founding element of human settlements, the book also includes consideration of such settlements within the scope of the chapters. The details of the two parts of the book are now discussed.

Part 1: The workers of the built environment

Part 1 starts with Chapter 3, also by Ofori, which analyses the characteristics of the construction industry and implications for labour. The conditions of work, further development, and general welfare depend on the features of the process of construction as well as on the policies of the government and practices of the companies. This chapter addresses questions related to the characteristics of the construction industry, process and product, the response of construction companies to these characteristics, the effects on labour, responses to the challenges, and future actions. Based on a review of the relevant literature, the chapter argues that the nature of construction and how the industry responds to it have major implications for workers, both in the construction industry and other sectors of the economy. Ofori suggests that actions are required to enable society to derive optimum benefit from the economic and social potential of construction.

It is a basic fact that humans need employment. Therefore, after the aforementioned overall analysis of Chapter 3, Part 1 goes into job creation in construction in Chapter 4 by Mitra. The chapter analyses the challenges related to access to employment in the sector and the details of the process, including employment elasticity and the relationship of the construction industry with the aggregate GDP per capita; the recruitment processes in the industry are seen to be very different from those in other sectors of the economy; and the relationships with the process of urban development. Mitra concentrates his analysis in India, but the contents of Chapter 4 are also of value elsewhere.

To be employable, workers need training. Chapter 5, by Clarke, Sahin-Dikmen, Winch, Price, Calvert, Bilodeau, and Dionne, addresses this topic. It is based on research in two different parts of the world (Europe and North America) but, again, it is also of value elsewhere. This is especially so because the chapter analyses training in the context of low energy construction (LEC). Training which takes LEC into consideration is increasingly a must in all countries. Construction activity is responsible for a large share of environmental problems, and there is an urgent need for changes in its production process. The chapter reveals sharp differences in the importance attached to vocational training for LEC in different countries. The conclusion is that equity in, and valuing of, labour is key to combating climate change.

Chapter 6, by Werna, Haddad, Ritter, and Wurtenberger, is also about training and the context is also climate change. It adds a different value to the discussion in Chapter 5, which deals with unionised workers and vocational education and training. Chapter 6 focuses on training and jobs for vulnerable populations. The authors argue that efforts to mitigate climate change should strive for climate justice to combat the current situation in which extreme weather events are becoming more frequent and have serious consequences for vulnerable communities. The built environment contributes considerably to greenhouse gas emissions, and it is imperative for cities to implement changes to reduce and aim to eliminate carbon pollution. Such changes can consist in making green jobs available to those who need them the most. With examples from the United States, the chapter is based on a systematic literature review considering how local governments in various cities are acting on climate change mitigation and climate justice, green jobs that also improve the built environment, workforce development, and training for vulnerable residents. The findings are used to provide recommendations for future research and local action.

Receiving adequate training and getting a job are the first steps for a worker. Naturally, the worker has to be paid. In Chapter 7, Wells shows that there are major problems in this respect. She argues that wage violations (late payment, under-payment, and non-payment of wages) are common in the construction industry in many countries. These violations may also be increasing in extent due to the changes in the business models adopted by the industry in many parts of the world over the past few decades. Chapter 7 includes examples from different countries, including Qatar, China, India, the United States, and some European countries. Whereas the links between wage violations and migration are important, it is argued by Wells that it is the change in business models that underlies both. Actions by governments to address the issue are assessed.

Jobs not only need to be available, and workers not only need to be remunerated appropriately for doing them, but also jobs need to be of good quality. Accordingly, Chapter 8, by Mella, provides an overview of quality employment, examining different topics such as poor conditions on site, occupational safety and health, child labour, and more – without the pretence of including every possible challenge related to work quality. The chapter concludes with suggestions for policy options.

Chapter 8 paves the way for the remaining chapters of Part 1, which deal with specific issues including human resource management (HRM), culture, gender, and migration. Such chapters are important and again there is no pretence that they encompass all the possible topics.

The first topic noted above, HRM, is considered in Chapter 9, written by Debrah, Jabaru, Nyuur, Ellis, and Banoeng-Yakubo. The authors note that the construction industry presents particular difficulties that might make it challenging to apply critical components of the fundamental HRM strategies that appear to have been successful in other, more stable, sectors of the economy. The authors conclude by noting that to avoid endangering the competitiveness of the construction industry in the future, stakeholders should develop practical measures and mechanisms that will help minimise the numerous unique present and future problems of the industry. It is also necessary to highlight the particular difficulties associated with some of the prevalent practices of HRM applied in the construction industry. This will act as a roadmap for taking the appropriate actions to address human resource concerns in the industry.

Chapter 10, by Tijhuis, can be seen as a counterpoint to Chapter 9. Tijhuis also deals with organisation in construction, but from the point of view of culture. The chapter is about organisation-related aspects of human behaviour within the construction processes. It uses a model, symbolising culture's influence on labour situations, which should lead to better working circumstances for individuals and organisations by stimulating a better understanding of each other, resulting in the improved delivery of the built environment to its owners and users.

It is noted throughout this book that women and migrants are groups of workers with specific challenges in the construction industry. Chapter 11, by Johansson and Johansson, addresses the first group, with an interface with Chapter 10, because the challenges of women in construction have a strong cultural component. A gender perspective in the consideration of construction work makes it evident that it is a gender-segregated arena dominated by men, which has cultures and notions associated with masculinity. A review of research on gender and equality in construction work points out norms and values connected to a masculine culture, as well as inequalities in structures and in the organising of work as hindrances to increased female participation. For example, there are reports on situations where women are met by gender-biased recruitment practices, discrimination, and harassment. Further, Johansson and Johansson argue that some stereotypical notions of male construction workers as being brave, strong, and risk-taking lead to exclusion of women from construction work. The chapter ends with a discussion of the imperatives for change, including some examples of related practical measurements.

Chapter 12, by Abdul-Aziz, Abdullateef Ashola, and Bala, addresses migration. The chapter presents a historical view of labour migration in construction, exposing its problems for the foreign workers involved. It argues that widely reported mistreatment of migrant construction workers is likely to persist into the future. Structural changes in the modern-day construction industry reinforced by discriminatory legislation and judicial systems of host countries have further entrenched undesirable treatment of migrants. Central to this is the perception of these workers as being no more than disposable construction resources. Given these hurdles, the chapter ends by asking whether fair, safe, and decent employment will ever be available worldwide for migrant construction workers. Abdul-Aziz, Abdullateef Ashola, and Bala suggest that the relevant international organisations and non-government organisations should combine their efforts to achieve greater impact, as demonstrated by the Global Compact for Safe, Orderly and Regular Migration which was endorsed by the United Nations General Assembly on 19 December 2018. This chapter concludes Part 1 of the book.

Part 2: The impact of the built environment on workers

Part 2 of the book begins with chapters which approach the topic by considering human settlements in general (territorial development and housing markets), followed by specific (although also broad) themes including the use of public spaces by workers, working from home, workplace safety, and the connection between construction sites and the urban economy.

Chapter 13, by Klink and Tepassê, argues that research and policy making regarding the linkages between territorial development, livelihoods, and labour should be revisited critically considering the increasing influence of what is called 'platform capitalism' on the way contemporary cities are built, managed, and financed. The chapter is organised around three complementary sections. The first section discusses the relevant science and technology literature and its links with debates on smart cities, platform capitalism, and platform urbanism. The following section provides an illustration of indirect impacts of platform urbanism on labour through the transformations of living costs and the structure of housing provision. More specifically, it elaborates upon how real estate and housing platforms, in combination with institutional investors and investment funds, have gradually emerged as new actors in the finance, building, and operation of the built environment. This consideration includes a discussion on real estate platforms and their connections to urban livelihoods, with highlights on the specific situation of Brazil. The chapter concludes by suggesting elements for a research agenda on platform urbanism, cities, and labour, with particular attention to the Global South.

Klink and Tepassê's Chapter 13 makes important references to housing. Next, Chapter 14, by Werna, Keivani, and Cho, interfaces with some of the aforementioned analysis as it concentrates on the relationship between the housing and labour markets. Together, these markets can strongly influence the development of a city. The chapter argues that current analyses that link both markets do not capture their diversities and, as such, provide only partial information for policy making. The chapter advocates an inclusive approach which needs to take on board the pluralism of both markets and how they operate to provide a better standard of living to increasingly diverse types of workers who need different options to be able to access decent housing. It concludes with the proposal of a broad research agenda, including a pluralistic approach for each market; housing as an asset in the labour market, with particular attention to home-based work; the impact of housing construction in the housing-labour market nexus; and the contribution of housing to the wellbeing of workers.

Following on from Chapter 14, which includes the consideration of urban policies related to buildings (specifically housing) and the relationship to labour, the book goes on to consider urban policies related to public spaces and labour, in Chapter 15. According to the authors, Were, Mukiibi, Majale, and Nawangwe, increasingly greater numbers of workers resort to working in public spaces as they do not have other options. The chapter discusses public policies such as the use of public markets, access to sanitary facilities, building of precarious shops on public land (which can be considered as the workplace equivalent of building slum accommodation on public land), issues of support versus eviction, and so on. The authors concentrate their analysis on Kampala, Uganda, but, as is the case of other chapters, the contents of Chapter 15 are also of value elsewhere. In addition, while many workers go to public spaces to work, many others resort to working from home due to the difficulties they face in renting an office or because they become a 'one person enterprise' and many enterprises now operate online (see also Chapter 14 by Werna, Keivani, and Cho). Chapter 15 also addresses such a phenomenon, but just generically. The focus is mainly on work in public spaces.

After such an analysis of work in open areas in Chapter 15, the book goes back to buildings. Chapter 16, written by Katz, Harwood, and Nielsen, focuses on the resilience and security of workplaces. Resilient workplaces are crucial not only for protecting the lives of workers, but also for maintaining economic stability in the face of disasters. As the world continues to grapple with the impacts of climate change, the safety of the built environment in general and workplaces in particular is becoming an increasingly urgent concern. From rising sea levels to extreme weather events, the effects are putting the infrastructure and the people who rely on it at risk. After analysing practical examples, the authors of Chapter 16 argue that if the world is not prepared to address the disasters of today, it will be ill-prepared to tackle those of tomorrow. Indeed, climate change will strengthen weather events and require the adoption of stronger adaptation measures across all industries. The workforce must be protected both at work and at home in order to safeguard lives and incomes and maintain strong economies. Katz, Harwood, and Nielsen conclude by calling for good practices and recommending that such solutions must now be adopted more broadly, and they suggest that policy makers need to put incentives and regulations in place to ensure that big businesses are not the only ones which can afford adaptive measures; small- and medium-sized enterprises make up the backbone of the global economy and their workforce must also be protected against the calamities which might come.

All the chapters in Part 2 analysed, in different ways, the products of the built environment. Chapter 17, focuses on the process of construction, making a link with Part 1, from an opposite angle (Part 1 is about the workers engaged in the construction process). Chapter 17, written by Aigbavboa and Akinradewo, explores the interface between local workers and businesses on the one side and construction sites on the other. The study identified factors influencing the impact of local labour and businesses in construction sites and vice-versa. It also presents the approaches to managing the impacts. Chapter 17 proposes a framework for the effective management of community stakeholders on construction project sites. The study recommends that on each project, construction professionals should develop and implement a stakeholder management plan, including a community engagement plan, to identify and prioritise stakeholders, assess their level of influence, interest, and power, and communicate effectively with each of them.

Filling the knowledge gap

Prospects for the reduction of poverty are intrinsically linked to the creation of good quality employment (ILO, 2015; Mella and Savage, 2017). In addition, reduction of poverty needs elements of the built environment, such as workplace facilities, other items of infrastructure, and other elements to be in place. The large existing literature on each of the bodies of knowledge on the subjects of labour, the built environment, and human settlement development has addressed the connections among these subjects only in a limited and fragmented manner. The book addresses the gap in knowledge which is represented by the lack of analysis of the inter-connections among the three subjects.

It is strategic to have a comprehensive approach to labour in the built environment and human settlements, as labour issues are inter-related and there are synergies to be derived from considering them together. It is evident that the current treatment of the subjects in silos, one or a few issues at a time, has a negative impact on the ability to develop effective policies and initiatives in the others, as well as even in the subject considered. The book was conceived as a platform to enable a critical examination of the complex issues in the three fields and the relationships among them to be undertaken. It also provides the opportunity for some new issues and under-explored ones in the various fields to be discussed. It is appropriate for the subject of Labour, the Built Environment, and Human Settlements to become a distinct area of research to guide policy making and the development of programmes and initiatives for each of the three component areas and also jointly for all of them.

Many chapters in the book do make references to issues included in other chapters, therefore providing some degree of integration. However, the coverage is not necessarily totally integrated – this was not the aim. The book contains a set of chapters about different topics. The discussion adds value by providing the state-of-the-art in each topic. Also, some of the topics are under-researched, as is shown throughout the book.

Aim and objectives of the book

The aim of the book is to present a state-of-the-art review of the complex and dynamic relationship between labour and the built environment and indicate possible new directions for the subject, also including human settlements. It is intended to be an authoritative reference text on the new subject of Labour, the Built Environment, and Human Settlements.

The objectives of the book are to:

1 present a comprehensive review of the literature on the main topics of the relationship between labour and the built environment, focusing on construction, the largest segment of the sector which produces the built environment;
2 explore the fundamental issues and underpinnings of the subject such as the critical nature of the built environment, prevailing and recurrent deficits, changes in technology, employment relations, and conditions of work and of living; and
3 propose new topics and sub-topics to be explored with the view to enabling the subject to grow.

Nature of the book and of each chapter

The book comprises original chapters authored by persons who are globally recognised experts on the particular topics of labour and the built environment. Each chapter is an authoritative and enduring piece on the topic. This first chapter of the book provides an introduction to the work and presents a synthesis of the other chapters.

Each chapter presents a thorough scholarly analysis of the existing knowledge on the topic it covers, compares the various views on it, and presents a reference point for further and advanced research leading to a contribution to the development of the subject. Each chapter presents and discusses current definitions and the main concept of the topic it covers; considers its development and current state, covering also issues relating to the topic; discusses likely future development of the topic; and proposes aspects for further research of the field of knowledge, as well as recommendations for action by government, industry, individual professionals and workers, and other relevant stakeholders.

As the subjects covered in the book constitute combinations of issues in new ways as envisaged by the editors of the book, the authors were guided by not only a topic, but also a brief outline of the possible contents, prepared by the editors. However, the authors were also given the flexibility to suggest possible fine-tuning of the topic, the approach, or the proposed contents.

How long will the book remain up to date?

The creation of quality employment and reduction of poverty remain major challenges throughout the world and are related to the expected outcomes of current global and national development agendas. This will continue to be so for some decades. The built environment plays an important role in such a context. It will be important that the construction industry, which creates it, is continuously developed to improve its performance. The construction industry is changing rapidly, owing to the pace of technological development brought about by the Fourth Industrial Revolution and the accelerated need to protect the environment, especially to arrest the impact of human activities and patterns of consumption on the climate. These factors should be considered together with the prospect of an increase in the world population by 50 percent (an additional 3.5 billion people) by 2050, which will require a massive expansion of cities, towns, and other human settlements.

Moreover, as noted on the first page of this review, the traditional view which implies a sharp division between the developed and developing countries is changing. The division is now more nuanced and complex. For example, some developing countries have made socio-economic advances in some areas, while lacking progress in others. It is not sufficient to simply use new terms such as 'emerging economies' or 'least developed countries'. It will remain important to study the dynamics of socio-economic development and the factors which contribute to progress in it. At the same time, issues which were deemed to be existing only in developing countries are now becoming evident in the developed nations. Examples include the informal sector of the economies and particular sectors and housing shortages coupled with issues of affordability which are leading to significant levels of homelessness and reductions in local services leading to 'slums'.

The need to consider the subjects of labour, the built environment, and human settlements is evident owing to the synergies such an approach affords. Thus, it is clear that the issues addressed in this book will remain of interest in the future.

How the book differs from the competition

There are no apparent competing titles that bring together the issues analysed in the book in such a specific manner. The latest academic book on labour in construction is Belman et al. (2021). While there are interfaces with the present book, there are also many differences, which shows that both books add value in different ways. While the subjects of some chapters coincide, the chapters of Belman et al. focus on specific countries. The chapters of the present book by Ofori and Werna include vast material about regions and the whole world, although a few chapters concentrate on a given locality but with implications elsewhere. Moreover, Ofori and Werna also include an analysis of the impact of the built environment on labour in other sectors beyond construction. This is beyond the scope of Belman et al. (2021).

There is also grey literature on labour in construction from the International Labour Office (ILO) and the then UK Department for International Development (Mella and Savage, 2017). Both are reports prepared for the benefit of administrators and do not involve deep academic analysis. There are many books which focus on specific issues covered in the present book but they do so by considering each of the issues on its own. Thus, there is a lack of interlinkages and comprehensiveness. For example, the ILO and related institutions have produced many reports on several issues related to labour. Yet, this has been done in a fragmented way, and not many of the reports are related to construction or the built environment, and there have been none related to human settlements. In addition, ILO literature tends to be politically loaded, with limited margin for academic discoveries. The publications have to be previously approved by workers' and employers' representatives and governments at the same time.

Moreover, the academic literature on labour (such as Standing, 2011) tends to focus on generic trends and not on a specific value-chain such as the built environment. Finally, there is also a body of literature on livelihoods (such as the seminal book by Lloyd-Jones and Rakodi, 2002) with some connection to construction, although it addresses other sectors. While interfacing, livelihoods and employment are not synonyms.

References

Belman, D., Druker, J. and White, G. (2021). Work and Labor Relations in the Construction Industry. An International Perspective. London: Routledge.

International Labour Organization (ILO) (2001). The Construction Industry in the Twenty-First Century: Its Image, Employment Prospects and Skill Requirements. Report for Tripartite Meeting. Sectoral Activities Department. Geneva: ILO.

International Labour Organization (ILO) (2015). Good Practices and Challenges in Promoting Decent Work in Construction and Infrastructure Projects. Issues Paper for Discussion at the Global Dialogue Forum on Good Practices and Challenges in Promoting Decent Work in Construction and Infrastructure Projects. Geneva: ILO.

Lloyd-Jones, T. and Rakodi, C. (2002). Urban Livelihoods: A People-Centred Approach to Reducing Poverty. London: Routledge.

Mella, A. and Savage, M. (2017). Construction Sector Employment in Low Income Countries. ICED (Infrastructure and Cities for Economic Development) Report, DfID (Department of International Development, UK). London: DfID.

Ofori, G. (2022). Construction economics: its origins, significance, current status and need for development. In Ofori, G. (Ed.). *Research Companion to Construction Economics*. Cheltenham: Elgar Publishing Ltd, pp. 18–40.

Standing, G. (2011). The Precariat: The New Dangerous Class. London: Bloomsbury.

Turin, D. A. (1973). *The Construction Industry: Its Economic Significance and Its Role in Development. Building Economics Research Unit*. University College London.

United Nations Environment Programme (2019). Global Environment Outlook 6: Healthy Planet, Healthy People. Report. Nairobi: UN Environment.

Wells, J. (2006). Labour subcontracting in the construction industries of developing countries: an assessment from two perspectives. *Journal of Construction in Developing Countries*, Vol. 11, No. 1, pp. 17–36.

Wells, J. (2013). *Relieving Chronic Poverty among Construction Workers: An Exploration of Possibilities to Improve the Quantity and Quality of Jobs*. Engineers Against Poverty, London, http://engineersagainstpoverty.org/resource/relieving-chronic-poverty-among-construction-workers-an-exploration-of-possibilities-to-improve-the-quantity-and-quality-of-jobs/

Werna, E. and Klink, J. (2022). The builders of cities: prospects for synergy between labour and the built environment. In Ofori, G. (Ed.). The Construction Economics Reader. Aldershot: Edward Elgar Publishers.

World Economic Forum (WEF) (2019). Transforming Infrastructure: Frameworks for Bringing the Fourth Industrial Revolution to Infrastructure. Community Paper. Shaping the Future of Cities. Infrastructure and Services Initiative. Geneva: WEF. http://www3.weforum.org/docs/WEF_Technology_in_Infrastructure.pdf. downloaded 19 March 2023.

2

THE BUILT ENVIRONMENT, CONSTRUCTION, DEVELOPMENT AND THEIR IMPLICATIONS FOR LABOUR

George Ofori

London South Bank University

Introduction

The aim of the chapter is to explain the nature of the built environment and construction, and explore the role of both of them in national development and how this impacts the construction workforce, labour in other sectors and communities.

The objectives are to:

- explain the nature and importance of the built environment and the construction industry and the difference and relationships between them.
- consider the role construction plays in the economy and in various aspects of national development and their implications for labour in construction and other sectors.
- consider and analyse labour as the key input into the construction process and the implications of construction for labour in other sectors.
- propose actions for addressing the labour challenges in construction and enabling the industry to attain its potential in its delivery of built items.

The built environment and its importance

The built environment: A definition

The built environment was introduced in Chapter 1 of this book. It is pertinent to expand its consideration. Elsevier (2023) refers to the built environment as: "the human-made structures and networks that surrounds us". Authors define it from the perspective of the field they are writing on. For example, in a paper on energy and environmental security, Thompson and Kent (2017) considered the built environment as that "which encompasses the places where we live and work and the ways we travel". Among the definitions used by Kaklauskas and Gudauskas (2016) in writing about smart cities is that the built environment "refers to the human-made surroundings that provide the setting for human activity,

DOI: 10.1201/9781003262671-2

ranging in scale from buildings and parks or green space to neighbourhoods and cities that can often include their supporting infrastructure, such as water supply or energy networks". Bekchanov et al. (2022), considering the circular economy, noted that "the built environment sector comprises … the surroundings modified by man and include all buildings and infrastructural objects such as transportation and telecommunication systems, energy and water supply, and waste management systems". Habash (2022), writing on urban sustainability, suggested that the built environment may be described as "the human-made or modified physical spaces of the environment where people live and perform their activities. These include physical spaces (buildings, open spaces, and infrastructures), human society spaces shaped and sustained by the culture and social interactions of urban residents, and cyberspace, which is comprised of computers, internet access, and the data flowing through these systems". Thus, for the purpose of discussion in this chapter, the built environment comprises the efforts made by humans to create the space and facilities for all activities.

Importance of the built environment

The built environment offers many important benefits apart from the physical space and installations in its products, and the connections between them. This is discussed in Chapter 16 of this book. It is briefly considered here. Kaklauskas and Gudauskas (2016) referred to the built environment as a material, spatial, and cultural product of human labour that combines physical elements and energy in forms for living, working, and playing. On the social drivers of health, Best and Erickson (2023) noted that "the built environment includes all of the components of where people live, including access to clean air and water, safe housing, green space, proximity to fresh food markets, and availability of walking trails or bike lanes". Thompson and Kent (2017) considered the built environment as important for human health, as the chronic diseases which people face can occur in places where it is easy to be physically active, socially connect with family and community, and eat nutritious food. They show that action for a health-supportive environment is also good for environmental sustainability. Pinter-Wollman et al. (2018) noted that the built environment affects health directly and indirectly. Perdue et al. (2003) reviewed the connection between public health and the built environment and suggested that public health advocates can help shape the design of cities and suburbs in ways that improve people's health. Similarly, studies show that buildings can be designed to positively influence human health (Croucher et al. 2012); and recovery (Glasgow Centre for Population Health, 2013). Fu et al. (2021) demonstrated that green building features can positively affect occupant well-being in various ways, such as improving indoor air quality and reducing noise pollution. Thus, development should take an occupant-oriented perspective to maximise the benefits.

Some sections of society have special needs. As observed by Maisel and Ranahan (2022), the concepts of universal design have gone beyond considering usability by the physically challenged to the approach Steinfeld and Maisel (2012) defined as: "a process that enables and empowers a diverse population by improving human performance, health and wellness, and social participation". These are required by building regulations in most countries, and form part of design considerations today (Maisel et al., 2018). Madanipour (2023) reconsiders public space in the context of contemporary global health and economic crises, and technological, political and cultural change, examining the potential for public spaces to contribute to inclusive social practices. Ji et al. (2023) noted that research has shown that the immediate environmental factors of senior housing, such as the design of the housing features and facilities, have a direct bearing on the satisfaction and quality of life of older people.

Much work exists on planning cities and towns to foster residents' safety (UN Interregional Crime and Justice Research Institute (UNICRI) and Massachusetts Institute of Technology Senseable City Lab, 2011; Schuilenburg and Peeters, 2018; Kaya-Malin and van Soomeren, 2021). Cui et al. (2023) undertook a safety perception evaluation of streets from the female perspective in Guangzhou, China and found differences in safety perceptions across genders; women gave lower safety scores; about 11 percent of the streets in the study area showed weak perceived safety; three of them had high pedestrian flows and required priority improvements.

Possible adverse impacts of the built environment

The built environment can have negative impacts on safety, as evinced by the occasional collapse of buildings in use (Windapo and Rotimi, 2012); as well as on health and well-being, including the effect of poorly designed facilities (Condihoto et al., 2009; Marchand et al., 2014). (This is discussed in more detail in Chapters 15 and 16 of this book.) For example, Best and Erickson (2023) suggested that stressful environmental exposures can have a negative impact on child health and wellness. Thus, care in planning, design and construction of the built environment is key. Shen et al. (2023) noted that cities pose many environmental and social challenges and it was urgent to develop urban strategies to create liveable, healthy, and inclusive cities that enhance the well-being of residents. UNCTAD (undated) noted that "around the world poorly designed transport systems contribute to environmental degradation, urban congestion, loss of human health and loss of lives" (p. 17). It reported that, in large cities in the Asia and Pacific region, one million people die annually in road accidents, and that, while the region has 4 percent of the world's motor vehicles, it accounts for 42 percent of global transport-related deaths. Smith (1999) found that, in India, indoor air pollution from the use of traditional biomass fuels for cooking was responsible for 500,000 premature deaths each year among women and children (5 percent of all deaths in India). The chapters in Martin-Ortega and Treviño-Lozano (2023) addressed links between sustainability and human rights in infrastructure projects and found a human rights gap in every stage of the public procurement processes.

Work also exists on the influence of indoor air quality on productivity and human health and well-being (Wang et al., 2022). Tham (2016) analysed research on the impact of indoor air quality on humans over a 30-year period (1986–2016). The findings showed how contaminants interact to accentuate adverse effects on humans and their performance. Rising affluence of the middle class, increased population density in cities and new synthetic materials have led to greater intensity of exposure to indoor air pollution. Innovations in air distribution, air cleaning, modularisation of indoor environmental devices/systems, leveraging on smart technologies and engagement of the user, as a sensor and indicator of preferences, are advancements towards a holistic solution. The COVID-19 pandemic underlined the importance of the linkage between the built environment and human health (see, for example, the work of an international group of experts, Morawska et al., 2020). Navaratnam et al. (2022) outlined the engineering control measures to mitigate COVID-19 spread and healthy building design as including modified heating, ventilation, and air conditioning systems; and ultraviolet germicidal irradiation, bipolar ionisation, vertical gardening, indoor plants, microbial paints and smart technologies.

Having defined the built environment and considered its importance and impacts, the discussion in the chapter now focuses on the construction industry, which produces the built environment, in its relation to the process of national development, and labour. The construction industry is next defined and its importance is considered.

Construction industry and its importance

The construction industry

The most appropriate definition for the construction industry has long been debated (Ofori, 2022a). The disagreements have been on: (a) what economic activities the industry includes or excludes, i.e., delineation of the boundaries of construction, for example, whether to include clients, or manufacturers and suppliers of materials (Hillebrandt, 2000); (b) what stages of the activities should be considered, mainly, whether to include the operation and facility management of the built item (Carassus, 2000); (c) whether construction is a discernible industry, as there are so many types of built items requiring different inputs (Meikle, 2019); (d) whether construction is a sector comprising many industries rather than one industry (de Valence, 2019); and (e) whether there is, in fact such a thing as a construction industry or a construction sector, owing to the complexities involved in trying to define it (Groak, 1994). Ive and Gruneberg (2000) referred to a "construction industry cluster". Carassus et al. (2006, p. 169) suggested that it is more appropriate to consider the "built environment cluster" which delivers a variety of services and improvements to the human environment rather than a building or item of infrastructure.

It is pertinent to consider what the data in national statistics cover. The definition in the North American Classification System is (US Bureau of Labor Statistics, 2023):

> The construction sector comprises establishments primarily engaged in the construction of buildings or engineering projects (e.g., highways and utility systems). Establishments primarily engaged in the preparation of sites for new construction and establishments primarily engaged in subdividing land for sale as building sites also are included in this sector.
>
> Construction work done may include new work, additions, alterations, or maintenance and repairs. Activities of these establishments generally are managed at a fixed place of business, but they usually perform construction activities at multiple project sites.

The classification adopted in the United Kingdom is (ONA, undated):

> This section includes general construction and specialised construction activities for buildings and civil engineering works. It includes new work, repair, additions and alterations, the erection of prefabricated buildings or structures on the site and also construction of a temporary nature.
>
> General construction is the construction of entire dwellings, office buildings, stores and other public and utility buildings, farm buildings etc., or the construction of civil

engineering works such as motorways, streets, bridges, tunnels, railways, airfields, harbours and other water projects, irrigation systems, sewerage systems, industrial facilities, pipelines and electric lines, sports facilities etc. (p. 149).

To deal with the issue of the variety of types of products and necessary inputs of construction activity, Turin (1973) proposed that the construction industry should be viewed as comprising distinct segments; he provided a matrix of the industry in developing countries with different segments, each with its own determinants of demand and operating constraints, and the relationships between them. Ofori (1989) built on Turin's (1973) matrix to propose a construction industry which has two main divisions: formal and informal, differing mainly in terms of regulation, which comprised six segments existing along a continuum of technological sophistication:

- formal sectors: international, conventional-large, conventional-small/medium, self-help projects
- informal sectors: monetary-traditional, subsistence.

Revisiting the matrix, Ofori (2022a) noted that there had been developments in construction since he presented the matrix in 1989. As the size and complexity of construction items have increased, the *International* segment has become ubiquitous, with stiff competition among international and large local firms (Yan et al., 2022; Turner & Townsend 2022). The *Conventional-Large* segment, which used to comprise only local firms, now includes many foreign companies wishing to operate in developing countries, increasing competition locally. Countries have introduced measures such as compulsory joint ventures (see, for example, Nippa and Reuer, 2019) and local content requirements such as the *Buy Uganda Build Uganda Policy* (Ministry of Trade, Industry and Co-operatives, 2014), although some argue against such interventions (see OECD, 2019). It is now more widely recognised among administrators and researchers that it is useful to find ways to formalise, improve and grow (not replace) the methods in the *traditional* segment; and the *informal* segment should be nurtured, to harness its entrepreneurship and innovation. Ofori (2022b) suggested that, despite these developments, Ofori's (1989) matrix is still relevant as a framework of analysis for the construction industry in developing countries, if not the whole world.

In the context of this book, it is necessary to agree on a definition of the construction industry, to provide the basis for discussing its importance in the economy and development, and the issues relating to its labour and that of other sectors. As Ofori (2022b) notes, the nature of these issues will depend on the way the industry is defined. For the purpose of this chapter, it is suggested that there is a construction industry which can be defined as the sector of the economy which plans, designs, builds, maintains, repairs or refurbishes as necessary, operates, and eventually demolishes buildings and other works. Thus, the entire lifecycle of the built item is considered. The items built by the industry can be classified into (a) economic infrastructure – which is used to produce directly (such as factories, hotels and offices) or to support production (such as roads, airports, harbours, power stations and water purification and distribution systems); and (b) social infrastructure which contributes to improve the quality of life of the people (such as schools, hospitals, housing). From this portrayal, it is clear that construction provides the physical basis for the long-term national development, as it builds the basis for production of goods and services, and social activities. It is also evident that the items which constitute the built environment are the result of construction activity. Thus, it is important for each country to have a construction industry which is capable of creating this critical foundation for development.

This chapter follows the three themes of this book: *how labour and the built environment relate to development; workers of the built environment; and the impact of the built environment on labour in other sectors.* In the next section, the role of the construction industry in development is discussed.

Construction in national development

As noted in Chapter 1, there has been much research on the role of construction in development (Turin, 1973; Kirchberger, 2020; Ofori, 2022a). The focus was initially on the role of infrastructure in development; this had been a key part of the considerations under all major economic development paradigms. *World Development Report 1994* (World Bank, 1994) was devoted to an examination of the link between infrastructure and development. It noted that developing countries had made substantial investments in infrastructure in previous decades, achieving dramatic gains for households and producers by expanding their access to services such as safe water, sanitation, electric power, telecommunications, and transport. However, even more infrastructure investment and expansion were needed to extend the reach of services, especially to people in rural areas and the poor. Quantity and quality improvements would modernise and diversify production, help countries compete internationally, and accommodate rapid urbanisation. This role of infrastructure was stressed under the Millennium Development Goals (MDGs). Similarly, construction must play a major role towards realising the Sustainable Development Goals (SDGs) (Opoku, 2022). Ofori (2023) suggests that all the SDGs are relevant to construction. They relate to what the industry produces; the inputs it needs; how it works; and its impacts.

It is useful to explore the role of infrastructure in development further. UNCTAD (undated) made these points about "infrastructure as an agent for economic development" (p. 1). First, development of infrastructure networks is intimately connected with the process of economic growth – studies show that a 1 percent increase in infrastructure stocks is associated with 1 percent increase in GDP (World Bank, 1994); the relationship is in both directions, infrastructure leads to growth which then leads to further expansion of infrastructure. Second, infrastructure helps create macro-economic conditions for poverty reduction – World Bank (2000b) estimated that on average, every additional percentage point growth in average household consumption reduces the share of the population living in poverty by 2 percent. Third, at the micro-level, access to infrastructure provides vulnerable households with opportunities to escape poverty. Access to safe water and modern energy helps to raise household productivity by improving health and enabling households to deploy power-assisted tools; transport and communication networks connect communities to markets for their products leading to higher income. However, access to infrastructure services is inequitable in the poorest countries, for example, in Chad, Mozambique and Uganda, only the wealthiest 20 percent of the population has access to electricity. Finally, transport infrastructure facilitates the integration of the global economy – World Bank (2000b) estimates that a reduction of 10 percent in transport costs leads to increase in trade flows by 25 percent.

Governments recognise the importance of infrastructure. For example, the United Kingdom government includes infrastructure among the five foundations of the national industrial strategy (HM Government, 2017, p. 4); and as one of the three main thrusts of the latest strategy in which it noted that: "High quality infrastructure is crucial for economic growth, boosting productivity and competitiveness" (HM Treasury, 2021, p. 8). Mulaudzi and Lancaster (2021) noted that in 2020, President Cyril Ramaphosa of South Africa said at the Sustainable Infrastructure Development Symposium that: "Meaningful infrastructure [investment] would have the potential to strengthen the economy, bringing the country a step closer to achieving the [National Development Plan's] commitment for the economy to attain 5.5% year-on-year economic growth, a 6% unemployment rate and 30% gross fixed capital formation to GDP".

Deriving the synergies

It is necessary to consider the synergies among key aspects of development. UNCTAD (undated) noted that infrastructure tends to improve the efficiency of investments made in other sectors. The World Bank (2000a) asserted the many synergies between infrastructure assets and other critical services such as health and education. For example, in the Philippines, electrification was estimated to increase returns to education by 15 percent. It also cited the African Development Bank reporting that in some countries, some 30 percent of perishable foodstuff are prone to damage owing to the lack of all-season roads to transport them to the market centres. Table 2.1 shows linkages between infrastructure and broader development goals. Thus, the sole focus on the extension of physical networks without considering the wider impacts should change. UNCTAD (undated) advised developing countries to improve the development impact of infrastructure by: designing sector reform processes to ensure that the benefits reach the poor; considering the potential role of regional solutions to national infrastructure problems; requiring infrastructure projects to explicitly consider the needs of the social sector; and requiring social sector

Table 2.1 Linkages between infrastructure and broader development goals

	Water, sanitation, drainage	*Electricity and modern fuels*	*Telecommunications*	*Transport*
Households	Improving hygiene and household productivity	Improving hygiene and household productivity and quality of life	Building social capital	Building social capital
Enterprises	Improving enterprise productivity	Improving enterprise productivity	Providing links with markets	Providing links with markets
Health sector	Reducing incidence of water-borne diseases	Reducing indoor air pollution	Improving access to medical expertise and information	Improving access to health facilities
Education sector	Improving hygiene	Increasing length of day and scope of curriculum	Improving access to information	Providing access to schools
Environment	Reducing stress on water resources and contamination of the water environment	Increasing the efficiency of energy use and the extent of deforestation	Reducing the need for travel	Reducing local air pollution and congestion

Source: UNCTAD (undated).

projects to assess the potential contribution of the infrastructure sector. Also, industrialised countries should take a more holistic view of how sectoral interventions they support in developing countries can contribute to wider development goals.

The discussion here points to the impact of the creation of infrastructure on labour in other sectors of the economy, and demonstrates the possible synergies that can be realised from a multi-sectoral and multi-dimension approach to development when considering investments in infrastructure.

Construction in the economy: Contribution to capital formation and value added; and output generator and multiplier

The construction industry plays an important role in the economy. Hillebrandt (2000) pointed out that construction is the only industry which appears twice in the national accounts, in the gross domestic product (GDP) tables, and in the Gross Domestic Capital Formation tables which indicate savings made in the form of capital which can be used for economic activity in growing GDP. Thus, buildings and other infrastructure items constitute much of the nation's wealth and capital.

The construction industry is an important contributor to GDP. Studies find differences in sectoral distributions across countries, which change over time (Lopes et al., 2017). Ernst et al. (2015) studied 40 countries and found that construction comprised between 4.5 percent (Bulgaria) and 10.4 percent (Spain) of overall gross output in 1995 and between 4.2 percent (Taiwan) and 14.6 percent (Spain) in 2009. In terms of total value added, construction represented between 4.2 percent (the United States) and 10.1 percent (Korea) in 1995 and between 2.2 percent (Taiwan) and 10.8 percent (Spain) in 2009. It should be noted that the figures depend on how the "construction industry" is defined. For example, in Australia, Hampson and Brandon (2004) found that construction accounted for 14.4 percent of GDP (instead of the 6.3 percent in the national accounts data). Ruddock and Ruddock (2008) noted that there has been concern that data on the construction industry are not as valid, reliable, comprehensive in their coverage and useful as for other sectors of the economy. Thus, the role of construction in the economy is not yet fully understood.

The construction industry also has complex backward and forward linkage effects with the rest of the economy (as it obtains inputs from the rest of the economy; and builds the premises which are used for further production, respectively) (Hillebrandt, 2000). Ernst et al. (2015) found that intermediate inputs, such as materials and finance, the construction industry demands are relatively higher in comparison with aggregated industries: in most of the 40 countries studied, these inputs explained between 8 and 14 percent of the total intermediate demands in 1995, increasing to 18 percent in 2009. Moreover, the total output multiplier effects in the construction industry were higher than the overall level, for each group of countries. The absolute gap between overall and construction industry multiplier effects varied among countries, being 0.29 points above the overall figure in high-income countries and 0.11 in middle-income countries.

Yousaf et al. (2019) used Asian Development Bank input-output data for Bangladesh, Nepal and Sri Lanka to analyse and compare the performance of their construction industries. The results indicated that construction has strong backward and weak forward linkages for all the three economies; and that the "pull effect" is very significant in these countries while the "push effect" is very insignificant. Wibowo et al. (2008) analysed the relationship between the construction and other sectors of the economy in Indonesia (focusing on the trade, chemical, mining and finance industries), applying the country's input-output tables (for 1990, 1995, 2000 and 2003). Construction's contribution to GDP from 1995 to 2005 was about 6–8 percent. On the backward linkages (coefficients indicating required amounts from other industries per unit of construction output), construction was one of the industries with the highest direct backward linkage indicator. Considering forward linkages (coefficients representing how much construction delivers to other producers and final demand users per unit of output), construction is the highest among the industries in most years. Finally, in terms of the output multipliers which represent the effect of one monetary unit change in final demand of the construction industry on total output of all other industries, construction ranked highest compared to the other industries considered.

In the next section, the role construction plays in generating employment in the industry and in the wider economy is considered.

Construction as employment generator

Studies have highlighted the employment generation potential of construction activity (Ofori, 1993). This potential is of interest to governments. Chapter 4 of this book discusses this subject in detail. It is briefly considered here to present one of the major elements of the role of construction in national development. In the United States, Ball (1981) estimated that US$ 1 billion spent on new construction in 1980 generated 24,000 full-time jobs for one year, mostly in supporting industries such as manufacturing, trade and transport. Anderson et al. (2001) estimated direct and indirect employment generated by expenditures on federal-aid highway improvement projects, with data from the Federal Highway Administration database and input-output data. They generated employment estimates for 14 categories of highway improvement. A broader study by the International Labour Organisation (ILO) (Ernst et al., 2015) analysed output and employment multiplier effects between 1995 and 2009 after constructing a closed input-output model from a global database. It was found that: (i) total output multipliers in construction are higher than for the whole economy but total employment multipliers in construction do not follow simple patterns; they are usually higher in high-income countries; and (ii) there are trends of declining total employment and output multiplier effects at the overall level and in construction.

Schwartz et al. (2009) estimated the potential effects on direct, indirect and induced employment for different types of infrastructure projects in Latin American countries. They found that the direct and indirect short-term employment generation potential of infrastructure investment projects averages around 40,000 annual jobs per US$ 1 billion in Latin American countries, depending upon variables such as the mix of sub-sectors in the investment programme; technologies deployed; local wages for labour; and degrees of leakages to imported inputs. The study suggested that rural road maintenance projects might provide 200,000 to 500,000 annualised direct jobs for every US$ 1 billion spent. Boakye-Gyasi and Li (2015) found that flows of foreign direct investment from China to Ghana have a positive impact on employment growth in construction and its development, leading to an efficient workforce which results in growth in household incomes and benefits the economy from high productivity. Ball and Wood (1995) found only a weak link between quarterly increases in total construction output and construction employment on public projects in the United Kingdom, and no significant relationship between housebuilding and employment. They attributed these results to poor-quality data and like Ruddock and Ruddock (2008), suggested that there is a need for more accurate construction data.

Many countries have sought to generate employment with construction programmes, either with stimulus packages to avert recessions as occurred in the 2008 global economic crisis and the COVID-19 pandemic (Ofori, 2022), or to address unemployment. McCutcheon (2001) suggested that the theory, and experience and research in Africa give reasons to advocate carefully considered and well-designed long-term programmes using employment-intensive methods for the construction and maintenance of infrastructure. McCord (2012) outlined objectives of public works projects undertaken to foster employment as being to: alleviate poverty; create short-term work opportunities to alleviate transient unemployment, intensify labour in the production of infrastructure to increase aggregate demand, or develop small, medium and micro enterprises to promote labour intensification or promote aggregate demand; deliver skills development; improve physical and social infrastructure provision; or derive macro-economic stimulation by introducing capital into the economy through the wages to stimulate demand. McCord (2012) suggested that, to be successful, the objectives should align with the country's particular circumstances. She cited the inappropriateness of short-term relief projects such as in South Africa, Malawi, and Bangladesh, where chronic poverty and unemployment persist.

Bentall et al. (1999) noted that construction strategies adopted to utilise the relationships between employment opportunities, available skills, entrepreneurship and the use of small-scale enterprises in creating and maintaining infrastructure assets can facilitate the economic empowerment of marginalised groups in communities and micro- or small enterprises. Thus (as discussed above, and shown in Table 2.1), the process of constructing assets can be just as important to the communities and the economy as the provision of the assets themselves. Wells (2013, p. 8) noted:

> It may be concluded that planning and implementing an employment intensive construction programme is not an easy option. It requires a long time horizon, a lot of capacity building, a guaranteed and regular flow of funds and serious commitment on the part of all involved. But when all of these are present, the evidence strongly suggests that the adoption of employment-intensive approaches can generate additional employment opportunities for construction workers. If not spread too thinly these opportunities can make a significant impact on chronic poverty amongst both rural and urban populations.

Impact of construction on employment in other sectors

The activities which construction generates in the rest of the economy begin with the stimulation of its supply chain and have a wide influence on the economy and the community. Spence et al. (1993) noted that the role of the construction industry, building materials producers and the informal sector in the building of housing creates both wealth and income. Housing and community building, such as in urban regeneration through the improvement of informal settlements, can create employment in the production of building materials which has multiplier effects on the local economy. MIT (undated) outlined strategies that promote the construction and materials supply sector toward local income enhancement and broader development, offering proposals to governments and aid agencies. It suggests that residents in the informal sector are best served by the formal private sector, with a materials and tools "supermarket" as an interface, an approach that worked well in Brazil. Crowley and Goethert (1995) outlined these reasons why focusing on the improvement of informal sector housing positively impacts urban development: consolidation of dwellings with better materials and construction improves family amenities and the cityscape; such consolidation allows the addition of rental rooms which increases family income, reduces urban sprawl, mitigates squatter development and reduces inner-city overcrowding; there is increased employment for low-income families; and there are increased tax opportunities for cities from the construction and the enhanced property values.

In a report on housing in Ghana, UNHSP (2011) noted that for every dollar in wages in construction, another dollar is earned as the workers spend the money locally. Moreover, for every job created in construction, another one is created in the industries providing materials, transportation, vehicle spare parts and so on. The completed building generates more jobs in internal decoration, fitting out, furnishing, and repair, maintenance and renovation. The multiplier effect differs among types of items constructed, being higher in house building than large civil engineering projects. It was estimated that if 200,000 dwellings are built in Ghana annually, this would generate between 75,000 and 96,000 jobs in construction; and the backward linkages of construction would be twice the direct construction jobs, and would mean an additional 150,000 to 200,000 jobs (UNHSP, 2011, p. xxv).

Thus, investment in development projects of all types increases employment in construction and also in the rest of the economy, starting in the industry's supply chain such as materials manufacture and supply. The next section considers the nature, challenges and prospects of the construction workforce.

The construction workforce

Nature of the construction workforce

Considering the unique, strategic, indispensable role that construction and the built environment play in the economy, in long-term national development and in the enhancement of the quality of life for all people in all countries, the construction worker is a national asset. They should be enabled to grow, enhance their skills, and improve their performance. ICED Facility (undated) estimated that the construction, renovation, maintenance and demolition of buildings and civil engineering projects accounted for over 273 million (part-time and full-time) jobs worldwide in 2014 (about 8.6 percent of total global employment). An average of about 9.5 percent were female. Mella and Savage (2017) put the number of workers engaged in the production of the built environment around the world at about 330 million (formal and informal). This figure increases if those working in the production and supply of construction materials and equipment and others undertaking other activities in the value chain and those providing services to construction companies such as accountants and lawyers, are taken into account.

Workers in the construction industry range in expertise and responsibilities from professionals of various backgrounds who plan and design, and manage the execution of the works on site; through the technicians who assist the professionals and also form part of the design and site management teams; the supervisors of work on site; the tradespersons who undertake the work on site and their apprentices; and the unskilled labourers on site. Thus, the construction labour force is a complex array of specialities. It should be noted that the distribution of professions and specialisations differs among countries. For example, the Building Surveyor and Quantity Surveyor only exist in the United Kingdom and some Commonwealth countries. The size, profile and quality of this workforce determine the capacity and capability, as well as the performance of the construction industry in any country. Figure 2.1 presents the various

Direct employment in construction	*Other sub-sectors in construction value chain*	*Organisations in construction firms' operating environment*	*Users of constructed items*
Professionals 1. Architects 2. Planners 3. Engineers ... of many types (eg. Structural, Mechanical and Electrical) 4. Surveyors (Land, Building, Quantity) 5. other specialists 5a. in types of buildings, eg. airports, hospitals 5b. in aspects of buildings, eg. acoustics	**Materials producers** 1. producing for only construction industry, eg. cement 2. producing for construction and other uses, eg. steel, glass 3. producing items for general use (including in construction) *Prefabrication establishments* 1. producing components (eg. beams, slabs) 2. making volumetric units (eg. bathrooms) ...can replace site work; being pursued in some countries, eg. Singapore, Malaysia	**Public sector** 1. local authorities applying planning regulations and building control 2. industry development agency 3. industry regulatory agencies 4. procuring entities 5. civil defence organisations, eg. fire service	**Productive facilities (private)** 1. industrial buildings 2. offices 3. retail and wholesale 4. hotels 5. food and beverage businesses 6. leisure and entertainment ...leading to additional production in these premises
Other professionals ... 1. Landscape Designers 2. Interior Designers 3. Facilities Managers 4. Valuers and Investment Advisors		**Private sector** 1. clients, initiating buildings 2. end purchasers of buildings 3. users and tenants of buildings	**Social facilities** 1. educational buildings 2. health care facilities 3. community, recreation centres 4. religious buildings ...enhancing the quality of life of citizens
Technicians ... of various professional areas, eg. architectural, engineering, surveying ...becoming fewer as degree programmes are upgraded, but pyramid structure required	**Producers of components and installations for buildings** 1. ventilation and air-conditioning 2. vertical transportation equipment 3. laboratory equipment 4. production machinery	**Public-Private Partnerships (PPP)** ...for commissioning major projects ...for industry development	**Infrastructure** 1. utilities: power, water, gas ...production and distribution; telecommunication; digital networks 2. distributive: roads, highways, bridges; railways, airports, harbours 3. part of production, such as, eg. irrigation systems
Supervisors ... in the various trades ...increasingly being formalised	**Suppliers of materials and components** ...differ by items they specialise in, and services they provide (discounts; loan periods and so on)	**Financial institutions** 1. loans for contractors (with high expense of construction and contractor being paid in arrears, a key area) 2. loans for developers 3. mortgages, loans for end purchasers ... built items can be used as collateral to raise funds for more investments	**Interior fitting and installation companies** a. hospital and laboratory installations b. hotel, shop and leisure facility fitouts c. kitchen assembly
Tradespersons 1. Brick- and Block-Layers 2. Carpenters and Joiners 3. Electricians 4. Equipment Operators (of many specialisations) 5. Plasterers and Tilers 6. Plumbers 7 Steel Benders and Fixers; and Structural Steel Workers	**Plant-hire companies** 1. providing only specialist plant, equipment 2. providing a whole range of equipment ...increasingly challenging owing to enhancing performance requirements, eg. energy; changing technology	**Insurance companies** 1. contractors' all-risk insurance 2. insurance for workers 3. insurance of the built asset **Accountants** 1. corporate accounts 2. leading novel projects such as PPP projects	**Facilities management of constructed items** ...whole range of skills in construction and other fields
Unskilled workers ...importance of location specificity, and use of construction to provide jobs, 1. possibility of diversity, eg. women workers in India 2. possibility of additional earnings for non-construction personnel	**Haulage and transportation companies** **Manufacturers of equipment and tools**	**Lawyers** 1. drafting bespoke contracts 2. dispute settlement and arbitration **Entities providing services to construction workers on sites** eg. food and beverages; dormitories; transportation; employment agencies	
Sectors 'patronised' by construction workers			
Entire economy, including construction. Construction workers at all levels spend their earnings, enabling further employment in all parts of the economy. ...all workers in the economy, including those in construction, invest in construction ...companies in the rest of the economy expand their facilities or build new items. Cyclical nature of construction activity as demand is influenced by developments and prospects in the other sectors also has an impact on construction workers' role in the economy.			

Figure 2.1 Range of competences in construction, importance of construction in job creation and parallel competences which construction relies on.

roles in construction and other jobs in related sectors of the economy. The figure illustrates the employment-generating potential of the construction industry.

Construction workforce around the world

Profile of construction workforce and its challenges

Aspects of the profile of the construction workforce are discussed in various chapters of this book, including Chapters 5, 7, 8 and 12. ICED Facility (undated) noted that construction is often considered an "easy entry" industry because it needs large numbers of non-skilled and semi-skilled workers. Wells (2013) noted that employment in the construction industry is often seen as a way out of poverty for those with little education or skill; it provides a point of entry into the urban labour market for internal migrant workers. These observations are reiterated in studies on the construction workforce in India (Johari and Jha, 2020). Data from Ghana's labour force survey 2017 indicate that 316,368 persons were employed in construction, forming 3.4 percent of the total number of persons 15 years and older who were employed. Some 18,832, or 6.0 percent were female (Ghana Statistical Service, 2017). There was a very high level of informality in construction. Of the total of 316,368 workers employed in the industry, 309,132 (98 percent) were in the "informal sector"; all the females employed in construction were with informal sector organisations.

ICED Facility (undated) noted that in construction industries in developing countries, wages are low (often due to the oversupply of informal labour); working hours may be long and workers are paid at piece rates; and health and safety conditions are poor. The lack of social protection benefits such as pension schemes, maternity leave, and unemployment benefits is often correlated with poverty and vulnerability. Roy and Koehn (2006) noted that in developing countries, whereas labour is available, they generally lack the requisite skills and technical knowledge. The workers are less aware of their rights and the benefits available to them, including social security, health insurance, safety, and the responsibility of the contractors to pay minimum wages and maintain a safe working environment. This may lead to exploitation by the contractor and, consequently, poor relations with the workers. In addition to the shortage of semi-skilled and skilled labour, construction companies also face shortages of construction materials, lack of capital, and poor people management skills. Workers are engaged on a project basis, and due to this instability; contractor-employee relationships are poor. Boadu et al. (2020) outlined these features of construction labour in developing countries: reliance of firms on a temporary and casual labour force, and as labour is abundant and cheap, contractors can easily dismiss employees deemed to be performing unsatisfactorily and replace them with others (thus, workers accept work in high-risk situations without demanding that employers provide appropriate safety measures); inadequate supply of skilled, educated, trained and experienced workforce (World Bank Skills, 2010), attributed to the limited capacity of technical and vocational training institutes in terms of class sizes, workshops, teaching and administrative personnel and currency of the curriculum. Johari and Jha (2020) noted that 50 million people were engaged in the Indian construction industry. Some 8 percent of India's 1.3 billion workforce is involved in jobs like carpentry, masonry and plumbing and possess knowledge and experience of the trades through non-formal vocational education and training. Only 3–4 percent of the workforce have a formal VET certificate mainly owing to the non-availability of appropriate skill-training facilities.

Box 2.1 presents a profile of the United States construction workforce by McAnaw Gallagher (2022) showing the importance, size, mix and dynamism of the workforce, and some of the challenges.

Box 2.1 Profile of the construction workforce in the United States

Total employment in construction in the United States peaked in 2007 at 11.9 million, before falling to 9.1 million in 2010 in the wake of the Great Recession (2007 to 2009). Employment in the industry grew after 2012, reaching 11.4 million in 2019. In 2020, as a result of the COVID-19 pandemic, employment dropped to 10.8 million. In 2020, 10 percent of construction workers were women.

In 2020, 30.0 percent of construction workers were Hispanic, higher than their 17.6 percent share of the total employed. Other racial minorities were underrepresented; non-Hispanic Blacks held 5.1 percent of construction jobs, compared with their 11.8 percent share of total employment. Non-Hispanic Asians' share of construction employment was 1.8 percent, compared with 6.2 percent of total employment.

Hispanics accounted for 46.7 percent of construction labourers and 52.5 percent of painters and paperhangers, but (higher than their share of those employed in the industry); but were relatively underrepresented among construction managers (14.3 percent) and other managers (14.4 percent). Non-Hispanic Whites accounted for 60.9 percent of all those employed in the construction industry, with higher shares among construction managers (78.9 percent) and other managers (78.5 percent). By contrast, 44.1 percent of construction labourers and 42.2 percent of painters and paperhangers were non-Hispanic White.

People working in the construction industry were more likely than the employed overall to be foreign-born. In 2020, 25.3 percent of construction workers were foreign-born, higher than the 16.8 percent for the total employed. Over two-thirds of Hispanics in the construction industry were foreign-born, while less than half (45.2 percent) of all employed Hispanics were foreign-born.

Among full-time wage and salary workers, the native-born earned more than the foreign-born, both for the construction industry and for the nation overall ($1,031 versus $786; and $1,000 versus $885, respectively). The differences in earnings reflect many factors, including variations in the distributions of foreign-born and native-born workers by educational attainment, occupation, industry, and geographic region.

Between 2003 and 2020, the percentage of construction workers aged 55 years and over nearly doubled, from 11.5 percent to 22.7 percent (also reflecting the aging of the population; the share of those aged 55 years and over increased from 15.4 percent to 23.9 percent over the same period).

In 2020, 56.4 percent of construction workers had a high school diploma or had not completed high school; nearly double the percentage for the employed overall (30.2 percent). Construction workers were less likely than the employed overall to have a bachelor's degree or higher (17.9 percent versus 44.1 percent). Some 38.4 percent of the

Hispanics in construction had less than a high school education (much higher than for non-Hispanic Asians, non-Hispanic Blacks, and non-Hispanic Whites). By contrast, 46.8 percent of non-Hispanic Asian construction workers had a bachelor's degree or higher (more than double the rate for all other race and ethnicity groups). (These data include only workers aged 25 or older.)

The number of self-employed construction workers peaked at 2.9 million in 2007, had dropped to 2.3 million by 2012 and in 2020, 2.5 million construction workers were self-employed (representing 23.3 percent of total employment in the industry and more than twice the rate of workers in all industries).

Source: McAnaw Gallagher (2022).

As discussed in Chapter 8 of this book, in general, construction work does not have a good image. Construction 21 Steering Committee (1999) highlighted how people in Singapore consider it to be demanding, dirty, dangerous and low paying. Similarly, Wells (2013) noted that construction work is dirty, difficult, dangerous, irregular, badly paid and not decent. It contains more than its fair share of working poor. The reasons include the increasing casualisation of labour in construction. She outlined the challenges and inadequacies of even policies that have been relatively successful in addressing the issue of chronic poverty among construction workers. In the United Kingdom, Construction Industry Training Board (CITB) (Wiseman et al., 2018) found that the appeal of construction as a career option had fallen to 4 out of 10 among 14- to 19-year-olds, although more firms were sponsoring apprentices, due to an image problem of what it is like to work in construction amongst younger people. HM Government (2018) puts the perceived low image down to lack of gender diversity, long hours, low pay and little job security. Health and Safety Executive (HSE) (2022) data show that construction leads with the highest number of health and safety-related accidents in the United Kingdom; the fatal injury rate of 1.63 per 100,000 workers is about 4 times the whole-industry rate. The reasons for this include lack of equipment compliance, training and safety processes and procedures. Australia Constructors Association (2022) presented this profile of the country's construction industry: in the eyes of the next generation of workers, construction is stuck in the past. More businesses fail in construction than any other industry and, as an industry that operates on thin margins, its workers are under pressure. They work long hours, suffer high stress levels and are six times more likely to die from suicide than a workplace incident. Diversity is low and women make up only 12 percent of the workforce. In a survey, less than one-third of Gen Z respondents said they would consider a built environment career. The industry must transform to, at least, maintain a workforce large enough to deliver the substantial pipeline of work.

ManpowerGroup (2020) identified these five challenges facing the construction industry in the United Kingdom: skills shortages; an ageing workforce; health and safety issues; lack of gender diversity in the workforce; issues with technology adoption; and low worker productivity. Fahy (2023) presented a summary of the current situation in the United Kingdom construction workforce. It is estimated that the industry needs to recruit an additional 45,000 workers per year, over and above the 190,000 required to replace those leaving an industry (which has an annual churn rate of 8 percent). It is also estimated by the Federation of Master Builders (2023) that the shortage, evident since 2004, is causing 64 percent of projects to be delayed, and hindering the industry's ability to grow. Total employment numbers in the construction industry have dropped by 11 percent since the start of 2019, with an 8 percent decline in the number of foreign workers. Over one-fifth of the construction workforce is now over 55 years old, compared with 11 percent in 2003. Another labour challenge is that, mainly owing to changes to off-payroll working rules, the number of self-employed workers has fallen by 23 percent since 2019, declining from 41 percent to 35 percent of the total.

Bateman et al. (2023) noted that skills shortages have hampered the productivity of the construction industry in North America and the United Kingdom for many years. This crisis has been exacerbated by the COVID-19 pandemic, with tens of thousands of workers seeking early retirement. There is also a mental health crisis in the industry. In the United Kingdom, one in four individuals has suffered from poor mental health, and the crisis is particularly acute in construction (McHugh, 2018). Some 25 percent of workers in construction contemplated suicide, and more than 1,400 construction workers died this way between 2011 and 2015. Reasons given for poor mental health include: lack of job security; long, unsociable hours; time away from family; and lack of employer support. Initiatives which are underway to tackle the problem are considered to be slow and piecemeal (Ward, 2022). Similarly, Bateman et al. (2023) observed that, in North America, construction faces a mental health problem. In Canada, construction workers are five times more likely to die by suicide than the average worker. Addiction issues (and prescription drugs related to anxiety, depression and sleeplessness) also affect many of them, leading to high overdose rates. In North America, with housing affordability challenges, rising immigration and an increasing number of infrastructure projects, shortages will be a challenge for several years. In the United States, there were 450,000 construction job openings in the last quarter of 2022, the highest level ever recorded. The $550 billion Infrastructure Investment and Jobs Act (Library of Congress, 2021) is expected to create 300,000 new construction jobs in 2027 and 2028. Similarly, construction in Canada reported record volumes of activity in 2022; construction employment reached a peak of 1.6 million in July 2022; its unemployment rate is an around historically low 3.4 percent. In Canada, 21 percent of the construction workforce is aged 55 years and older, an 8 percent increase since 2002; between 2023 and 2027, around 156,000 construction workers are expected to retire and another 16,000 will be needed to respond to rising demands, such as housebuilding, to achieve the government's goal of doubling new home construction over the next 10 years, the transition to a greener economy that will require retrofits to homes and businesses, and the electrification transition.

Changing attitudes to construction work

Some studies find the construction worker to be among the most satisfied of all the workers in the various sectors. For example, in the United Kingdom, CITB (2022) noted that the industry is described by those inside it as a place which offers generous pay, opportunities for progression and a varied working environment. They also reported the construction industry to be diverse, flexible and aligned to their values, offering the ability to make a positive impact, including its contribution to the Net Zero drive, the opportunity to specialise and progress one's career through training, such as apprenticeships, and job satisfaction from creating something – making a difference to other people's lives such as by building homes. However, many outside the industry are unaware that construction can offer them these, and there is a need to sell construction better.

Studying factors affecting the job satisfaction of construction industry employees in Ghana, Anin et al. (2015) found that non-wage-based factors such as recognition, the task itself, the work environment, supervision and job security appeared to influence job satisfaction more than the wages paid. They suggest that managers in the construction supply chain should formulate policies incorporating factors which allow employees to achieve their high-order individual goals to secure their loyalty and improve productivity. Alzubi et al. (2023) sought to determine the most significant factors affecting the overall job satisfaction of construction engineers in Jordan and found that the working environment, satisfaction with co-workers, pay and benefits, and satisfaction with supervision were the most significant factors affecting the overall job satisfaction of construction engineers. As discussed in Chapter 11 of this book, the construction industry is known to have a masculine culture. Lingard and Lin (2004) studied the career, family and work environments determining the organisational commitment of females working in the Australian construction industry and found that career choice management, career progression satisfaction and career/job commitment had predictive effects on the organisational commitment of the female employees.

On to the wider level, the construction industry has an unenviable reputation in most countries owing the poor performance on construction projects, in terms of cost, time, health and safety and productivity (see, for example, Ssegawa and Ngowi, 2009; Ribeirinho et al., 2020). Much of this unsatisfactory performance is blamed on inadequacies in the construction labour, including low skills, poor training and poor attitudes (Farmer, 2016). Dabirian et al. (2019) asserted that the success of projects is correlated with good human resource management. They add that correctly selecting talent for a project is a determinant of its successful conclusion and suitable performance. Romo et al. (2023) found that a firm's human resources have an influence on its competitiveness owing to the impact that labour has on project execution. They note that the human resource management can develop factors that are difficult to imitate from company to company, for example, values, which cannot be copied or distributed, but rather become implicit in workers and contribute to teamwork, motivation and commitment to the company. However, in construction, the practice of human resource development is centralised, not in the power of the project and site management (Druker and White, 1995). (This issue is discussed in Chapter 9 of this book.) With the prevalence of sub-contracting and self-employment in construction in many countries, as discussed in Chapter 3, employers do not invest in human resource development. Government steps in, in some countries, with a training levy and incentives (Debrah and Ofori, 2001). As noted by Boadu et al. (2020) above, technical and vocational education and training (TVET) in construction (discussed in Chapter 5 of this book) is not effective in many countries. For example, in the United Kingdom, Farmer (2016) referred to a dysfunctional training funding and delivery model. Special programmes are implemented in various countries to realise TVET (Debrah and Ofori, 2001; Ofori, 2017). Australia Constructors Association (2022) noted:

> Government, industry and unions must … collaborate … If they do, construction could be an industry where both projects and workers are free and able to productively work the hours that best suit them. Construction could be an industry of choice and equal opportunity for all genders, nationalities and ages. It could be an industry that constructs resilient infrastructure without damaging the environment. It could be an industry at the forefront of technological advancement. Construction could be a

> profitable industry that rewards collaboration over conflict and innovation over status quo and an industry prepared to take risks on new ideas... Australia's construction industry could be … delivering high value infrastructure at a greatly reduced cost, for the benefit of all Australians. What are we waiting for? (p. 8)

Implications for labour in construction and in other sectors

Stressing the strategic importance of construction labour

The construction industry and its workers are of strategic importance to every country. In Australia, construction contributes 7 percent of GDP and employs almost 1 in 10 of the working population (Australia Constructors Association, 2022). It is "the industry called upon to rebuild Australia's economy following the COVID pandemic. It is vital to the health of the economy and, importantly, the quality-of-life Australians enjoy" (Australia Constructors Association, 2022, p. xx). FIEC (2023) noted that "the construction sector is a fundamental component of Europe's economic growth and a major source of employment. It generates about 10% of gross domestic product (GDP) in the EU and provides almost 13 million direct jobs". FIEC (2023) noted:

> The sustainable Europe of tomorrow cannot be achieved without the direct involvement of the construction industry. In fact, the construction industry is at the heart of our life … . Without the construction industry the EU cannot respond to its main challenges: competitiveness, youth unemployment, digital economy, urban regeneration, energy efficiency and energy poverty, circular economy, affordable housing, climate change, mobility and connected infrastructure, etc.

Considering technology

It has been suggested that technology is the future of construction (Goetjen et al., 2016). The construction industry has been shedding its image as a technology laggard in the last decade or so. It is necessary to briefly discuss this technological shift and consider its implications for labour. Sawhney et al. (2020) propose Construction 4.0 (based on the Fourth Industrial Revolution (Industry 4.0 or 4IR)), a framework based on a combination of trends and technologies likely to reshape the way built environment assets are designed, constructed, and operated. It covers these themes: Industrial production (prefabrication, 3D printing and assembly, offsite manufacture); Cyber-physical systems (actuators, sensors, IoT, robots, cobots, drones); and Digital and computing technologies (BIM, video and laser scanning, artificial intelligence and cloud computing, big data and data analytics, reality capture, blockchain, simulation, augmented reality, data standards and interoperability, and vertical and horizontal integration).

The rate of technology adoption differs among countries. Moreover, countries are taking different paths. Some governments are leading the use of the new technologies, such as adopting it on their own projects (such as the United Kingdom; HM Government, 2012); making its usage mandatory (such as Singapore; *Build Smart,* 2013); or providing guidance and incentives. Gayne (2023) notes that, in the United Kingdom, much money has been spent on modern methods of construction (MMC) since a report (Farmer, 2016) admonished the construction industry to "Modernise or Die". According to Make UK, a total of £1 billion of private finance has been invested in research and development and new modular factories since 2016, bringing total United Kingdom capacity to 5,000 units per year in 2022; and Homes England's affordable homes programme requires developers signing strategic partnership deals to ensure 25 percent of homes are built using MMC. Whereas technological innovation is proceeding relatively fast, construction is still behind other industries; its productivity levels lag behind every other sector in the United Kingdom (Plangrid, 2018). Government initiatives such as the Infrastructure Transformation Programme (IPA, 2017) are helping the industry to adopt new technologies and increase productivity. However, a report from McKinsey Global Institute identified construction as the second least digitalised sector in the world (Gandhi et al., 2016).

The implications of the new technology orientation in construction are different for the various groups of countries. For the industrialised countries, it is suggested that the new range of advanced technologies provides the opportunity to partially address the shortages of labour, rising wages and the poor productivity (Katara, 2022; Bateman et al., 2023). Developing countries have to balance their need for technological advancement with construction's employment generation possibilities. It is worth mentioning that the construction industry is a matrix of segments with different inputs (Ofori, 1989). Some types of projects lend themselves well to labour-intensive technologies, which might involve some high-technology inputs. For example, the ILO developed technologies and tools to enable roads to be constructed in rural areas and for municipal works and upgrading of low-income settlements. Thus, it is necessary to develop imaginative packages and programmes for application of technologies to foster growth and development while optimising the benefits of technological advancement in construction.

With each new technology, it is necessary to have the personnel who are trained in using it. McKinsey & Company (2022) pointed out that technology is only half of the Industry 4.0 equation; workers should be properly equipped through upskilling and reskilling. Similarly, Mansour et al. (2023) measured the strategic readiness of Malaysian construction firms in Industry 4.0 implementation. The results indicated that the human capital component is the most critical factor affecting the success of Industry 4.0 implementation, specifically, the intellectual agility, knowledge, skills, and competencies of the construction professionals.

Decent work

If there is an industry in which a focus on green and decent jobs is needed, it is construction. ILO (2008) defined decent work as "productive work for women and men in conditions of freedom, equity, security and human dignity" (p. vi). It noted that "decent work involves opportunities for work that: is productive and delivers a fair income; provides security in the workplace and social protection for workers and their families; offers prospects for personal development and encourages social integration; gives people the freedom to express their concerns, to organize and to participate in decisions that affect their lives; and guarantees equal opportunities and equal treatment for all" (p. vi).

Most construction workers in the developing countries will never be able to afford a built unit of their own. In many countries, they are also effectively excluded from using many of their products. Thus, the inequalities in the societies are mirrored in the construction industry. In construction, there is a divide, between the owner and the producer; between those who build and those who own or (in many cases) use. A key question is how to close this gap. An example of such an effort is the Artisans Association of Ghana, a non-profit organisation, which provides mentorship, and entrepreneurship and skills training to artisans, especially females in male-dominated trades and marginalised youth. In an interview with this author in 2017, the CEO disclosed that the association was formed to address the situation he knew from his personal experience, that no artisan in construction he had met or knew had become well-to-do. Many had been contractors, subcontractors or self-employed workers, building many houses, but almost all of them remained 'renters' of houses (Ofori, 2017). Addressing the question of how the construction worker can become a consumer of what the industry produces will require a range of measures including decent and especially well-paying jobs in construction; opportunities for training, upskilling and development; social protection for the workers; and access to funding for purchasing housing.

The future

Bateman et al. (2023) suggest that the construction industry must consider new approaches to resolve its labour challenges: step up efforts to recruit from underrepresented groups: women, Indigenous People, and newcomers to Canada; and note that part of the answer to shortages in North America will come from the increased adoption of digital technologies (BIM, IoT devices, robotics, immersive collaboration and drones), and expanded adoption of prefabrication, modular construction and digital twin capabilities. Outram et al. (2022) noted increased demand for construction across the globe. To capitalise on this, construction companies will have to invest more in people, processes and culture initiatives in order to acquire, retain, develop and motivate the talent they need. The top five human resource priorities in 2022 were: Redesigning work to improve agility; Improving total rewards packages; Putting environmental, social and governance (ESG) and sustainability at the heart of the transformation agenda; Investing in workforce upskilling/reskilling; and Redesigning HR operations. Companies leading the way were: constantly resetting for relevance, seeking to work in partnership with talent, working hard to deliver on total well-being and building for employability to harness the collective energy of their people.

Concluding remarks

Research agenda

It is useful to understand what the construction industry includes, in terms of the activities so that plans to develop or recruit more workers are appropriately informed. There is a need for better information on the construction workforce. While the casualisation of labour in construction will make this difficult, it is necessary to obtain a good picture of the challenge in its full complexity in order to address it.

Construction workers stimulate activities in the rest of the economy from their spending. It would be instructive to isolate or simulate this spending and its patterns to determine the real impact of investment in construction. In their spending, the construction workers fuel the other sectors which then leads to the derived demand for construction, continuing the circular flow. Similarly, workers in some of the other sectors express effective demand for and acquire built items. So, there is the dynamic symbiosis. The links between construction and the built environment need to be established and the synergies between the two segments explored.

With the introduction of advanced technology, the redundant workers and specialist companies such as subcontractors, will need to be retrained and repurposed. The design of multi-trade training courses will also have to be rethought as combinations of some skills might be lost. The path technology is taking in construction will enable construction to have a different, more progressive image. It is relevant to consider how technology can help address shortages while enabling workers to upgrade their skills to be better equipped to work with it.

The adoption of advanced technology in combination with global concern about environmental issues will improve the image of the construction industry and make it attractive to talent. It is necessary to explore how construction can do a better job in marketing the opportunities.

In construction, the changes in technology and materials imply a need to enhance the skills of the workforce; as well as to create some trades which are new in the countries such as those required for installing complex components; operating specialised construction equipment; and facilities management of premises with advanced equipment. Developing competence in changing technologies in educational and training programmes will require planning and continuous monitoring.

Conclusion

The workers in the construction industry who create the built environment contribute greatly to society, i.e., beyond their counterparts, labour in other sectors, and the community directly from the physical facilities they produce and indirectly from the wider benefits the items bring including education and training, health and well-being. The facilities should deliberately be designed and delivered to meet these benefits.

How to help the construction industry to consider how it attracts, trains, deploys, continuously develops and fosters the welfare of its future workforce is an uphill task owing to the industry's poor social image. It is necessary to address the incorrect stereotypes of construction work. It is key to celebrate what is good about it and to enhance the knowledge and understanding of those outside of the industry. It should build their enthusiasm and make construction the career of choice for more potential workers. Some of the messages are that the built environment provides services directly to all workers and their families and community, and improves their quality of life. The construction worker also stands to benefit from a well-performing construction industry and constructed items, including the direct work of the worker.

Finally, given its impact,maximising the potential of construction should be a key consideration in national development plans, and those for cities and regions. Developing its workers to operate to their potential should feature prominently in such plans.

References

Alzubi, K.M., Alkhateeb, A.M. and Hiyassat, M.A. (2023) Factors affecting the job satisfaction of construction engineers: evidence from Jordan. *International Journal of Construction Management*, 23(2):319–328. 10.1080/15623599.2020.1867945

Anderson, W.P., Lakshmanan, T.R. and Kuhl, B. (2001) Estimating employment generation by federal-aid highway construction projects. *Transportation Research Record*, 1777(1):93–104. 10.3141/1777-10

Anin, E.K., Ofori, I. and Okyere, S. (2015) Factors affecting job satisfaction of employees in the construction supply chain in the Ashanti Region of Ghana. *European Journal of Business and Management*, 7(6):72–91. https://citeseerx.ist.psu.edu/document?repid=rep1&type=pdf&doi=3036b156d16f730ddcbe736fbae309817adc3327

Australia Constructors Association (2022) *Disrupt or die: Transforming Australia's construction industry*. Sydney.

Ball, M. and Wood, A. (1995) How many jobs does construction expenditure generate? *Construction Management and Economics*, 13(4):307–318. 10.1080/01446199500000036

Ball, R. (1981). Employment created by construction expenditures. *Monthly Labor Review*, 104(12):38–44. http://www.jstor.org/stable/41841400

Bateman, J., Bray, M., Ferreira, B., Gabert, M. and Wilson, R. (2023) What key issues does construction face in 2023? *Construction Journal*, 12 January, ww3.rics.org/uk/en/journals/construction-journal/construction-2023-trends-predictions.html

Bekchanov, M., Wijayasundara, M. and de Alwis, A. (2022) Circular economy—A treasure trove of opportunities for enhancing resource efficiency and reducing greenhouse gas emissions. In M. Asif (Ed.) *Handbook of energy and environmental security*. Academic Press, pp. 481–499. 10.1016/B978-0-12-824084-7.00016-3

Bentall, P., Beusch, A. and de Veen, J. (1999) Employment-Intensive Infrastructure Programmes: Capacity Building for Contracting in the Construction Sector – Guidelines. International Labour Office, Geneva.

Best, D.L. and Erickson, E. (2023) Social drivers of health. In B. Halpern-Felsher (Ed.) *Encyclopedia of child and adolescent health*. Academic Press, pp. 366–377. 10.1016/B978-0-12-818872-9.00116-3.

Boakye-Gyasi, K. and Li, Y. (2015) The impact of Chinese FDI on employment generation in the building and construction sector in Ghana. *Eurasian Journal of Social Sciences*, 3(2):115.

Boadu, E.F., Wang, C.C. and Sunindijo, R.Y. (2020) Characteristics of the construction industry in developing countries and its implications for health and safety: an exploratory study in Ghana. *International Journal of Environmental Research and Public Health*. Jun 9;17(11):4110. 10.3390/ijerph17114110.

Build Smart (2013) BIM – The way forward. Singapore, https://www.bca.gov.sg/Publications/BuildSmart/others/buildsmart_13issue18.pdf

Carassus, J. (2000) A mesoeconomic analysis of the construction sector: applied to the French construction industry. CIB W55-W65 Joint Meeting Proceedings. The University of Reading. http://www.construct.rdg.ac.uk/bon/2000/index.htm

Carassus, J., Andersson, N., Kaklauskas, A., Lopes, J., Manseau, A., Ruddock, L. and de Valence, G. (2006) Moving from production to services: a built environment cluster framework. *International Journal of Strategic Property Management*, 19:169–184.

Çelik, G.T. and Oral, E.L. (2019) Mediating effect of job satisfaction on the organizational commitment of civil engineers and architects. *International Journal of Construction Management*, 21(10):969–986. 10.1080/15623599.2019.1602578

Condihoto, R., Tzortzopoulos, P., Kagioglu, M., Aouad, G. and Cooper, R. (2009) The impacts of the built environment on health outcomes. *Facilities*, 27(3/4):138–151. 10.1108/02632770910933152

CITB (2022) Construction needs to rethink how it attracts its workforce as war for talent hots up, https://www.citb.co.uk/about-citb/news-events-and-blogs/construction-needs-to-rethink-how-it-attracts-its-workforce-as-war-for-talent-hots-up/

Croucher, K., Wallace A. and Duffy, S. (2012) *The influence of land use mix, density and urban design on health: a critical literature review*. The University of York, York.

Crowley, J. and Goethert, R. (1995) The hidden consumer: a new paradigm using materials distribution and manufacturers as a bridge to improved housing in emerging markets. Research Report, MIT Department of Architecture.

Cui, Q., Gong, P., Yang, G., Zhang, S., Huang, Y., Shen, S., Wei, B. and Chen, Y. (2023) Women-oriented evaluation of perceived safety of walking routes between home and mass transit: a case study and methodology test in Guangzhou. *Buildings*, 13(3):715. 10.3390/buildings13030715

Dabirian, S., Abbaspour, S., Khanzadi, M. and Ahmadi, M. (2019) Dynamic modelling of human resource allocation in construction projects. *International Journal of Construction Management*, 22(2):182–191. 10.1080/15623599.2019.1616411

Debrah, Y.A. and Ofori, G. (2001) The State, skill formation and productivity enhancement in the construction industry: the case of Singapore. *The International Journal of Human Resource Management*, 12(2):184–202. 10.1080/713769601
de Valence, G. (2019) Accounting for the built environment. In R. Best and J. Meikle (Eds.) *Accounting for construction: frameworks, productivity, cost and performance*. Routledge, pp. 14–31.
Druker, J. and White, G. (1995) Misunderstood and undervalued? Personnel management in construction. *Human Resource Management Journal*, 5(3):77–91.
Elsevier (2023) Built environment, https://www.sciencedirect.com/topics/engineering/built-environment#:~:text=The%20term%20built%20environment%20refers,water%20supply%20or%20energy%20networks.
Ernst, C. and Marianela Sarabia, M. (2015) *The role of construction as an employment provider: a world-wide input-output analysis*, Employment Working Paper No. 186. International Labour Office, Geneva.
European Construction Industry Federation (2023) Construction is the solution industry, https://www.fiec.eu/construction-industry/solution-industry
Fahy, M. (2023) Construction industry calls for overseas labour help. *Investors' Chronicle.* March 26, https://www.investorschronicle.co.uk/news/2023/03/16/construction-industry-calls-for-overseas-labour-help/
Farmer, M. (2016) *The farmer review of the UK construction labour model – modernise or die: time to decide the industry's future*. Cast Consultancy, https://www.cast-consultancy.com/wp-content/uploads/2021/03/Farmer-Review-1-1.pdf
Federation of Master Builders (2023) *State of trade survey Q4 2022*. London. https://www.fmb.org.uk/resource/state-of-trade-survey-q4-2022.html
Fu, Y., Wang, H., Sun, W. and Zhang, X. (2021) New dimension to green buildings: turning green into occupant well-being. *Buildings*, 11:534. 10.3390/ buildings11110534
Gandhi, P., Khanna, S. and Ramaswamy, S. (2016) Which industries are the most digital. *Harvard Business Review*, April 1, https://hbr.org/2016/04/a-chart-that-shows-which-industries-are-the-most-digital-and-why
Gayne, D. (2023) Modular without the factory: modulous' plan to turn offsite building on its head. *Building*, 16 January, https://www.building.co.uk/building-the-future-commission/modular-without-the-factory-modulous-plan-to-turn-offsite-building-on-its-head/5121141.article
Ghana Statistical Service (2017) *Ghana's labour force survey 2017*. Accra.
Glasgow Centre for Population Health (2013) The built environment and health: an evidence review, Briefing Paper 11, Concepts Series. Glasgow, https://www.gcph.co.uk/assets/0000/4174/BP_11_-_Built_environment_and_health_-_updated.pdf
Goetjen, H.K., Dann, C. and Van den Berg, J. (2016) Engineering and construction industry trends – E&C companies need to break out of the commoditization trap. PwC, https://www.strategyand.pwc.com/media/file/2016-Engineering-and-Construction-Trends.pdf
Groak, S. (1994) Is construction an industry? *Construction Management and Economics*, 12(4), 287–293. 10.1080/01446199400000038
Habash, R. (2022) Urbanization as an intelligent system. In R. Habash (Ed.) *Woodhead publishing series in civil and structural engineering, sustainability and health in intelligent buildings*. Woodhead Publishing, pp. 239–257, 10.1016/B978-0-323-98826-1.00009-0.
Hampson, K. and Brandon, P. (2004) *Construction 2020: a vision for Australia's property and construction industries*. Cooperative Research Centre for Construction Innovation, Brisbane.
Health and Safety Executive (HSE) (2022) *Construction statistics in Great Britain, 2022*. Bootle, https://www.hse.gov.uk/statistics/industry/construction.pdf
Health and Safety Executive (2022) *Construction statistics in Great Britain, 2022*. https://www.hse.gov.uk/statistics/industry/construction.pdf
Hillebrandt, P.M. (1975) *Economic theory and the construction industry*. Macmillan.
Hillebrandt, P.M. (2000) *Economic theory and the construction industry*. Macmillan.
HM Government (2012) *Industrial strategy: government and industry in partnership: building information modelling*. London, https://www.gov.uk/government/uploads/system/uploads/attachment_data/file/34710/12-1327-building-information-modelling.pdf
HM Government (2017) *Industrial Strategy: Construction sector deal*. London.
HM Government (2018) *The industrial strategy: building a Britain fit for the future*. London.
HM Treasury (2021) *Build back better: our plan for growth*. London.

ICED Facility (undated) ICED evidence library – construction sector employment in low-income countries: size of the sector. Infrastructure and Cities for Economic Development, DfID, London, http://icedfacility.org/resource/construction-sector-employment-low-income-countries-size-sector/

Infrastructure and Projects Authority (2017) Infrastructure transformation programme. London, https://chrome-extension://efaidnbmnnnibpcajpcglclefindmkaj/Infrastructure Transformation Programme

International Labour Office (ILO) (2008) *Toolkit for mainstreaming employment and decent work – country level application*. Geneva. https://www.ilo.org/wcmsp5/groups/public/—dgreports/—exrel/documents/publication/wcms_172612.pdf

Ive, G. and Gruneberg, S.L. (2000) *The economics of the modern construction sector*. Palgrave-Macmillan.

Ji, T-T., Wei, H-H., Sun, Y., Seo, J. and Chen, J-H. (2023) An explorative study of the political, economic, and social factors influencing the development of senior housing: a case study of Hong Kong. *Buildings*. 2023; 13(3):617. 10.3390/buildings13030617

Johari, S. and Jha, K.N. (2020) Challenges of attracting construction workers to skill development and training programmes. *Engineering, Construction and Architectural Management*, 27(2):321–340. 10.1108/ECAM-02-2019-0108

Kaklauskas, A. and Gudauskas, R. (2016) Intelligent decision-support systems and the Internet of Things for the smart built environment. In F. Pacheco-Torgal, E. Rasmussen, C.-G. Granqvist, V. Ivanov, A. Kaklauskas and S. Makonin (Eds.) *Start-up creation*. Woodhead Publishing, pp. 413–449, 10.1016/B978-0-08-100546-0.00017-0.

Katara, S. (2022) Replenishing the construction labor shortfall. *Forbes Innovation*, 18 August, https://www.forbes.com/sites/forbestechcouncil/2022/08/18/replenishing-the-construction-labor-shortfall/?sh=164f2bfe47a4

Kaya-Malin, F. and van Soomeren, P. (2021) *Security by design: SecureCity –10 rules of thumb – Urban agenda for the EU partnership on security in public spaces*. European Commission, Amsterdam.

Kirchberger, M. (2020) The construction sector in developing countries: some key issues. In J. Page and F. Tarp (Eds.) Mining for change: natural resources and industry in Africa (Oxford, 2020; online edn., Oxford Academic, 19 March) 10.1093/oso/9780198851172.003.0003, accessed 1 May 2023.

Library of Congress (2021) H.R.3684 – infrastructure investment and jobs act, 117th Congress (2021–2022). Washington, D.C., https://www.congress.gov/bill/117th-congress/house-bill/3684

Lingard, H. and Lin, J. (2004) Career, family and work environment determinants of organizational commitment among women in the Australian construction industry. *Construction Management and Economics*, 22(4):409–420.

Lopes, J., Oliveira, R. and Abreu, M.I. (2017) The sustainability of the construction industry in Sub-saharan Africa: some new evidence from recent data. *Procedia Engineering*, 172(2017):657–664, 10.1016/j.proeng.2017.02.077

Madanipour, A. (2023) *Rethinking public space*. Elgar Publishing Ltd.

Maisel, J., Steinfeld, E., Basnak, M., Smith, K. and Tauke, M.B. (2018) *Inclusive design: implementation and evaluation*. Routledge.

Maisel, J.L. and Ranahan, M. (2022) Beyond accessibility to universal design. Whole building design guide accessible committee, Washington, D.C., https://wbdg.org/design-objectives/accessible/beyond-accessibility-universal-design

ManpowerGroup (2020) 5 major challenges in the construction industry, 4 September, https://www.manpowergroup.co.uk/the-word-on-work/manpower-tech-expertise-5-major-challenges-construction-industry/#:~:text=5%20major%20challenges%20in%20the%20construction%20industry%201,Technology%20adoption%20...%205%205%29%20Worker%20productivity%20

Mansour, H., Aminudin, E. and Mansour, T. (2023) Implementing industry 4.0 in the construction industry – strategic readiness perspective. *International Journal of Construction Management*, 23(9):1457–1470. 10.1080/15623599.2021.1975351

Marchand, G.C., Nardi, N.M., Reynolds, D. and Pamoukov, S. (2014) The impact of the classroom built environment on student perceptions and learning. *Journal of Environmental Psychology*, 40:187–197. 10.1016/j.jenvp.2014.06.009.

Martin-Ortega, O. and Treviño-Lozano, L. (Eds.) (2023) *Sustainable public procurement of infrastructure and human rights: beyond building green*. Elgar Publishing Ltd.

Massachusetts Institute of Technology (MIT) (undated) Carrying it out: does the project/program support local initiatives in construction? https://web.mit.edu/urbanupgrading/upgrading/issues-tools/issues/support-contruction.html

McAnaw Gallagher, C. (2022) The construction industry: characteristics of the employed, 2003–20, US Bureau of Labor Statistics, https://www.bls.gov/spotlight/2022/the-construction-industry-labor-force-2003-to-2020/home.htm

McCord, A. (2012) *Public works and social protection in sub-Saharan Africa: do public works work for the poor?* United Nations University Press, New York.

McCutcheon, R.T. (2001) Employment generation in public works: recent South African experience. *Construction Management and Economics*, 19(3):275–284. 10.1080/01446190010020381

McHugh, C. (2018) Building mental health: construction industry training board mental health first aiders' programme 2018–2022 social value report. CITB and Lighthouse Club, Ipswich, https://www.lighthouseclub.org/wp-content/uploads/Building-Mental-Health-2022-v3.pdf

McKinsey & Co. (2022) What are industry 4.0, the fourth industrial revolution, and 4IR? 22 August, https://www.mckinsey.com/featured-insights/mckinsey-explainers/what-are-industry-4-0-the-fourth-industrial-revolution-and-4ir#/

Meikle, J. (2019) A response to George Ofori's special note. *Journal of Construction in Developing Countries*, 24(2):207–208. 10.21315/jcdc2019.24.1.10

Mella, A. and Savage, M. (2017) Construction sector employment in low income countries. Infrastructure and cities for economic development (ICED) report. Department for International Development, London.

Ministry of Trade, Industry and Cooperatives (2014) *Buy Uganda Build Uganda Policy*. Kampala.

Morawska, L., Tang, J.W., Bahnfleth, W., Bluyssen, P.M., Boerstra, A., Buonanno, G., Cao, J., Dancer, S., Floto, A., Franchimon, F., Haworth, C., Hogeling, J., Isaxon, C., Jimenez, J.L., Kurnitski, J., Li, Y., Loomans, M., Marks, G., Marr, L.C., Mazzarella, L., Melikov, A.K., Miller, S., Milton, D.K., Nazaroff, W., Nielsen, P.V., Noakes, C., Peccia, J., Querol, X., Sekhar, C., Seppänen, O., Tanabe, S-I., Tellier, R., Tham, K.W., Wargocki, P., Wierzbicka, A. and Yao, M. (2020) How can airborne transmission of COVID-19 indoors be minimised? *Environment International*, 142:105832, 10.1016/j.envint.2020.105832.

Mulaudzi, G. and Lancaster, L. (2021) South Africa's construction mafia trains its sights on local government, ISS Today, 25 October, Institute of Security Studies, Pretoria. https://issafrica.org/iss-today/south-africas-construction-mafia-trains-its-sights-on-local-government

Navaratnam, S., Nguyen, K., Selvaranjan, K., Zhang, G., Mendis, P. and Aye, L. (2022) Designing post COVID-19 buildings: approaches for achieving healthy buildings. *Buildings*, 12(1):74. 10.3390/buildings12010074

Nippa, M. and Reuer, J.J. (2019) On the future of international joint venture research. *Journal of International Business Studies*, 50:555–597. 10.1057/s41267-019-00212-0

Office for National Statistics (2009) *UK standard industrial classification of economic activities 2007 (SIC 2007) – structure and explanatory notes*. Palgrave Macmillan, Basingstoke.

Ofori, G. (1989) A matrix for the construction industries of developing countries. *Habitat International*, 13(3):111–123.

Ofori, G. (1990) *The construction industry: aspects of its economics and management*. Singapore University Press, Singapore.

Ofori, G. (1993) *Managing construction industry development: the case of Singapore*. Singapore University Press, Singapore.

Ofori, G. (1992) The environment: the fourth construction project objective? *Construction Management and Economics*, 10(5):369–395. 10.1080/01446199200000037

Ofori, G. (2016) Construction in developing countries: current imperatives and potential. K. Kahkonen and M. Keinanen (Eds.) *CIB World Building Congress 2016 on Intelligent Built Environment for Life*. Tampere 30 May – 3 July 2016 Tampere Department of Civil Engineering, Construction Management and Economics, Tampere University of Technology.

Ofori, G. (2017) Study on Skills and Market Assessment of the Construction Sector in Ghana. Report prepared for the State Secretariat for Economic Affairs of the Federal Department of Economic Affairs, Education and Research, Switzerland, Berne.

Ofori, G. (2022a) Construction economics: its origins, significance, current status and need for development. In G. Ofori (Ed.) *Research companion to construction economics*. Elgar Publishing Ltd., pp. 18–40.

Ofori, G. (2022b) Defining and describing construction in developing countries. In R. Best and J. Meikle (Eds.) *Describing construction: industries, projects and firms*. Routledge, pp. 171–196.

Ofori, G. (2023) From the MDGs to the SDGs: the role of construction. In Opoku, A. (Ed.), *Research Handbook on Construction and the Sustainable Development Goals*. Edward Elgar: Cheltenham, UK.
Olugboyega, O. and Abimbola Windapo, A. (2023) Modelling the indicators of a reduction in BIM adoption barriers in a developing country. *International Journal of Construction Management*, 23:9, 1581–1591. 10.1080/15623599.2021.1988196
Opoku, A. (2022) Construction industry and the sustainable development goals (SDGs). In Ofori, G. (Ed.), *Research Companion to Construction Economics*. Edward Elgar: Cheltenham, UK, pp. 199–214.
Organisation for Economic Cooperation and Development (OECD) (2019) Trade Policy Brief – local content requirements. Paris. https://www.oecd.org/trade/topics/local-content-requirements/
Outram, K., Harvanek, D., Chalmers, B. and Juay, R. (2022) Outlook for the construction industry in 2022: cautious optimism for continued recovery and growth, with investments in people and culture essential for success. Marc & McLennan Companies, Inc., https://www.marshmclennan.com/content/dam/mmc-web/insights/publications/2022/april/gl2022globaltalenttrendsindconstruction.pdf
Perdue, W.C., Stone, L.A. and Gostin, L.O. (2003) The built environment and its relationship to the public's health: the legal framework. *American Journal of Public Health*, 93(9):1390–1394. 10.2105/ajph.93.9.1390
Pinter-Wollman, N., Jelic, A. and Wells, N.M. (2018) The impact of the built environment on health behaviours and disease transmission in social systems. *Philosophical Transactions of the Royal Society B*, 373: 2017024520170245, 10.1098/rstb.2017.0245
Plangrid (2018) *The digital groundwork: beyond construction's productivity gap*. Autodesk, London.
Ribeirinho, M.J., Mischke, J., Strube, G., Sjödin, E., Luis Blanco, J.L., Palter, R., Biörck, J., Rockhill, D. and Andersson T. (2020) *The next normal in construction: how disruption is reshaping the world's largest ecosystem*. McKinsey & Company.
Rietveld, P. (1992) Housing and employment in Indonesia: prospects for employment generation in the construction materials sector. *Bulletin of Indonesian Economic Studies*, 28(2):55–73. 10.1080/00074919212331336214
Romo, R., Orozco, F., Forcael, E. and Moreno, F. (2023) Towards a model that sees human resources as a key element for competitiveness in construction management. *Buildings*, 13(3):774. 10.3390/buildings13030774
Roy, S.K. and Koehn, E.E. (2006) Construction labor requirements in developing countries. Proceedings of the 2006 ASEE Gulf-Southwest Annual Conference Southern University and A & M College – Baton Rouge. American Society for Engineering Education, https://peer.asee.org/construction-labor-requirements-in-developing-countries.pdf
Ruddock, L. and Ruddock, S. (2008) The scope of the construction sector: determining its value. In L. Ruddock (Ed.) *Economics for the modern built environment*. Routledge, Abingdon.
Sawhney, A., Riley, M. and Irizarry, J. (Eds.) (2020) *Construction 4.0: an innovation platform for the built environment*. Routledge, Abingdon. 10.1201/9780429398100
Schuilenburg, M. and Peeters, R. (2018) Smart cities and the architecture of security: pastoral power and the scripted design of public space. *City, Territory and Architecture*, 5(2018):13. 10.1186/s40410-018-0090-8
Schwartz, J., Andres, L. and Dragoiu, G. (2009) Crisis in Latin America: infrastructure investment, employment and the expectations of stimulus. *Journal of Infrastructure Development*, 1(2):111–131. 10.1177/097493060900100202
Sepasgozar, S.M.E., Costin, A.M., Karimi, R., Shirowzhan, S., Abbasian, E. and Li, J. (2022) BIM and digital tools for state-of-the-art construction cost management. *Buildings*, 12(4):396. 10.3390/buildings12040396
Shen, L., Ochoa, J.J. and Bao, H. (2023) Strategies for sustainable urban development—exploring innovative approaches for a liveable future. *Buildings*, 13(3):764. 10.3390/buildings13030764
Smith, K. (1999) *Indoor air pollution: pollution management in focus, discussion note no. 4*. World Bank, Washington, D.C.
Spence, R., Wells, J. and Dudley, E. (1993) Jobs from housing: employment, building materials, and enabling strategies for urban development. Prepared by Cambridge Architectural Research, Ltd. for the Overseas Development Administration of the United Kingdom, London.
Ssegawa J.K. and Ngowi A.B. (2009) *Challenges in delivering public construction projects: a case of Botswana*. Lambert, Saarbrücken.

Steinfeld, E. and Maisel, J. (2012) *Universal design: creating inclusive environments*. Wiley, Hoboken, NJ.
Tham, K.W. (2016) Indoor air quality and its effects on humans—a review of challenges and developments in the last 30 years. *Energy and Buildings*, 130:637–650. 10.1016/j.enbuild.2016.08.071.
Thompson, M. and Kent, J.L. (2017) Human health and a sustainable built environment. In M.A. Abraham (Ed.) *Encyclopedia of sustainable technologies*. Elsevier, pp. 71–80. 10.1016/B978-0-12-409548-9.10178-2.
Turin, D.A. (1973) Construction and Development. Building Economics Research Unit, University College Research Group, University College London.
Turner & Townsend (2022) International Construction Market Survey 2022, https://www.turnerandtownsend.com/en/perspectives/international-construction-market-survey-2022/
UNCTAD (undated) Infrastructure Report, https://www.oecd.org/sweden/2083051.pdf
United Nations Human Settlements Programme (2011) *Ghana: housing profile*. UNHSP, Nairobi.
United Nations Interregional Crime and Justice Research Institute (UNICRI) and Massachusetts Institute of Technology Sensible City Lab (2011) *Improving urban security through green environmental design new energy for urban security*. Turin.
US Bureau of Labor Statistics (2023) Industries at a glance: construction: NAICS 23, https://www.bls.gov/iag/tgs/iag23.htm#workforce
Wang, M., Li, L., Hou, C., Guo, X. and Fu, H. (2022) Building and health: mapping the knowledge development of sick building syndrome. *Buildings*, 12(3):287. 10.3390/buildings12030287
Ward, B. (2022) CITB-funded initiative tackles mental health epidemic. *Construction Journal*, 30 August, https://ww3.rics.org/uk/en/journals/construction-journal/mental-health-citb-funded-training-addresses-stigma.html
Wells, J. (2013) *Relieving chronic poverty among construction workers: an exploration of possibilities to improve the quantity and quality of jobs*. Engineers Against Poverty, London, http://engineersagainstpoverty.org/resource/relieving-chronic-poverty-among-construction-workers-an-exploration-of-possibilities-to-improve-the-quantity-and-quality-of-jobs/
Wibowo, M.A., Sugiyanto, F.X., Firmansyah, and Amoudi, O. (2008) The effects of the Indonesian construction industry on the economy: a series of input-output table analysis. In K. Carter (Ed.) *Proceedings of the CIB W065/055 Joint Symposium on "Transformation through Construction"*, Edinburgh, 15–17 November, https://www.irbnet.de/daten/iconda/CIB17562.pdf
Windapo, A.P. and Rotimi, J.O. (2012) Contemporary issues in building collapse and its implication for sustainable developments. *Buildings*, 2(3):283–299. 10.3390/buildings2030283.
Wiseman, J. , Roe, P. and Parry, E. (2018) *Skills and Training in the Construction Industry 2018*. CITB, Peterborough.
World Bank (1994) *World development report 1994: infrastructure for development*. Oxford University Press, New York. http://hdl.handle.net/10986/5977
World Bank (2000a) *Can Africa claim the 21st century?* Washington, D.C.
World Bank (2000b) *World development report 2000*. Washington, D.C.
World Bank Skills (2010) *Development strategies to improve employability and productivity: taking stock and looking ahead*. World Bank, Washington, D.C.
Yan, P., Liu, J., Zhao, X. and Skitmore, M. (2022). Risk response incorporating risk preferences in international construction projects. *Engineering, Construction and Architectural Management*, 29(9):3499–3519. 10.1108/ECAM-03-2019-0132
Yousaf, A., Sabir, M. and Muhammad, N. (2019) A comparative input-output analysis of the construction sector in three developing economies of South Asia. *Construction Management and Economics*, 37(11):643–658. 10.1080/01446193.2019.1571214

3

CHARACTERISTICS OF THE CONSTRUCTION INDUSTRY AND CONSTRUCTION PROCESS, AND IMPLICATIONS FOR LABOUR POLICIES AND PRACTICES

George Ofori

London South Bank University

Introduction

Aim and objectives of the chapter

The aim of this chapter is to consider the characteristics of the construction industry and the construction process, and their implications for labour in the industry and other sectors of the economy, and the cities and towns they build with the view to proposing appropriate interventions to address the prevailing and expected future challenges. The objectives are to:

- consider the main characteristics of the construction industry and explore how these features influence the practices of construction companies and government legislation and control
- discuss the implications of the characteristics of the construction industry for labour in the industry, as well as in labour in other sectors of the economy
- consider the current situation, challenges and problems of labour in construction as a result of the characteristics, and actions being taken to address them
- propose appropriate future actions.

Features of the construction industry and implications for construction and labour

Some of the features of the construction industry, the construction process and the constructed product, and their implications for employment and training are now discussed. In each case, the *feature of construction* is outlined. Then, an *explanation of the feature* is presented. This is followed by consideration of the *implications for labour in construction and its development*, and then *implications for labour in other sectors and the community.*

DOI: 10.1201/9781003262671-3

Location specificity

Explanation of the feature

The product of the construction process is fixed in its location. This means that the work on a project must take place in the location of the planned built item. This has many implications. First, the construction companies involved in the project must marshal their resources to operate in the locality of the project. This makes construction different from manufacturing (with which it is often compared with respect to productivity levels and growth), in which the production takes place in permanent premises. This need for geography-bound mobilisation lies at the root of many of the implications outlined in this subsection. Second, the location specificity makes logistics a critical consideration, and it has a potential impact on productivity as several factors come into play where logistics are concerned (Wegelius-Lehtonen, 2001; Sullivan et al., 2011; Vidalakis et al., 2011). Third, the area within which a company can operate effectively is influenced by geography, depending on its ability to handle the logistics, as well as the cost of transportation and other considerations of the bulky material inputs (Hillebrandt, 2000). For this reason, *Construction is described as a local industry.* Segerstedt and Olofsson (2010) assert that the market of the construction company is mostly local and highly volatile. Thus, local businesses can also be competitive on some projects, especially the small and medium-sized ones.

Fourth, this geographical limitation on operational effectiveness has determined the structure of the construction industry in all countries; there is a small number of large companies which can operate on a national or regional basis, and a multiplicity of small firms. For example, in 2022, small and medium enterprises (SMEs) accounted for at least 99 percent of the overall population in each of the main industry sectors in the United Kingdom, and the largest number of SMEs (914,000 or 17 percent of the total SMEs in the economy) were operating in Construction (BEIS, 2022). Construction SMEs were responsible for 14 percent of total SME turnover and 12 percent of total SME employment.

Fifth, the project has direct economic impact on the local area where the work is undertaken. It is suggested that large projects can have a major socio-economic impact on a large area. Ribeirinho et al. (2020) of McKinsey & Company describe construction and its value chain as “the world’s greatest ecosystem”. Construction activity distributes income as workers spend their wages locally. Makenya and Mng’ong’o (2018) noted that studies in Tanzania showed that labour-based road construction methods could have a cost advantage compared to conventional equipment-based methods; the methods promote small-scale contractor development; studies showed that up to 60 percent of the costs can be retained by the community; and the approach could foster a sense of ownership and skills transfer by participation of local people in the road works.

Sixth, aspects of the operating environment of the construction project are also defined by its location. The factors here include the direct availability of resources locally, including labour, key items of materials, plant and equipment and efficient suppliers; subcontractors; and utilities. The indirect, but also influential factors include local regulations, politics, business dynamics and practices and social issues (Lizarralde et al., 2013). These factors have implications for the management of the project. Issues of culture and sensitivities between workers and local communities (discussed below) are other considerations.

Seventh, the location specificity also means that construction projects and the methods which are applied can have an environmental impact, such as encroachment on sensitive areas. The environmental impact of construction activity on site includes (Ofori, 1992;

Green Alliance, 2023): various forms of pollution (air, noise and water) from the activities and the wastes; reduction of greenery and localised soil degradation; hygiene, health and safety of workers, residents and users, including the occurrence of accidents caused by vehicles and equipment, and around uncovered excavations; and possible disturbance of historic and archaeological assets (buildings or other remains).

Eighth, it is also necessary to highlight that there can be adverse impacts from the presence of construction workers in local communities. This issue is discussed in Chapter 17 of this book. It is briefly considered here. While it occurs temporarily over the duration of the project, this can have deep, long-lasting effects. For example, the HIV Aids impact in such a situation is well researched (Klunklin and Greenwood, 2005; Ditangco et al., 2008; Bowen, 2014). It is evident that the project should include a programme to prepare the local community to receive the incoming workers. There can also be potential conflicts if there are ethnic and cultural differences between the workers and the community living where the project is based. Finally, there is a possibility of interface conflict with the communities in the environs of the project; this has generated a significant volume of literature (see, for example, Awakul and Ogunlana, 2002a, 2002b; Mahato and Ogunlana, 2011). Awakul and Ogunlana (2002b) identify two categories of conflicts in large construction projects: internal conflict (among the participants in the project); and interface conflict (between the construction project and groups outside the project such as the "project-affected" people, civil society organisations and so on).

The ninth point is that of community participation in the project. This is recognised as desirable and it is stressed in the national development plans and public projects of some countries, and indeed, it is enshrined in South Africa's constitution (Williams, 2006). However, it does not always happen, and a large volume of literature has developed on it (Hollnsteiner, 1976; Dudley, 1993; Raco, 2000). Mahato and Ogunlana (2011) found that interface conflicts at the construction stage of projects are caused mainly by lack of effective Environmental Impact Assessment, public participation and mutual consultation, on a timely basis, and accurate information from the early stages of projects. They proposed a model to help avoid and minimise interface conflict in the construction stage of a dam. Di Maddaloni and Sabini (2022) concluded in a study that although project managers recognise the moral obligations and benefits of including local communities in the decision-making process, project organisations are ill-equipped to embrace this inclusive approach in practice, and project managers seldom do it. Vosloo (2021) reviews and compares three public projects that included community participation to determine the total value added. These projects were examples of the government's initiatives to address some of the most pressing and challenging problems facing South Africa: unemployment, poverty, urban redress, infrastructural decay, undereducation and the transformation of the landscape left by apartheid through building and construction projects, which require the participation by, and employment of local community members. Some of these projects were flagship projects that were lauded by the architectural profession and attracted wide publicity. The socio-economic benefits to the community and local area, the extent of skills transfer to the community participants, and the long-term benefits they brought to the community participants are less obvious. He recommended that infrastructural development programmes such as the Extended Public Works Programme in South Africa should prioritise the socio-economic upliftment and sustainable empowerment of people and have these as their aim in configuring projects.

Now, the tenth point, interface conflict, is also discussed in Chapter 17 of this book, and briefly considered here. Researchers report many cases of actions by the local communities against construction projects in many countries. Teo and Loosemore (2010) found that collective action against projects is maintained by a high degree of interconnectivity and relational multiplexity between the participating individuals and groups; the protective role of hidden social networks; overlapping protest group memberships; the plurality of protest issues faced; and the quality and nature of social ties, experiences and emotions that link activists in collective action over the protest movement's lifetime. In South Africa, there are reports of many major construction projects being disrupted by well-organised groups (which call themselves "business forums" but are often referred to as the "construction mafia") which appear to be interpreting a government provision for a proportion of projects to be subcontracted to the local community. Mulaudzi and Lancaster (2021) note that by March 2019, the South African Forum of Civil Engineering Contractors (SAFCEC) estimated that construction projects worth about R25.5 billion had been violently disrupted and halted. Possibly misinterpreting the proportional subcontracting requirement in the South African government's Preferential Procurement Regulations of 2017 (Government Gazette, 2017), the groups typically demand from local business owners a portion of the cost of the project, or that specific individuals are employed on the site. If this is refused, the groups damage and disrupt their operations or intimidate staff.

Social Value by legislation

Drawing on the feature of location specificity, governments (national and local) have endeavoured to deepen and strengthen the social value of projects. Laws such as the UK's Social Value Act and South Africa's Preferential Procurement Regulations of 2017the seek to optimise the benefits to the local community and reduce the adverse impacts. In the United Kingdom, the Public Services (Social Value) Act, 2012 (HM Government, 2012) requires people who commission public services to think about how they can also secure wider social, economic and environmental benefits for their area or stakeholders before they start the procurement process. The Act is to help commissioners get more value for money out of procurement and encourages them to talk to their provider market or community to design better services and find new and innovative approaches. In 2018, the central government announced that it would henceforth explicitly evaluate social value when awarding most major contracts. Government departments will be expected to report on the social impact of their major contracts (Cabinet Office, 2021).

In South Africa, the government's 2017 Preferential Procurement Regulations directives aim to transform the economy by empowering historically disadvantaged individuals and small, medium and micro enterprises. The regulations require that winner of state tenders over R30 million subcontract at least 30 percent of the contract's value to small local developers to undertake construction work. Murphy and Eadie (2019) investigate the use of what they termed "socially responsible procurement" by the government as a means of generating social value from construction activities in Northern Ireland. They found that socially responsible procurement is being driven by social legislation and being delivered by contractors as part of their contractual obligations and they also found that it is generating social benefits through employment creation and feedback from employees is largely positive. They suggest that there is a need for a more holistic measurement of the impacts and outcomes of socially responsible procurement to ensure social targets are appropriate for the communities in which projects are being constructed.

In some countries, in addition to complying with regulatory requirements in some countries, the private sector is also increasingly extending its corporate social responsibility to enhance the social value of its projects. In the United Kingdom, Ahluwalia (2023) suggested that real estate developers and investors should aim to sustainably deliver social value for all the communities within their ecosystem through development and beyond, looking at it from more than a monetary perspective. To truly deliver change, it would require the industry to work in partnership, share best practices and deliver long lasting, sustainable social value. Examples of social value projects of a UK developer's company given by Ahluwalia (2023) are presented in Box 3.1.

Box 3.1 Social value in Dominvs group, UK

"It's tempting to think that the section 106 agreement and a community infrastructure levy embody our obligations. But a retail offering on the ground floor, or a few new additions to the street scene, are secondary to the importance of engaging with community needs to generate long-term social outcomes. Creating meaningful places that preserve and enhance community spirit must flow from the earliest stages of acquisition to operational use. The underlying principle of social value has to be about involving stakeholders through connecting with the local community. This enables direct engagement with, and a better understanding of, the needs and wants of residents. Too often, a top-down approach reinforces the status quo, leading to suboptimal uses and underutilised spaces. At the same time, engaging and collaborating with communities can lead to increased support for schemes at planning committees, and build genuine trust. At Dominvs we believe that to truly deliver social value, it needs to be embedded from the start of the decision-making process and considered beyond just the design of the scheme. Our in-house social value unit works in tandem with the development managers and our board.

At our scheme in Stratford, we'll provide a non-profit community pub to be run by the Made Up Collective and affordable workspace targeted at creatives. At Holborn we'll provide a public roof terrace and cultural amenity space. At Crutched Friars, we are discussing with the Migration Museum how we can provide three floors that will reflect the key role migration has played in the shaping of modern Britain.

But while plans embedded with schemes can deliver for the long-term, complex planning applications may wait more than 47 weeks at the appeal stage until a decision is made. This offers the opportunity to be creative and incorporate meanwhile uses at a site that is currently sitting vacant or unused. Our meanwhile use scheme, Gaia's Garden, attracted more than 1,300 young people from underrepresented backgrounds to the City of London to learn about sustainability and wellbeing through workshops and events.

Social value for the real estate industry has to be about more than just what we can do at the sites we own, too. It's about a widening of opportunities, encouraging a diverse workforce within the sector and inspiring the next generation of changemakers – by creating new opportunities, but also through collaboration with charities, local organisations, schools, businesses and local authorities to start bridging inequalities across our communities.

> There are currently 45,000 vacancies in construction and 16,000 in real estate activity. ... It is within our industry's own interest to widen participation and attract the next generation of talent. ... The industry is getting better at opening pathways to those from diverse backgrounds, but there is still work to be done.
>
> *Jay Ahluwalia is the principal at Dominvs Group*
>
> *Source*: Ahluwalia (2023)

Community safeguarding

The issue of safeguarding the community in the environs of the project is discussed in Chapter 17 of this book. It is briefly considered here. Major financial institutions such as the World Bank have instituted policies to ensure the attainment of environmental and social attributes on the projects they fund (World Bank, 1999 presents the main policy which has been updated a number of times). Di Leva (2021) discusses the World Bank's regime. These include a safeguarding policy for residents in the environs of projects. This nature of construction implies that the construction company should extend its consideration of safeguarding beyond its workers and on its site (see World Bank, 1999; on the website, the Bank advises readers that "Questions on the operational guidelines may be addressed to the Safeguard Policies Helpdesk in OPCS (safeguards@worldbank.org)".

Major projects endeavour, and declare their commitment, to be responsible. For example, HS2 describes itself as: "HS2 is more than a railway; it is the most important economic regeneration project in Britain for decades. We are joining up Britain to help build a more united country where every region of the UK can reach its economic potential, creating jobs and opportunities for millions of people. It will improve living standards, increase productivity and help Britain compete on the global stage" (HS2, 2022). Its approach to responsible business focuses on: People and Workplace – Helping businesses and people to reach their full potential as we build HS2, for now and the future; Communities – Supporting communities to thrive and build for the future; and Environment – Protecting and enhancing the environment for future generations.

In the United Kingdom, the voluntary Considerate Constructors Scheme supports and guides positive change in the construction industry (Considerate Constructors Scheme, 2023). To help constructors improve their behaviours and the impact of construction activity, the Code of Considerate Practice focuses on: Respecting the community; Caring for the environment; and Valuing the workforce. Those registered with the Scheme make a commitment to conform to the Code and work to raise their standards above and beyond it. A dedicated Public Support team helps if one has a concern or feedback about construction activity. Clients of construction projects which register agree that their projects are registered with the Scheme and commit to the Code.

Implications for labour in construction

The location specificity of construction projects has major implications for labour. First, construction jobs are generated in the locality. However, there have to be sufficient numbers of skilled persons for the companies.

In addition to the training of workers for the industry, in a remote area, training of the local people in the environs of the project can equip them to take up jobs on the project, in the locality or elsewhere, and increases their social mobility although they might work on these other jobs on an off-season or casual basis. The labour-intensive rural roads programme of the International Labour Office (ILO), discussed in Chapter 2, sets out to take direct advantage of the features of the construction industry (Bentall et al., 1999).

The need for the project to take place where the item is needed generates service jobs in the locality, to provide for the workers on the project. This issue is discussed in Chapter 17 of this book. It is briefly considered here. Companies can be created to provide non-construction services to the companies and participants in the project. Local small enterprises can be energetic and can thrive for the duration of the project. Apart from the opportunities for citizens in the district to fill the job positions created on the site, there are derived opportunities for jobs in businesses which are set up locally to provide services to the construction project (Adato and Haddad, 2001; Vosloo, 2021). Thus, this feature of construction has direct implications for labour in other sectors of the economy. For example, with regard to location specificity, there might be a need for temporary accommodation for the workers on infrastructure projects away from the cities such as highways, bridges and dams.

Uniqueness of each project

Feature of construction

Construction is a project-based industry. A project is defined as a specific unique endeavour to realise particular objectives under defined constraints. For example, to APM (2022): "A project is a unique, transient endeavour, undertaken to achieve planned objectives, which could be defined in terms of outputs, outcomes or benefits". Thus, each project is considered to be unique. Construction projects demonstrate this uniqueness even more clearly. They differ from each other in ways that other projects do, for example, in terms of the client commissioning the project, the design and material inputs and the participants involved. However, in addition, in construction, the location for each project is different; each site has its own physical features (such as soil conditions) and accessibility; and social dynamics in the environs might differ.

The uniqueness of the project sets construction activities apart from other works. It implies that first, construction firms might not be able to transfer much of the surplus material and component inputs acquired for one project to the next. Second, participants in a project cannot transfer all the knowledge directly to the next project, especially as the team is disbanded after each project. A knowledge management system is necessary to propagate lessons to help enhance performance on future projects. Third, it is likely that the companies which work on any project will not work together again. This makes it necessary for appropriate business and management practices to be adopted. The project-based nature of construction underlines the discontinuities in the business of construction companies. A related feature of construction is that the design team and consultants have limited influence in the decision making on the project. They only come into the picture when the client has decided to express effective demand for the item. They are engaged in projects when they win bids. This makes construction risky, and its companies susceptible to failure; these points are now further discussed.

Considering the nature of the construction business (as discussed in Chapter 2), employment in construction might be project based. Thus, there is the spectre of discontinuity and uncertainty. This makes it difficult to attract good people (an issue discussed further in Chapter 2). For the construction firm, the project-based nature of construction means there is discontinuity in jobs. This leads to the firms being reluctant to invest in any significant manner. The construction industry is well known for casualisation of labour, as discussed in Chapter 7. Firms are not willing to invest in the training and development of their workforce. Çelik and Oral (2019) referred to the idiosyncratic conditions of the construction industry and suggested that the industry has an erratic structure owing to the uniqueness of each project and the different resource combinations in the sector. Thus, it is one of the sectors with the highest labour turnover.

Moreover, coupled with the discontinuity (and perhaps because of it), there is intense competition in construction. There is a common practice to award contracts to the lowest bidder. Thus, the construction business is characterised by low profit margins. For example, in the Turner & Townsend 2021 global construction report, it asked a sample of firms in 90 world-wide markets to provide typical profit margins on a medium-sized commercial job. These were the average margins: the United Kingdom, 3.9 percent; Australia and New Zealand at 4.5 percent; North America, 4.6 percent; and Continental Europe, 6.1 percent (Smulian, 2021). As another example, in 2021, although only 14 firms in the Top 100 UK contractors registered a loss in their latest accounts and the total profits increased (perhaps because of comparison with a COVID-19 affected 2020), 3 of the reporting top 10 firms made losses (The Construction Index, 2022). From its recent annual global survey, Turner & Townsend (2023) concluded that these are times of increasing complexity and risk for large organisations managing diverse capital programmes. Proactive clients are tackling these challenges directly by: forming closer supplier partnerships; Sharing risk; and Embracing innovation.

Studies show that the main contractors tend to delay payments due to subcontractors as part of their financial management (Construction Leadership Council, 2018; Jones, undated). This exacerbates the financial difficulties of the smaller firms. (This issue is discussed in detail in Chapter 7 of this book.) This has led to a number of governments such as the states in Australia (Government of Victoria, 2013), Singapore (Government of Singapore, 2020; Building and Construction Authority, undated) and the United Kingdom (Government of UK, 1996), to pass security of payments regulations. The Subcontract Act of Japan (Act No. 120 of June 1, 1956) has the same objective (Japan Free Trade Commission, 2023). Clause 1 reads:

> The purpose of this Act is, by preventing a delay in payment of subcontract proceeds, etc., to ensure that transactions between main subcontracting entrepreneurs and subcontractors are fair and, at the same time, to protect the interests of the subcontractors, thereby contributing to the sound development of the national economy (p. 1).

The construction industry has a high level of mortality among companies (Construction Leadership Council, 2018; Jones, undated). Decker (2021) suggested that such business exit can be healthy and productive because it is productivity enhancing as lower-productivity establishments are selected and replaced. On the other hand, exit can have adverse productivity consequences if selection does not function or if it is not matched with business creation (for example, in such a pandemic environment, exit selection may not operate productively); it permanently destroys some jobs, as well as intangible firm-specific capital. Moreover, displacement causes long-term harm; for example, it might destroy the proprietor's wealth. Most important in the context of this chapter, it alters the economic geography of local communities, and can reshape communities. Ofori (2022a) noted that, for the construction industries in developing countries, such exits mean the loss of capacity built over a period of time.

Given the forward linkage effects of construction (as discussed in Chapter 2), *the demand for construction is derived demand* (Hillebrandt, 2000). The demand for facilities for operations of all types of business is based on the perceptions of demand for the goods and services to be produced in them. Considering the long period of gestation of construction projects (see below), it is difficult to estimate the demand for construction. Thus, the construction market is characterised by instability; there can be acute shortages of certain types of construction, such as houses and office space, and also, overbuilding (see below). This makes it difficult for firms to plan for the future and invest appropriately (as discussed below). Some governments have recently been publishing infrastructure plans with indications of projects to be launched in future. These include Australia (Infrastructure Australia, 2021), New Zealand (NZIC, 2022) and the United Kingdom (IPA, 2017). The BCA (2023a) in Singapore publishes forecasts of construction demand into the medium term, forming a rolling database to guide developers and contractors.

Implications for labour

With project uniqueness, there are discontinuities in work opportunities and in the work itself. The construction worker faces a situation where the working environment constantly changes, in terms of what is built, what role they might play, the technology they might adopt, the location of the operations, and the team they will work with. (This issue is discussed in detail in Chapter 8 of this book.) This has an impact on income. This characteristic also influences the ability of the construction industry to attract talent. Çelik and Oral (2019) noted that the project-based production of the construction industry and the different combinations of construction site, project, production methods and labour create an erratic atmosphere that affects its workers.

One way in which some countries have addressed the discontinuities is to design and deliver multi-skills training where the worker learns a number of closely related trades. This is beneficial to both the worker and the employer. For the former, it provides involvement on projects for a wider range of activities, and therefore, for a longer period, increasing their income earning potential. For the company, apart from reducing the number of workers and the amount of downtime on each project, it reduces the recruitment effort and expenses. In Singapore, the traditional construction skills have been consolidated into a few, such as: Structural Trades – bricklaying, steel bending, concrete; and Finishing Trades – plastering, tiling and painting. As the Building and Construction Authority (BCA) (2023b) notes: "Multi-Skilling workers offer greater flexibility in deployment on site and hence help to enhance productivity by reducing productivity downtime. This will benefit firms as they will be able to deliver their construction work with lesser workers". For the country, there is a higher, strategic benefit: it reduces the number of foreign workers to be employed.

Corporate strategies with respect to human resource development and deployment should address this subject of discontinuities and job insecurity. Leadership styles in construction should reflect this discontinuity. Finally, leaders in construction should consider how to attract the talent which will improve performance on the projects, in the companies, and at the industry level.

Involvement of many companies and persons

Feature of construction

A construction project involves many participants from different companies and other organisations. The individual practitioners are from different professional backgrounds, and get involved in the project during various stages, when the work involving their expertise comes up (RIBA, 2020). The issues this raises include, first, how to co-ordinate the contributions of the different parties. Çelik and Oral (2019) noted that, as the processes they involve require long-term collaborations among a large number of different teams, the dynamic structure of construction projects renders their management difficult. Studies show that the inefficiencies in construction are due to both inadequacies in the work at the various stages as well as the inability to manage the interfaces well. For example, Agbaxode et al. (2023) study how the poor quality of design documentation impacts construction project delivery. They found that it leads to project delays, project abandonment and cost overruns, and they suggest improving collaboration between design disciplines should be enhanced and specialists should be involved during design to improve quality.

Each project involves the physical planner, a design team comprising an architect, various types of engineers, cost consultants and possibly, a design specialist in the type of building; a project manager; the main contractor and subcontractors; and suppliers. Each of these organisations employs a range of professionals and technicians as they require. The contractors and subcontractors employ workers with various trades and unskilled workers. There are variations in the roles of these players around the world. The procurement approach and the nature of the contract have implications for the workers on the project. Construction is an activity in which many persons and groups have legitimate stakes. This might be because of the benefits they bring, and the adverse impacts they can have. Other stakeholders include the end purchasers (of private-sector houses or commercial buildings) or beneficiaries (of public projects); the community in the environs of the project; the professional institutions; the trade unions; and the environment. Government agencies are involved in development and building control, and health and safety enforcement and inspection. In some countries, including Hong Kong, SAR (Buildings Department, 2018), Malaysia (Government of Sarawak, 2019) and the United States (see, for example, Durham County, 2020) the completed item has to be inspected and certified by the public agencies before it is occupied.

Construction firms build their capacity through subcontracting (see Debrah and Ofori, 2001; Raidén et al., 2007). (This issue is discussed in detail in Chapter 7.) There has been a phenomenon where many construction firms have tended to employ very few direct workers and rely on subcontractors. Wells (2006) notes that the practice of employing labour through subcontractors, an old practice known by different names in various countries, such as *Kepala* in Malaysia and Singapore, and *mistri* or *jamadar* in India is increasing in both developing and developed countries. She assesses the advantages of the practice from the perspective of the contractors and of the workers and concludes that subcontracting is likely to endure and that "interventions may be needed to deal with some of the negative repercussions, but they have to accept and build on current labour practices" (p. 17). The responsibility for the welfare of the workers on sites is a major issue. In the United Kingdom, many studies have examined the "hollowed out firm" which has followed the rapid and extensive growth of "self employed labour" in the construction industry (Harvey, 2002). Ofori and Debrah (1998) argued that the conditions for the increasing use of peripheral workers in construction in Singapore are different from those suggested to be underlying the quest for "flexibility" in industrialised countries (see Atkinson, 1984). They noted that the practice has considerable adverse effects, although firms derive some of the benefits relating to labour market flexibility. They proposed possible measures for improving construction labour use strategies in Singapore. The practice of labour subcontracting is blamed for the poor quality, low productivity, and high level of wastage of materials in construction (Winch, 1994; Construction 21 Steering Committee, 1999).

The benefits subcontracting offers to contractors include (Ofori and Debrah, 1998; Wells, 2006; Raidén et al., 2007): flexibility in the recruitment of labour which makes it a variable cost, which is important in construction due to the firms' fluctuating workload and changing product mix and thus, the variety of construction skills; to further reduce their labour costs by avoiding overhead costs associated with the employment of labour and the "on-costs" associated with legal employment including social security coverage, paid vacations, etc.; the delegation of responsibility for supervising the labour force and the risk in getting the work done (which is referred to by many authors over the years such as Bresnen et al., 1985; Ofori and Debrah, 1998; Leonard, 2000); it simplifies the estimating process, as the subcontractors bid for specific packages of the work, saving further on overhead costs.

There have been fewer studies of subcontracting from the perspective of the worker. Wells (2006) noted that in the United Kingdom, while a survey found that at least a proportion of the construction workforce prefers the status of self-employment, which can provide higher wages but at the expense of regular work and social security benefits (Nisbet and Thomas, 2000) others argue that the self-employed workers are disadvantaged in terms of both income and employment security (Harvey, 2002). However, for the majority of construction workers in developing countries, the choice, in practice, is between employment with a labour subcontractor on a regular basis or trying to find work (with a main or subcontractor) as an individual worker in the casual labour market. Wells (2006) discusses some of the social dynamics among workers of subcontractors which offer benefits to workers, and opportunities for further training and development.

Ocean (2021) highlighted several characteristics of construction. The one related to the discussion here is the problem of "pointing fingers or the blame game" – when something goes wrong, "the contractor points fingers at subcontractors, the client blames the contractor, and gets reprimanded by the project manager". He suggested taking up a builders' risk policy, a type of property insurance that covers the majority of possible problems that may or may not happen during the construction process (from vandalism and theft to natural disasters and extreme weather conditions). The second related characteristic is "Forgetful" clients – for example, where a client requests a change mid-project and then pretends to "forget" it when paying for the project upon completion. He suggests that, to avoid these situations, contractors should get a signed change order for each change in the project.

Impact on labour

For the worker, the location specificity of construction projects means that their skills are transferable to countries where similar items are to be built, and they are permitted to operate.

In some countries, such as Singapore and the Middle Eastern countries, the foreign workers form the bulk of the workforce. This issue is discussed in detail in Chapter 12 of this book. The United Kingdom government, in response to calls by the industry organisation Build UK (Fahy, 2023) is considering including construction as a key sector on the government's "shortage occupation list" to address the shortage (Boycott-Owen, 2023). This would allow employers to bring in staff on a lower salary than the current skilled worker threshold of £25,600 and reduce the visa fees. The Migration Advisory Committee recommends that bricklayers, carpenters and plasterers be included. It is suggested that foreign workers tend to be exploited and have to work in harsh physical conditions, live in poor conditions and earn low wages (see, for example, ILO (1995) a global report on the main destinations of such workers; and ILO (1996) on their low wages). Wells (2018) addresses the issue of late or non-payment of foreign workers. (This subject is discussed in Chapter 7 of this book.) Countries have different control measures. They include the need to apply for permits; quotas on their numbers; being tied to the employer who brought the worker into the country; and various other restrictions (ILO, 1995, 2010; OHCHR, 2013; Faraday, 2021). There are many protectionist measures using labour. Some countries limit the entry of foreign companies through various means such as stipulating types of qualifications and language requirements for professionals (see, for example, Nah, 2012). The control is intended to dissuade the contractors from relying on cheap labour and ignoring investment in labour-saving technologies. Foreign workers are blamed for poor productivity and quality of work and difficulties in managing them because of the language barrier. They also do not have the right attitude to safety. In Singapore, the foreign worker must be trained and certified before being granted a visa and work permit, and must have passed a safety course before starting work on-site.

Another aspect which needs addressing is multi-level subcontracting which has negative effects on the quality of work as it squeezes the margins, leaving little for investment in training and development of workers; and in equipment, or research. The national reviews of the construction industries in Hong Kong, SAR and Singapore sought to eliminate it (Construction 21 Steering Committee, 1999; Construction Industry Review Committee, 2001). Subcontracting can be better performing if actions are taken. These include improved and more continuous relationships between main and subcontractors both recognising the potential benefits of the symbiosis. Practice in Japan offers an example of such practices (Reeves, 2002; Matsumoto, 2007; An and Watanabe, 2011). The workers of the subcontractors benefit from the greater continuity of work.

Multiplicity and variety of activities

Feature of construction

The construction project comprises a very large number of separate but closely interrelated and interdependent activities. The activities are of different sizes and duration. Some are done in conjunction with other tasks. Each of these activities should be undertaken by a team with the right mix of skills. A number of gangs are put together to do the same kind of work, and this might take place simultaneously on different floors of the building. Each of these teams is on the site only to undertake their task. It is possible that one crew will work on more than one site, on any given day. Thus, special action is needed to monitor who enters and leaves the site each day. Determining the labour requirements of the project in order to complete it within the period set out in the contract is a challenge. The supervisors are faced with the problem of human resource optimisation on the project. In Singapore, the workforce required for each project is determined using a formula put together by the main industry stakeholders, to form the basis for calculating the Man-Year Entitlement (MYE) which indicates the number of foreign workers that can be recruited for the job; the formula and approach are outlined at: Ministry of Manpower (2023a).

Again, the coordination of the multiplicity of discrete operations on a single project is a major challenge, owing to the large volume of the activities and the fact that they take place in different isolated places on the site.

Implications for labour

Construction companies have sought to apply both technology and management approaches in a bid to find ways to reduce the duration of the project. Apart from the application of mechanisation and now the application of advanced ICT including robotisation, as discussed in Chapter 2, there are novel complete approaches. An example is top-down construction that enables the superstructure to be built without having completed the substructure.

The industry's difficulty in attracting potential workers is further amplified by this feature which brings up the issue of decent jobs. Construction work on site is characterised by hard physical work for low pay (Wells, 2013). The range of skills is very wide, as discussed in Chapter 2. (Singapore has attained a workforce comprising only skilled workers (Ministry of Manpower, 2023b)). There are major differences in wages among the skilled workers in any country (McAnaw Gallagher, 2022); and also among countries (Turner & Townsend, 2023). The factors which determine wages have not been explored. What is clear is that it does not necessarily depend on skills and experience. For example, ethnicity is a possible factor, as noted by McAnaw Gallagher (2022) in the US construction industry.

Given the possibility of setting out particular tasks requiring particular types of skills, subcontracting and self-employment are possible in construction, the latter even for the workers of the subcontractor. Moreover, the interface of construction activity with other sectors brings up the issue of the interface with relevant public agencies. It can be suggested that construction activity supports the existence of certain public agencies and their employees, such as the planning, and building control authorities.

As discussed in Chapter 2, planning, designing and delivering projects intended to promote personal safety and security in towns and cities has a wide impact on the community. Another is designing for health and recuperation. For example, open areas can help to promote healthy living. Parks can provide facilities such as outdoor exercise equipment which help to make what might only be available in private gymnasiums accessible to all, towards promoting health among the population.

The involvement of foreign workers brings up the issue of communications and differences in culture among the workers, and managing these factors appropriately to ensure that they do not have a negative impact on the work.

Role of government

Explanation of the feature

Government plays a role in the construction industry which has many dimensions. This has been highlighted since the emergence of Construction Economics in the late 1960s

(Turin, 1969; Hillebrandt, 2000). Government is a major client of the construction industry as it is the sole investor in public goods such as roads, airports, harbours and power, gas and water services. Although, with Public-Private Partnership (PPP) projects, the government's involvement has been decreasing, this is still a dominant role in all countries. Government is said to use construction as "a balance wheel of the economy" by increasing or reducing its investment. Second, the government can use its bargaining power to bring about change in the industry. Moreover, the government is a beneficiary of a well-performing construction industry delivering the infrastructure on time, below cost, to high levels of quality and so on. Thus, the government, through its own procurement arrangements, can influence change in the construction industry (Ofori, 2022a). For example, in the United Kingdom, Infrastructure and Projects Authority (2016) noted that: "We need to ensure that taxpayers' money is spent carefully; that projects are delivered on time and to budget; and that government uses its muscle to help our economy become more competitive – by helping SMEs bid for contracts or improving our workforce's skills". The *Government Construction Strategy 2016–20* set out how this would be done (HM Government, 2015). It is also pertinent to note that government sets the PPP ground rules, and manages the programme.

Through regulation, the government can dictate the technologies adopted in construction, such as industrialised building (as in Malaysia, CIDB, 2023), sustainable construction (as in Singapore, BCA, 2023c), and application of information and communication technology (ICT) (as in Singapore, BCA, 2023d). Fourth, government regulation can have a big influence on procedures and practices in the industry, especially in health and safety (Health and Safety Executive, 2022) and operating costs.

Government is a regulator of construction, given its public health and safety implications. Construction is referred to as *one of the most regulated sector of the economy* (Ofori, 2022a). For a typical construction project, the government's regulatory involvement starts at the (physical) planning stage and continues through the design and construction stages to the commissioning and in some cases, certification, of the built item and throughout the entire life of the built item. It involves almost all the activities on the project, the material inputs and in some cases, the methods applied. In all countries, the main reasons for formulating and enforcing building regulations are safety and health. Again, the government can use this to introduce, demand or encourage change in construction.

Government is a facilitator of the construction process and enterprises. Through incentives, the government can spur investment in certain areas of the country, such as in enterprise zones. It can also guide the adoption of technologies, through its own procurement, and through the provision of incentives (BCA, 2023e). Governments also provide incentives to nudge companies and individuals towards desirable technologies and practices (see, for example, BCA, 2023c).

Government is responsible for managing the development of the industry. Most governments recognise the clear benefits of improving construction industry as mentioned above and have formulated appropriate policies and strategies, often after major reviews of the industries (see, for example, Construction 21 Steering Committee, 1999; CIDB, 2015; Development Bureau, 2018). Few governments have established dedicated agencies and set policies for developing the industry. The examples are the Construction Industry Council in Hong Kong, SAR; Construction Industry Development Boards in Malaysia, Mauritius and South Africa; BCA in Singapore (established in 1984); National Construction Council in Tanzania (the first such agency, which started operating in 1981; and National Construction Industry

Councils in Malawi and Zambia. As suggested by Ofori (2016), the main task of such an agency is to manage the expansion and continuous development of the construction industry with the view to building up its capacity and capability to improve its performance in order to deliver the infrastructure required by the nation for its development.

Implications for labour

Developing human resources for construction is an important component of industry development (Ofori, 2022b). As Debrah and Ofori (2001) noted, it is a task that all governments tackle, through their further and higher education, and technical, vocational education and training (TVET). This forms the subject of Chapter 5 and is also discussed in Chapter 2. It can be suggested here that government should consider the education and training needs in construction when formulating laws, regulations and policies in a wide range of areas. Government should take the lead in planning for, funding and delivering and monitoring education and training in construction as it is necessary if it (the government) is to realise its policy intentions in many areas.

Government can influence the way projects are designed and built through its own procurement approaches. Also, by regulating some elements of types of work such as buildability and constructability (to facilitate construction and boost productivity) as in Singapore and Malaysia with the minimum buildability and constructability scores; and industrial building system proportions respectively; and performance such as in quality and environmental considerations, again, as in Singapore with the minimum environmental performance stipulated in building control legislation (BCA, 2023c, 2023d). Government can also use its investment in construction to create employment directly in desired parts of the country, as noted above.

The regulations in health and safety are among the most commonly applied. This is helping to address the poor safety records in construction which are also commonly seen in most countries, as discussed in Chapter 8. Increasingly, safeguarding and worker welfare are key considerations.

High cost and indivisibility

Explanation of the feature

Constructed items are very expensive. The uncompleted item cannot be used well, or often, not for any purpose. Owing to the high value of each project, there is an increased risk, as a failure on any project can put the whole company at risk. This characteristic of a construction project also limits the number of jobs a firm can take on at any one time. The difficulty of the decision on the optimal volume of work the company can have at any time is exacerbated by the fact that the firm will be working on projects at different stages of completion and each of them would have its different set of circumstances.

The clientele of significant projects in the construction industry is a small proportion of the population. For example, for individuals, a house is considered to be the largest investment a person would make in their life. Access to mortgage finance enables the effective demand to be executed. Such access is conditional upon a number of factors including salary. In many countries, the overwhelming majority of construction workers are excluded. The high cost also means that companies need to borrow working capital as, in most countries, there is no advance payment and the client pays the contractor in arrears, usually monthly, upon inspection and certification of the work done. Contractors can use what they borrow to manage their finances appropriately. Construction is said to offer high gearing potential.

As projects are expensive and last for a number of years, on public projects, there is a mismatch with an annual basis of budgeting of the government. This causes delays in payment. The delays are due to the bureaucracy in disbursing the funds to the agencies to which it is allocated and in the payment procedure. This is commonly cited in research as one of the reasons for both time and cost overruns. The medium-term national budgeting approach developed and promoted by the World Bank has not taken off (World Bank, 2012).

Implications for labour

Labour costs are a significant component of the total cost of a project. Johari and Jha (2020) found that they could be 50–60 percent of the total cost. Thus, firms endeavour to minimise it. This explains not only the low level of wages in construction, but also, the practice of subcontracting, engagement of self-employed tradespersons and casualisation of labour to avoid oncosts and other employment costs.

Educating and training to enhance the quality of the workforce is key to realising higher efficiency and productivity in construction. As discussed elsewhere in this chapter and others in the book, this calls for the commitment of governments and industry to enhance the quality of the workforce.

Long period of gestation

Explanation of the feature

Constructed items take a long period to complete. This is because of the many tasks that have to be accomplished, the relative labour intensity of the work; the need to wait for concrete to set; delays caused by the bureaucratic processes in projects; and the relative low productivity of construction work. The construction project is also well known for time overruns.

The long period makes the project susceptible to any adverse effects of many possible occurrences during the period, including (Hillebrandt, 2000; Ofori, 2022a): changes in the property market which might affect the sales or values of the completed items; increases in the prices of inputs – materials, wages, increase in interest rates (which affect the client); increases in the rate of inflation; and changes in regulation which might have consequences for aspects of the project. Also likely are changes in government and impact of political patronage and continuing flow of funding for the project. Political or social instability might also affect the project.

Implications for labour

The long period of gestation means that, whereas each construction project is a temporary endeavour, it can provide sound employment for reasonably long periods (for the trades which are involved in many tasks on the project). On the other hand, developments in the operating environment might affect the promptness of payments to the contractor, and have an impact on the regularity of wages. This issue is considered in detail in Chapter 7. Again, this brings up the importance of training and development to enhance the skills in construction, as it will help to increase the productivity of the workers and minimise possible delays to projects.

Long life of the item

Another aspect of longevity in the construction industry is that construction products have a long period of life. Buildings and other items of infrastructure are designed to have a service life of a minimum of around 30 years. This has a number of related issues:

- the completed building, especially commercial or health buildings, creates, by itself, a set of operations in Facility Management, a distinct segment of the built environment sector
- the building generates additional work on occasion to maintain, repair, alter or refurbish it which might enhance the value and utility of the item, and/or lengthen its potential service life
- the building forms the basis of more additional work to extend it. In these regards, the added parts of the building might be intended to perform a task different from that of the original
- the building can be redeveloped if the planning regulations change, and it is possible to develop the same site to a higher level of intensity.

Segerstedt and Olofsson (2010) assert that the long durability of the construction "product" contributes to the volatility of the construction market. However, buildings seldom reach their designed service life, or a state of physical obsolescence. This is because various forms of obsolescence set in during the life of the building. In real estate, three forms are usually identified: functional, economic and physical obsolescence (see, for example, Australian Property Institute, 2008). Reed and Warren-Myers (2010) argue that "sustainability" could be the fourth form of obsolescence. They stated: "concept of sustainable obsolescence applies to a broad range of considerations including the location of the property (locational sustainability), the amount of embodied energy in the building, the energy efficiency levels of the building and the degree of social sustainability (p. 1).

On sustainability, calls are being made for better design and specification, changes in procurement, materials used and in construction processes, practices and methods, and reuse of buildings and materials for the sector to achieve Net Zero Carbon Emissions by the year indicated by the country. For example, Green Alliance (2023) suggests that moving onto a more circular construction sector in the United Kingdom, prioritising reuse, re-manufacturing and recycling, could significantly reduce the environmental footprint of the industry while still providing the housing and infrastructure the country needs. There is an upsurge in the retrofitting of buildings, as it has become evident that the stock of buildings is a major drag on the desire of nations to attain net zero by 2050 (Green Alliance, 2023). This will lead to additional work for labour in construction. Indeed, as mentioned in Chapter 2, there are concerns that the non-availability of the required labour will make countries unable to meet their Net Zero commitments.

Concluding remarks

Construction has great potential to contribute to the effort towards the attainment of all the SDGs. In doing this, the most relevant issue in the context of this chapter is the fulfilment of the Sustainable Development Goals (SDGs) relating to decent work with respect to construction labour and labour in other sectors. The principles of the San Marino Declaration adopted in 2022 by the UN Economic Commission for Europe (Bureau of the Committee on Urban Development, Housing and Land Management, 2022) encapsulate what construction can do in respect of workers in other sectors, the community and society. It is instructive to outline the principles here (Kayatekin, 2022): social responsibility, inclusivity, and participatory mechanisms to anchor built-environmental work; projects to respect the local socio-cultural fabrics within which they are situated; the built environment to minimise its carbon, energy, and water footprints, limit material extraction, and become a more active player in urban food production; the promotion of multi-modal transport systems and mobility networks, increasing access to urban green areas, and enforcing fundamental standards of built environmental health and safety; limiting impacts on the local environment and biodiversity; minimising carbon, energy, and emissions footprints and adding to local energy production; promoting urban transparency, this helps the vulnerable, and curbs corruption anchored in the development and maintenance of the built environment; strengthening the resilience of the building stock and urban infrastructure with regard to natural disasters – with a specific emphasis placed upon durability, flexibility, and adaptability; assuring the affordability and accessibility of the built environment to all citizenry; supporting cohabitation, community participation, community engagement, and social cohesion; and promoting collaboration across stakeholder networks and to foster and support long-term built environmental research.

Among the characteristics of construction, the fact that the item is fixed where it is required to be built (location specificity) has the widest implications for labour in construction and in other sectors, as well as the community. Another feature is the project-based nature of construction. In the era of projectification of the economy, with all sectors adopting the project approach in many aspects of their activities, construction can offer lessons to other sectors of the economy with respect to working practices and the management of labour in such temporary organisations. This makes construction and its approaches worth exploring, in particular, teamwork, the matrix structure of work organisation. Also worth studying for possible lessons for other sectors is working in a situation of interdependent sequential activities, especially in an environment fraught with risk.

It is necessary to consider, in a holistic manner, how construction responds to its features, with its approach to labour. So far, the approach has been studied in fragments, such as subcontracting and self-employment of workers. The relationships among the approaches which have led to silos of considerations need to be studied.

The construction worker is an important factor in the socio-economic development of the nation as their efforts result in the provision of the physical facilities which enable all economic and social activities by businesses large, small and micro-sized, and individuals including the poor. Effectively, these forward linkages of construction make construction activities determinants of how quickly companies can respond to developments in their particular markets. The efficiency and productivity of the construction worker have implications downstream for companies in other sectors, and by implication, their labour. This is an important factor in industries such as electronics and computer manufacturing which require short lead times. It is necessary to ensure the continuous development of the skills and improvement in the performance of the worker. The construction workers need to be given continuous training and development to enhance their skills and capabilities, and to be able to earn higher wages. Education and training programmes should cover the sustainability impacts of construction and how they are to be avoided or addressed. New skills necessary for sustainable construction might be created.

Finally, it is pertinent to consider how to engender optimal economic impact of the construction project, in the context of the desire of the developing countries to progress beyond reliance on overseas development assistance, such as the "Ghana Beyond Aid" programme. For example, how the country can build a whole ecosystem of the supply chain locally, including materials and equipment manufacturing and supply.

References

Adato, M. and Haddad, L. (2001) Targeting poverty through community-based public works programmes: Experience from South Africa. *Journal of Development Studies*, 38(3), pp. 1–36. 10.1080/00220380412331322321

Agbaxode, P.D.K., Saghatforoush, E., and Dlamini, S. (2023) Assessment of the impact of design documentation quality on construction project delivery. *Journal of Engineering, Project, and Production Management*, 13(2), pp. 81–92.

Ahluwalia, J. (2023) How real estate can use social value to connect with communities, *Estates Gazette*, https://main-estatesgazettebolt-rbi.content.pugpig.com/2022/11/16/how-real-estate-can-use-social-value-to-connect-with-communities/content.html

An, T. and Watanabe, T. (2011) Towards the sustainable improvement in construction labour's employment in China: Cultivating subcontractor's role. In C. Egbu and E.C.W. Lou (Eds) *Proceedings, 27th Annual ARCOM Conference, 5-7 September*, Bristol, UK. Association of Researchers in Construction Management, 663–672.

APM (2022) What is a project? https://www.apm.org.uk/resources/what-is-project-management/

Atkinson, J. (1984) Manpower strategies for flexible organisations. *Personnel Management*, August: 28–31.

Australian Property Institute (API) (2008) *Glossary of Property Terms*. API, Canberra.

Awakul, P. and Ogunlana, S.O. (2002a) The effect of attitudinal differences on interface conflicts on large scale construction projects: The case of the Pak Mun Dam project. *Environmental Impact Assessment Review*, 22(4), pp. 311–335.

Awakul, P. and Ogunlana, S.O. (2002b) The effect of attitudinal differences on interface conflicts in large scale construction projects: A case study. *Construction Management and Economics*, No. 20, pp. 365–377.

Boycott-Owen, M. (2023) Foreign construction workers set to be brought in to fill labour shortages. *The Yorkshire Post*, 10 March, https://www.yorkshirepost.co.uk/news/politics/foreign-construction-workers-set-to-be-brought-in-to-fill-labour-shortages-4058313

Bresnen, M.J., Wray, K., Bryman, A., Beardsworth, A.D., Ford, J.R. and Keil, E.T. (1985) The flexibility of recruitment in the construction industry: Formalisation or re-casualisation? *Sociology*, 108–124.

Bureau of the Committee on Urban Development, Housing and Land Management (2022) *San Marino Declaration – Note by the Bureau of the Committee on Urban Development, Housing and Land Management*. Economic Commission for Europe, https://unece.org/sites/default/files/2022-12/ECE_HBP_2022_2_REV-E.pdf

Bentall, P., Beusch, A. and de Veen, J. (1999) *Employment-Intensive Infrastructure Programmes: Capacity Building for Contracting in the Construction Sector – Guidelines*. International Labour Office, Geneva.

Bowen, P.A., Govender, R., Edwards, P.J. and Cattell, K. (2014) An integrated model of HIV/AIDS testing behaviour in the construction industry. *Construction Management and Economics*, 32(11), pp. 1106–1129, DOI:10.1080/01446193.2014.958509

Buildings Department (2018) New Building Works – Roles of Buildings Department, https://www.bd.gov.hk/en/building-works/new-building-works/index.htm

Building and Construction Industry (BCA) (undated) *A Quick Guide: Security of Payment Act*. Singapore, https://www1.bca.gov.sg/docs/default-source/docs-corp-regulatory/sop_brochure.pdf?sfvrsn=3777e0ae_4

BCA (2023a) Media Release – Singapore's Construction Demand to Remain Strong in 2023, 12 January, https://www1.bca.gov.sg/about-us/news-and-publications/media-releases/2023/01/12/singapore's-construction-demand-to-remain-strong-in-2023
BCA (2023b) Registration of Multi-Skilling Personnel. https://www1.bca.gov.sg/buildsg/manpower/bca-approved-training-and-testing-centres/coretrade_faqs/registration-of-multi-skilling-personnel#:~:text=The%20Multi%2DSkilling%20scheme%20was,)'s%20Foreign%20Worker%20Levy
BCA (2023c) Green Mark Certification Scheme. https://www1.bca.gov.sg/buildsg/sustainability/green-mark-certification-scheme
BCA (2023d) Building Plan Submission. https://www1.bca.gov.sg/regulatory-info/building-control/building-plan-submission
BCA (2023e) Other Incentives Scheme, https://www1.bca.gov.sg/buildsg/productivity/other-incentives-scheme
BCA (2023f) Green Mark Incentive Schemes. https://www1.bca.gov.sg/buildsg/sustainability/green-mark-incentive-schemes
Cabinet Office (2021) Guidance – Social Value Act: Information and Resources, Updated 29 March. https://www.gov.uk/government/publications/social-value-act-information-and-resources/social-value-act-information-and-resources
Çelik, G.T. and Oral, E.L. (2019) Mediating effect of job satisfaction on the organizational commitment of civil engineers and architects. *International Journal of Construction Management*, 21(10), pp. 969–986, DOI: 10.1080/15623599.2019.1602578
Considerate Constructors Scheme (2023) About the Considerate Constructors Scheme, https://www.ccscheme.org.uk/
Construction 21 Steering Committee (1999) *Reinventing Construction*. Ministry of Manpower and Ministry of National Development, Singapore.
Construction Industry Review Committee (2001) *Construct for Excellence – Report of the Construction Industry Review Committee*. Hong Kong, SAR, https://www.legco.gov.hk/yr00–01/english/panels/plw/papers/plw0611-487e-scan.pdf
Construction Industry Development Board (CIDB) (2015) *Construction Industry Transformation Programme*. Kuala Lumpur.
CIDB (2023) Industrialised Building System (IBS). Kuala Lumpur, https://www.cidb.gov.my/eng/ibs/
Construction Leadership Council (2018) *Procuring for Value: Outcome based, transparent and efficient*. London. https://www.constructionleadershipcouncil.co.uk/wp-content/uploads/2018/07/RLB-Procuring-for-Value-18-July-.pdf
Debrah, Y.A. and Ofori, G. (2001) The state, skill formation and productivity enhancement in the construction industry: The case of Singapore. *The International Journal of Human Resource Management*, 12(2), pp. 184–202, DOI: 10.1080/713769601
Debrah, Y.A. and Ofori, G. (2001) Subcontracting, foreign workers and job safety in the Singapore construction industry. *Asia Pacific Business Review*, 8(1), pp. 145–166, DOI: 10.1080/713999129
Decker, R. (2021) Measuring business exit, Prepared for FESAC, December 10. US Bureau of Economic Analysis, https://apps.bea.gov/fesac/meetings/2021-12-10/Decker.pdf
Department of Business, Energy and Industrial Strategy (2022) Business Population Estimates for the UK and the Regions 2022. https://assets.publishing.service.gov.uk/government/uploads/system/uploads/attachment_data/file/1106039/2022_Business_Population_Estimates_for_the_UK_and_regions_Statistical_Release.pdf
Development Bureau (2018) *Construction 2.0 – Time to change*. Hong Kong, SAR.
Di Leva, E. (2021). The challenge and promise at the intersection of environmental and social policies: How the World Bank established a policy framework that fully integrates environmental and social concerns. *Global Social Policy*, 21(2), pp. 344–348. 10.1177/14680181211019170
Di Maddaloni, F. and Sabini, L. (2022) Very important, yet very neglected: Where do local communities stand when examining social sustainability in major construction projects? *International Journal of Project Management*, 40(7), pp. 778–797, 10.1016/j.ijproman.2022.08.007.
Ditangco, R., Kanda, K., Obayashi, Y., Matibag, G. and Tamashiro, H. (2008) Mainstreaming HIV/AIDS program into infrastructure development and community preparedness. *AIDS* – Correspondence, 22, pp. 167–171. DOI: 10.1097/QAD.0b013e3282f2c94b
Dudley, E. (1993) *The Critical Villager: Beyond Community Participation*. Routledge, London.

Durham County (2020) Home Occupation Permit – City-County Planning Department, Durham, NC, https://www.durhamnc.gov/DocumentCenter/View/1078/Home-Occupation-Application-PDF

Fahy, M. (2023) Construction industry calls for overseas labour help. *Investors' Choice*, March 16, 2023, https://www.investorschronicle.co.uk/news/2023/03/16/construction-industry-calls-for-overseas-labour-help/

Faraday, F. (2021) *The Empowerment of Migrant Workers in a Precarious Situation – Mapping Barriers to Migrant Workers' Rights Enforcement*, KNOMAD Paper No. 39. World Bank, Washington, DC. https://www.knomad.org/sites/default/files/2021-12/KNOMAD%20Working%20Paper%2039-Empowerment%20of%20Worker-Dec%202021.pdf

Government Gazette (2017) National Treasury: Preferential Procurement Policy Framework Act, 2000: Preferential Procurement Regulations, 2017. 20th January, http://www.thedtic.gov.za/wp-content/uploads/PPPFA_Regulation.pdf

Government of Singapore (2020) Building and Construction Industry Security of Payment Act 2004, 2020 Revised Edition. Singapore, https://sso.agc.gov.sg/act/bcispa2004

Government of Victoria (2013) Building and Construction Industry Security of Payment Act 2002, incorporating amendments as at 1 July 2013. Melbourne. https://content.legislation.vic.gov.au/sites/default/files/cc273c60-f6d4-3c6b-8059-b7da96d139ae_02–15aa012%20authorised.pdf

Government of UK (1996) *Housing Grants, Construction and Regeneration Act, 1996: Part II Construction Contacts*. https://www.legislation.gov.uk/ukpga/1996/53/contents

Green Alliance (2023) What's Stopping Construction from Becoming More Resource Efficient? https://green-alliance.org.uk/event/whats-stopping-construction-from-becoming-more-resource-efficient/

Government of Sarawak (2019) Laws of Sarawak – Chapter 76 Strata Management Ordinance, 2019. Kuching, https://jurusurveikonsult.com/img/about/Strata%20Management%20Ordinance%202019%20(Chapter%2076).pdf

Harvey, M. (2002) The United Kingdom: Privatization, fragmentation and inflexible flexibilization in the UK construction industry. In G. Bosch and P. Philips (Eds) *Building Chaos: An international comparison of deregulation in the construction industry*. Routledge, Abingdon, pp. 188–209. 10.4324/9780203166130

Health and Safety Executive (HSE) (2022) *Construction Statistics in Great Britain, 2022*. Bootle, https://www.hse.gov.uk/statistics/industry/construction.pdf

Hillebrandt, P.M. (2000) *Economic Theory and the Construction Industry*. Macmillan, Basingstoke.

HM Government (2012) Public Services (Social Value) Act, 2012. https://www.legislation.gov.uk/ukpga/2012/3/enacted

HM Government (2015) The Construction (Design and Management) Regulations 2015. https://www.legislation.gov.uk/uksi/2015/51/contents/made

Hollnsteiner, M.R. (1976) People power: Community participation in the planning and implementation of human settlements. *Philippine Studies*, 24(1), pp. 5–36. http://www.jstor.org/stable/42632307

HS2 (2022) Our responsible business, https://www.hs2.org.uk/about-us/responsible-business/

Infrastructure and Projects Authority (2016) *Government Construction Strategy 2016–20*. London.

Infrastructure Australia (2021) Reforms to meet Australia's future infrastructure needs: 2021 Australian Infrastructure Plan. Canberra, https://www.infrastructureaustralia.gov.au/sites/default/files/2021-09/2021%20Master%20Plan_1.pdf

Infrastructure and Projects Authority (2017) Analysis of the National Infrastructure and Construction Pipeline. London, https://assets.publishing.service.gov.uk/government/uploads/system/uploads/attachment_data/file/665332/Analysis_of_National_Infrastructure_and_Construction_Pipeline_2017.pdf

International Labour Office (ILO) (1996) Migrant Construction Workers Facing Pressure on Wages and Working Conditions, ILO/96/4, https://www.ilo.org/global/about-the-ilo/newsroom/news/WCMS_008082/lang–en/index.htm

International Labour Organisation (ILO) (1995) Social and Labour Issues Concerning Migrant Workers in the Construction Industry. International Labour Organisation; Tripartite Meeting on Social and Labour Issues Concerning Migrant Workers in the Construction Industry. Geneva.

International Labour Organisation (2010) *International Labour Migration: A Rights-Based Approach*. Geneva.

Japan Free Trade Commission (2023) Subcontractor Act – up to the revision in Act No. 87, 2005, https://www.jftc.go.jp/en/legislation_gls/subcontract.html

Johari, S. and Jha, K.N. (2020) Challenges of attracting construction workers to skill development and training programmes. *Engineering, Construction and Architectural Management*, 27(2), pp. 321–340. 10.1108/ECAM-02-2019-0108

Jones, P. (undated) How insolvency affects companies in the construction sector. The Gazette, https://www.thegazette.co.uk/insolvency/content/103418

Kayatekin, C. (2022) Reflections on the San Marino Declaration, Buildings & Cities, 15 December, www.buildingsandcities.org/insights/commentaries/san-marino-declaration.html

Klunklin, A. and Greenwood, J. (2005) Buddhism, the status of women and the spread of HIV/AIDS in Thailand. *Health Care for Women International*, 26(1), pp. 46–61, DOI: 10.1080/07399330590885777

Leonard, M. (2000) Coping strategies in developed and developing societies: The workings of the informal economy. *Journal of International Development*, 12, pp. 1069–1085.

Lizarralde, G., Tomiyoshi, S., Bourgault, M., Malo, J. and Cardosi, G. (2013) Understanding differences in construction project governance between developed and developing countries. *Construction Management and Economics*, 31(7), pp. 711–730, DOI: 10.1080/01446193.2013.825044

Mahato, K.B. and Ogunlana, S.O. (2011) Conflict dynamics in a dam construction project: A case study. *Built Environment Project and Asset Management*, 1(2), pp. 176–194. 10.1108/20441241111180424

Makenya, A.R. and Mng'ong'o, G.C. (2018) Analysis of participation of labour based road contractors in rural road works in Tanzania. *International Journal of Science and Research*, 7(2), DOI: 10.21275/ART20179739

Matsumoto, T. (2007) Main contractor and subcontractors relationship in international construction project: Japanese contractor's perspective. Doctoral thesis, University College London, https://discovery.ucl.ac.uk/id/eprint/1569427/

McAnaw Gallagher, C. (2022) The construction industry: Characteristics of the employed, 2003–20, US Bureau of Labor Statistics, https://www.bls.gov/spotlight/2022/the-construction-industry-labor-force-2003-to-2020/home.htm

Ministry of Manpower (2023a) Man-year entitlement (MYE). Singapore, https://www.mom.gov.sg/passes-and-permits/work-permit-for-foreign-worker/sector-specific-rules/man-year-entitlement

Ministry of Manpower (2023b) Basic-Skilled workers for construction sector. https://www.mom.gov.sg/passes-and-permits/work-permit-for-foreign-worker/sector-specific-rules/basic-skilled-workers

Mulaudzi, G. and Lancaster, L. (2021) South Africa's construction mafia trains its sights on local government, ISS Today, 25 October, Institute of Security Studies, Pretoria. https://issafrica.org/iss-today/south-africas-construction-mafia-trains-its-sights-on-local-government

Murphy, M. and Eadie, R. (2019) Socially responsible procurement: A service innovation for generating employment in construction. *Built Environment Project and Asset Management*, 9(1), pp. 138–152. 10.1108/BEPAM-02-2018-0049

Nah, A.M. (2012) Globalisation, sovereignty and immigration control: The hierarchy of rights for migrant workers in Malaysia. *Asian Journal of Social Science*, 40(4), pp. 486–508.

New Zealand Infrastructure Commission (2022) *Rautaki Hanganga o Aotearoa 2022–2052 New Zealand Infrastructure Strategy*. New Zealand Government, Wellington.

Nisbet, P. and Thomas, W. (2000) Attitudes, expectations and labour market behaviour: The case of self-employment in the UK construction industry. *Work Employment and Society*, 14(2), pp. 353–368.

Ocean, J. (2021) Glossary: Top 12 Construction Issues & Challenges, https://revizto.com/en/construction-issues-challenges/

Ofori, G. (1992) The environment: The fourth construction project objective? *Construction Management and Economics*, 10(5), pp. 369–395, DOI: 10.1080/01446199200000037

Ofori, G. and Debrah, Y.A. (1998) Flexible management of workers: Review of employment practices in the construction industry in Singapore. *Construction Management and Economics*, 16(4), pp. 397–408, DOI: 10.1080/014461998372187

Ofori, G. (2022a) Construction economics: Its origins, significance, current status and need for development. In G. Ofori (Ed), *Research Companion to Construction Economics*. Elgar Publishing Ltd.: Cheltenham, pp. 18–40.

Ofori, G. (2022b) Defining and describing construction in developing countries. In R. Best and J. Meikle (Eds), Describing Construction: Industries, projects and firms. Routledge, Abingdon, pp. 171–196.
Ofori, G. (2016) Construction in developing countries: current imperatives and potential. In K. Kahkonen and M. Keinanen (Eds), CIB World Building Congress 2016 on Intelligent Built Environment for Life. Tampere 30 May - 03 Jul 2016 Tampere Department of Civil Engineering, Construction Management and Economics, Tampere University of Technology.
Raco, M. (2000) Assessing community participation in local economic development — lessons for the new urban policy. *Political Geography*, 19(5), pp. 573–599, 10.1016/S0962-6298(00)00004-4.
Raidén, A., Pye, M. and Cullinane, J. (2007) The nature of the employment relationship in the UK construction industry: A flexible construct? In A. Dainty and B. Bagilhole (Eds) *People and Culture in Construction*. Taylor and Francis, Abingdon, pp. 39–55.
Reed, R. and Warren-Myers, G. (2010) Is sustainability the 4th form of obsolescence? Paper presented at the 16th Pacific Rim Real Estate Society (PRRES) Conference, 24–27 January, Wellington, New Zealand.
Reeves, K. (2002) Construction business systems in Japan: General contractors and subcontractors. *Building Research and Information*, 30(6), pp. 413–424, DOI: 10.1080/09613210210159857
Ribeirinho, M.J., Mischke, J., Strube, G., Sjödin, E., Luis Blanco, J.L., Palter, R., Biörck, J., Rockhill, D. and Andersson T. (2020) *The Next Normal in Construction How Disruption is Reshaping the World's Largest Ecosystem*. McKinsey & Company.
Royal Institute of British Architects (RIBA) (2020) *RIBA Plan of Work 2020*. London. https://www.architecture.com/knowledge-and-resources/resources-landing-page/riba-plan-of-work
Segerstedt, A. and Olofsson, T. (2010) Supply chains in the construction industry. *Supply Chain Management*, 15(5), pp. 347–353. 10.1108/13598541011068260
Smulian, M. (2021) UK ranked as 'worst major market for profit margins'. *Construction News*, 22 July, https://www.constructionnews.co.uk/financial/uk-ranked-as-worst-major-market-for-profit-margins-22-07-2021/
Sullivan, G., Barthorpe, S. and Robbins, S. (2011) *Managing Construction Logistics*. John Wiley & Sons.
Teo, M.M.M. and Loosemore, M. (2010) Community-based protest against construction projects: The social determinants of protest movement continuity. *International Journal of Managing Projects in Business*, 3(2):216–235. DOI:10.1108/17538371011036554
The Construction Index (2022) Top 100 Construction Firms, https://www.theconstructionindex.co.uk/market-data/top-100-construction-companies/2022
Turin, D.A. (1969) *The Construction Industry: Its Economic Significance and its Role in Development*. University College Environmental Research Group: London.
Turner & Townsend (2023) International construction market survey 2022. https://www.turnerandtownsend.com/en/perspectives/international-construction-market-survey-2022/
United Nations, Office of the High Commissioner of Human Rights (OHCHR) (2013) *Migration and Human Rights: Improving Human Rights-based Governance of International Migration*. Geneva. https://www.knomad.org/sites/default/files/2021-12/KNOMAD%20Working%20Paper%2039-Empowerment%20of%20Worker-Dec%202021.pdf
Vidalakis, C., Tookey, J.E. and Sommerville, J. (2011) The logistics of construction supply chains: The builders' merchant perspective. *Engineering, Construction and Architectural Management*, 18(1), pp. 66–81. 10.1108/09699981111098694
Vosloo, C. (2021) A comparison of three public projects that included community participation to determine the total value add. *Acta Structilia*, 28(2), pp. 170–207.
Wegelius-Lehtonen, T. (2001) Performance measurement in construction logistics. *International Journal of Production Economics*, 69(1), pp. 107–116, 10.1016/S0925-5273(00)00034-7.
Wells, J. (2006) Labour subcontracting in the construction industries of developing countries: An assessment from two perspectives. *Journal of Construction in Developing Countries*, 11(1), pp. 17–36.
Wells, J. (2018) *Protecting the Wages of Migrant Construction Workers*. Engineers Against Poverty, http://engineersagainstpoverty.org/wp-content/uploads/2018/07/Protecting-the-Wages-of-Migrant-Construction-Workers.pdf

Wells, J. (2013) *Relieving Chronic Poverty Among Construction Workers: An exploration of possibilities to improve the quantity and quality of jobs*. Engineers Against Poverty, London, http://engineersagainstpoverty.org/resource/relieving-chronic-poverty-among-construction-workers-an-exploration-of-possibilities-to-improve-the-quantity-and-quality-of-jobs/

Williams, J.J. (2006) Community participation: Lessons from post-apartheid South Africa. *Policy Studies*, 27(3), pp. 197–217, DOI: 10.1080/01442870600885982

Winch, G. (1994) The search for flexibility: The case of the construction industry. *Work, Employment and Society*, 8(4), pp. 593–606.

World Bank (1999) Operations Manual: OP 4.01 - Environmental Assessment, https://web.worldbank.org/archive/website01541/WEB/0__-2097.HTM

World Bank (2012) Beyond the annual budget: Global experience with medium term expenditure frameworks. Washington, DC. 10.1596/978-0-8213-9625-4

4

EMPLOYMENT GENERATION IN THE CONSTRUCTION SECTOR

Volatility and prospects

Arup Mitra

South Asian University, New Delhi, India

Introduction

The construction sector is believed to be labour intensive but at the same time, it is highly sensitive to the short- and long-term business ups and downs. With a huge deceleration in economic growth, the real estate sector and construction activities suffer the most. This chapter proposes to reflect on the employment growth in the construction sector and the relationship of this sector with the aggregate GDP per capita. Secondly, the focus would lie on the recruitment processes in this sector which are seen to be very different from those in other activities. The role of the labour intermediaries and the denial of minimum benefits to the workers, even when the parent organisations are willing to adhere to the norms, are some of the common and prevalent practices followed within the construction sector. Besides, the gender discrimination in terms of wage differentials and varying duration of employment will also be considered.

The agglomeration literature suggests that the large cities offer better opportunities both in terms of employment scope and earnings compared to the medium-sized and small towns (Mitra, 2013; Mitra, 2020; Mitra and Nagar, 2018; Ramaswamy and Agrawal, 2012). Construction activities, in particular, are supposed to spread fast in the large cities as both industrialisation and commercialisation create huge demand. With migrants flowing in from different areas add to the demand for residential units and other commercial buildings. Thus, as construction activities shoot up, labour demand rises and migration of labour from rural areas takes place on a large scale. However, the workers often do not have access to decent living conditions and they reside near the work place in temporary slums under precarious conditions. The labour contractors do not offer any facility to the workers as they are keen on raising their profit margin. This study proposes to reflect on some of these issues relating to the overall wellbeing of the workers. How do the wages within the construction activities compare with the wages in other activities and can construction in cities be used as a strategy for employment generation and poverty reduction? Can urbanisation have growth and development spill-overs so that rural and urban livelihood opportunities are created in the process and poverty is reduced significantly? This is a key question which has been bothering researchers as well as policy makers for a long time. Agglomeration

DOI: 10.1201/9781003262671-4

economies are supposed to contribute to productivity growth and levels of living significantly. Productivity growth may reduce the input utilisation for the same amount of production to take place but that happens only at the enterprise level. In order to take advantage of the agglomeration benefits, more firms flow into the city space which then due to the scale effect creates more job opportunities (Eichengreen and Gupta, 2013; Ghose, Undated; Gordon and Gupta, 2003; Harriss and Todaro, 1970; Kundu and Mohanan, 2009; Lall, Harris and Shalizi, 2006; Luke and Munshi, 2006).

In the absence of adequate industrialisation of the workforce, the informal sector is said to grow rapidly and this being a repository of residual employment, migrants do not benefit significantly, and thus, rural poverty gets transferred to its urban counterpart (Dandekar and Rath, 1971). However, an alternative view is that even the informal sector holds prospects for upward mobility, and thus, urbanisation is associated with a decline in poverty. Hence, whether urbanisation is sustainable (or not) is an empirical question which needs to be investigated in the context of a given country. The UN definition of sustainability is very wide. **Sustainable urban development** is considered to be the way forward for cities to mitigate climate change. Integrated **urban** places are "designed to bring people, activities, buildings, and public spaces together, with easy walking and cycling connection between them and excellent transit service to the rest of the city".[1] The UN-HABITAT views sustainability more from the point of view of environmental issues. It is suggested that demographic and economic changes have propelled cities and urban centres to become the principal habitat of humankind, and at the same time, the cities are increasingly the drivers of global prosperity. However, the planet's resources are fast depleting, and hence, ways will have to emerge to achieve economic and socially equitable growth without further cost to the environment. The solution partly lies in how cities are planned, governed, and provide services to their citizens as poorly managed urbanisation can be detrimental to sustainable development. So, with growing global population, efforts will have to be made to create jobs, reduce ecological footprint, and improve quality of life in a holistic sense. On the whole, if sustainable urbanisation can be prioritised, it can eventually address many critical development challenges related to energy, water consumption and production, biodiversity, disaster preparedness, and climate change adaptation. Whether urbanisation is able to contribute to productive employment creation and poverty reduction is a key question. A strong empirical fact is that with a residual absorption in low-productivity activities and meagre earnings, households are forced to reside in unhygienic and inhuman conditions which have a strong spill-over effect on the urban environment and health. Considering this as the backdrop, it is argued in this chapter that the construction sector can emerge as a major outlet for employment generation and equitable growth.

Even during the pre-COVID years, livelihood problems had been serious in India as the agriculture sector has not been in a position to absorb any additional labour, while the non-agriculture activities either in the rural or urban areas have not been a preamble to gainful opportunities on a large scale, particularly for the unskilled and semi-skilled variety of workers. In the urban context, sluggish industrialisation in many parts of the country associated with the adoption of capital-intensive technology reduced the pace of industrialisation of workforce. The services sector's share, which shot up much before the per capita income rose to comparable levels as per the historical experience of the present developed nations, involves a significant dualism in terms of a high productivity segment and a vast spectrum of residual jobs, overlapping with the urban informal economy. The lack of regular jobs for a sizeable proportion of the workforce forced the casual labour market to become significant and the own account enterprises to emerge as a livelihood source of last resort. The casual job market encompasses a great degree of volatility as labour demand and supplies interact almost on a daily basis. On the other hand, the self-employed households are characterised by a gross deficiency of financial and tangible capital in relation to the availability of the human resource.

Livelihood diversification has been presented as an effective strategy for reducing the income risks and the consumption fluctuations. In other words, by shifting to different economic activities across seasons/months or taking recourse to multiple activities at a given point in time, household incomes can be augmented, and the probability of facing major decline in consumption and moving below the poverty line can actually be reduced. However, there are a number of inflexibility issues in relation to occupational mobility and income augmentation due to the lack of capital, credit, information, and skill. In the urban context, the shortage of space and enforcement of regulations make diversification more difficult. Rural to urban migration is envisaged as a household strategy to mitigate the livelihood challenges. Given the rural–urban development disparity and particularly the growth of large cities with concentration of economic activities and investment in a few centres, rural–urban migration is seen to have a strong large city bias. Hence, the large cities in general face greater problems relating to livelihood in spite of the fact that they have more opportunities than the small- and medium-sized towns. In the backdrop of all this, the outbreak of COVID and lockdown aggravated the labour market outcomes, especially in the large urban spaces which also registered higher incidence of the disease (both positivity and mortality rates) due to heavy concentration of population and congestion.

Since the industrial sector has been stagnating in terms of employment creation, it becomes pertinent to examine the potential that the construction sector may possess. This sector is expected to be more flexible as it can grow and generate employment in relation to the income growth originating from either the industry or the services sector.

Data

The share of construction in the occupational structure is quite small (around 10 percent of the workers) in relative sense (Table 4.1), though there are around 64 million workers in this

Table 4.1 Workforce structure in India

	2017–2018		*2018–2019*		*2019–2020*		*2020–2021*	
Sector	*Rural*	*Urban*	*Rural*	*Urban*	*Rural*	*Urban*	*Rural*	*Urban*
Agriculture	59.4	6.14	57.8	5.49	61.59	5.71	60.76	6.49
Mining and Quarrying	0.39	0.48	0.36	0.54	0.21	0.43	0.29	0.4
Manufacturing	7.78	22.97	7.75	22.41	7.26	20.79	7.55	20.29
Electricity, Gas and Water Supply	0.36	1.16	0.34	1.08	0.36	1.24	0.36	1.21
Construction	12.27	10.17	13	10.01	12.15	10.34	12.46	11.1
Wholesale and Retail Trade	7.9	22.1	8.39	22.85	7.54	27.38	7.72	24.84
Transport, Storage and Communication	3.98	10.78	4.06	10.47	3.78	10.17	3.67	10.21
Other Services	7.93	26.2	8.3	27.16	7.11	23.94	7.19	25.47
Total	100	100	100	100	100	100	100	100

Source: Periodic Labour Force Survey (PLFS), Government of India.

sector as per the 2020–2021 periodic labour force survey (PLFS). Further, the share of construction in the total gross value added (in constant prices) is even lower than the employment share, indicating that labour productivity in certain other sectors is higher (Table 4.2). However, labour productivity in the construction sector is higher than that in the agriculture sector. Given the fact that agriculture is not in a position to absorb labour any further, construction is indeed a much better outlet. Value added originating from the construction sector has grown at a rate of around 4 percent per annum while the aggregate value added (GVA) grew at around 6.5 percent per annum over 2011–2012 through 2019–2020. The average elasticity of the construction sector value added with respect to the aggregate GDP turns out to be 0.62 based on the year-to-year observations over the period mentioned above. This is indicative of the fact that the construction sector is quite responsive to the overall economic growth of the economy. In the fiscal year 2020, the real gross value added in construction industry across India increased by around 1.3 percent. However, subsequently, there was a decline in the GVA growth due to the impacts of the coronavirus (COVID-19) pandemic: the GVA of the construction industry is estimated to have decreased by 12.6 percent in the financial year 2021.[2]

Table 4.3 gives the average daily wages in different activities over different quarters in urban India. The average daily wage in the urban construction sector is quite comparable with other activities. However, there are severe fluctuations from quarter to quarter. For example, an average worker who earned a sum of Rs 329/- in quarter 1 (Q1V1) received only Rs 284/- in quarter 2 (Q2V1). In quarter 3 though it improved, there is a major decline again in quarter 4. Such fluctuations are indicative of the instability of demand for labour in the construction activity. Massive government interventions in pursuing construction projects can reduce such volatility in demand. The male–female differences in wages are huge. In spite of regulations against gender discrimination, difficulties at the implementation level are clearly evident. In fact, the demand for female construction workers is believed to be much more volatile in comparison to the male workers.

In the rural context, NREGA encourages a variety of construction programmes, engaging labour. Irrigation, housing (PM Awas Yojana), and so on provide an impetus to livelihood diversification, asset creation, and employment generation. However, the lack of rural industrialisation and the lack of growth in the corporate reduce the pace of employment generation. The stringent regulations not encouraging conversion of agricultural land for non-agricultural purposes are also the major obstacles to construction. The rural hinterland of the big cities is undergoing rapid transformation but the regulations regarding green zone are strict, reducing the employment potential of the construction activities.

In the urban context, there is a severe space shortage. Slowly, the industries are witnessing death in the large cities because of anti-pollution rules. The business corporates and private housing are the only outlets. Multi-story housing and the vertical growth of the cities can be encouraged to reap the benefits of employment-rich growth.

However, the mechanisation that is followed reduces the pace of employment creation. There are certain stages in which work can be done manually without any environmental pollution, but the builders are preferring mechanised ways. Strict regulations can save both livelihood opportunities and the environment.

The rise in material cost is reducing the pace of construction activity. In many instances, speculative and hoarding activities are responsible for the rise in the cost of production. Import of materials which is expensive can be curtailed by encouraging domestic production in a cost-effective manner. Even the land market in India involves severe speculative

Table 4.2 Percentage share of agriculture and construction sectors in total value added (2011–2012 prices)

Sectors	*2011–2012*	*2012–2013*	*2013–2014*	*2014–2015*	*2015–2016*	*2016–2017*	*2017–2018*	*2018–2019*	*2019–2020*
Agriculture	18.5	17.8	17.8	16.5	15.4	15.2	15.3	14.8	14.8
Construction	9.6	9.1	8.8	8.6	8.2	8.1	8.0	8.0	7.8

Source: National Accounts Statistics, MOSPI, Government of India.

Table 4.3 Average daily wages of casual worker employed across the industries in urban India (in INR)

	Q1V1		
Industry groups	**M**	**F**	**P**
Agriculture	273	166	230
Manufacturing	309	174	282
Construction	329	200	315
Mining and Quarrying	320	224	311
Services	299	194	281
Total	309.75	189.50	289.76
SD	118.34	83.62	121.83
CV	38.20	44.12	42.04

	Q2V1			***Q2V2***		
Industry groups	**M**	**F**	**P**	**M**	**F**	**P**
Agriculture	274	164	239	295	167	245
Manufacturing	295	162	262	298	173	273
Construction	284	121	266	269	–	269
Mining and Quarrying	324	231	317	336	243	329
Services	310	230	296	287	186	272
Total	312.02	194.10	292.78	316.78	192.69	298.11
SD	105.91	89.66	112.24	112.65	84.28	117.55
CV	33.94	46.19	38.33	35.56	43.74	39.43

Industry groups	Q3V1			Q3V2			Q3V3		
	M	F	P	M	F	P	M	F	P
Agriculture	302	162	240	279	153	235	291	177	254
Manufacturing	302	189	267	285	151	247	317	162	288
Construction	342	207	329	334	276	327	362	–	362
Mining and Quarrying	352	270	346	329	241	320	342	245	334
Services	310	194	296	331	258	319	298	207	286
Total	333.649	201.55	312.86	319.52	197.58	297.46	326.63	201.63	309.20
SD	118.6666	87.91	124.07	120.65	100.64	126.33	126.97	99.09	130.83
CV	35.56	43.61	39.65	37.75	50.93	42.46	38.87	49.14	42.31

Industry groups	Q4V1			Q4V2			Q4V3			Q4V4		
	M	F	P	M	F	P	M	F	P	M	F	P
Agriculture	340	190	281	306	147	249	286	156	242	323	205	292
Manufacturing	294	173	265	305	184	277	309	158	262	322	179	302
Construction	247	173	240	287	204	285	295	123	274	305	–	305
Mining and Quarrying	356	243	342	364	272	357	344	268	339	360	250	351
Services	319	212	305	319	236	312	328	232	313	325	202	307
Total	336.68	207.44	314.48	342.12	212.19	325.47	330.79	193.89	309.20	343.86	214.71	328.46
SD	135.34	88.62	137.47	130.09	96.84	133.58	123.15	118.06	132.14	145.61	105.38	147.48
CV	40.19	42.72	43.71	38.02	45.64	41.04	37.22	60.89	42.73	42.34	49.08	44.90

Source: Periodic Labour Force Survey (PLFS) 2017–2018.

Note: Q stands for quarter and V for visits; SD=Standard Deviation and CV=Coefficient of Variation.

pursuits, raising the transaction costs and the operational costs. Ultimately, they hinder the pace of construction and the employment potential that this sector possesses.

Rent control acts are extremely conservative and they try to protect the tenants by extending undue advantages to them. On the flip side, rebuilding of the houses and pace of employment generation get severely affected. The overall business environment for construction is not conducive: it requires several clearances at various levels, which affects adversely the growth in this sector.

Are the wages major constraints? The wages could have been high to suit the demand and supply conditions, but the actual disbursements are much lower because of layers of subcontracting at various levels. So, laws will have to change if labour exploitation must be reduced, and employment is to grow. The Lewis turning point may not have occurred as India has enormous supplies of labour stuck in the agriculture and other informal activities. However, labour mobilisation from these activities to construction is an important problem, as the stories of labour exploitation by the contractors are very much in currency.

Singh (2017) concluded that wage rate in both agriculture and construction increased over time. It increased in the urban sector because of high demand for labour in construction, while wage rate in the rural agricultural sector increased due to MGNREGS. However, a high demand for labour in urban construction also creates shortage labour in the rural agricultural sector. In fact, the wage gap (difference in the wages of both sectors) remained the same in 2004–2005, 2011–2012, and 2019–2020. Until 2019–2020, India had been experiencing a shortage of skilled labour in the rural areas and also in the urban construction sector. This shortage compelled the rural agriculture sector to compete for workers, and this competition raised the wage rate in both rural and urban areas. There has been an increasing trend in wages in the agricultural and construction sectors, a decline in surplus labour in rural India, and a concomitant rise in the marginal product of labour. However, labour absorption and wage increase are very slow both in the agriculture and construction sectors, and there is still a surplus labour available in the rural areas.

The PLFS (2017–2018) which has information on the same households over different sub-rounds within a year shows that some workers do not remain within the same activity throughout. They keep depending on different activities in different seasons or keep shifting from one activity to another as sustainable livelihood sources are shrinking. Besides, at a given point in time, some workers have to depend on multiple sources of livelihood as one source is not sufficient to provide subsistence level of income. Some of the workers engage themselves in two or more subsidiary status activities simultaneously while some augment their income by working in a principal and a subsidiary status activity. The livelihood diversification strategy is beneficial for the households as it helps smoothen the consumption expenditure. Else, fluctuations in consumption can push some of the households, particularly those marginally above the poverty line, into the domain of poverty, and such fluctuations affect the productivity of the workers severely. However, on several occasions, households remain below the poverty line in spite of pursuing multiple sources of livelihood. Due to the lack of gainful employment, they strive hard engaging in a number of petty activities. In the backdrop of this, the role of the construction sector can be envisaged as a major respite, particularly in certain months of the year such as agricultural slack seasons and non-festive times.

The recent pandemic and the subsequent lockdown wiped away the sources of livelihood, particularly in the urban space and more so in large cities, which kept attracting the migrants on a significant scale. The return migration occurred, raising the vulnerability of the workers. The regions which had witnessed lower unemployment rates before the pandemic experienced a hike. Regions with higher levels of urbanisation and large cities and with higher rates of rural–urban migration are the ones which encountered massive increase in the unemployment rate. The complete close down of the informal sector forced millions of workers to become openly unemployed and could bring in considerable overlaps between unemployment and poverty, a situation quite opposite to what prevailed during normalcy, as unemployed were usually those who could afford it. There has been a severe stress in the rural sector: those who have returned to the rural areas cannot get absorbed in gainful activities as the rural non-farm sector hardly comprises demand-induced activities. On the other hand, the agricultural sector has already been in a state of excess supplies of labour. The return migration only meant the same activity and income being shared by many, which again adds to the pathos of the working poor.

From these nightmares, certain lessons can be drawn for the future. In fact, it is high time that the strategy of rural industrialisation must be pursued which can contribute to the overall industrial growth of the nation and generate employment opportunities for the rural population and the migrants who have returned to the rural areas. The rural infrastructure building programmes along with major irrigation projects can offer livelihood to many, and on completion of the construction activities, new opportunities will emerge for inclusive growth to take place. The entire rural economy needs massive investment for which the government will have to take a great initiative. This should also be the occasion to revive the handicrafts and cottage industries so that high-value products can be manufactured through labour-intensive methods and these products can be sold both in the domestic and international markets. For all this to happen again, the construction activities will have to be pursued on a large scale so that individuals have enough space to pursue their new ventures and for storage facilities.

What needs to be done

The "employment problem" in a developing country context cannot be gauged merely in terms of open unemployment rate. The phenomenon of working poor engaged in low-productivity activities with meagre earnings is a vast manifestation of the employment problem in a country like India. The PLFS (PLFS: 2017–2018) witnessed a rise in the unemployment rate which was also accompanied by improvements in educational attainment. Besides, the positive association between the rate of urbanisation and unemployment was evident. These patterns perhaps can be taken as a sign of maturing of the economy, indicating that people could afford to remain unemployed till they found a job of their desirable status. Interestingly, the relative size of the informal sector slithered which could further substantiate that people were not under compulsion to get residually absorbed in petty activities. With such positivity as the economy was aspiring for rapid economic growth, the pandemic struck the world and hit the Indian economy with utmost adversity in terms of widely prevalent livelihood loss at various levels.

There has been a close link between rural–urban migration and informal sector employment, indicating transfer of rural poverty to the urban space. Though parts of the informal sector have been growing in response to demand, the vast stretches of low-productivity activities cannot be ignored and they do unravel significant overlaps with urban poverty and slum dwelling. However, several of the workers have been residing in the urban space for more than a decade and it would be highly erroneous to interpret urban poverty purely in terms of the spill-over effect of rural poverty. In fact, the elasticity of urban poverty with respect to rural poverty is highly nominal in spite of being statistically significant. Hence, the importance of public provisioning of work opportunities in the urban context must be felt in policy circles. Though some of the jobs in the informal sector help reduce poverty, the recent pandemic has affected those processes miserably. While certain segments of the ınformal sector bear strong linkages with the formal sector through the processes of ancillarisation, business subcontracting, and a great deal of complimentary activities, possibilities of exploitation, particularly with the operation of the intermediaries and contractors, have been there on a large scale to reduce the pace of upward mobility. Further, informalisation processes adopted in the formal sector in order to experience efficient utilisation of labour have been introducing vulnerability from a different angle. In what way these issues can be tackled is an important policy question: while the flexibility angle is important, the exploitation side cannot be overlooked. How the construction sector can be used as a key strategy to provide decent employment is a pertinent question, and the future policies will have to work on these lines.

Poverty in the rural areas persists both in the agricultural sector and the non-farm sector. Mechanisation, water shortage, and climate change effects have rendered agricultural labour redundant, and disguised unemployment has been significant. The lack of demand-induced activities in the rural non-farm sector has led to the supply-push phenomenon. Marginal activities may raise the share of non-agricultural sector, and the process of urbanisation may have been underway, but that does not result in sustainable development and livelihood. The massive growth of the census towns between 2001–2011 is a witness to this fact. Rural transformation was not accompanied by the growth of demand-induced activities, and the emergence of the new towns (not recognised by the government as statutory towns) is rather a misleading process if urbanisation has to be taken as an indicator of development. The expansion in the city limit and the rural hinterland being considered as a part of the city agglomeration is interesting but whether it is able to create employment opportunities for the rural population is an important question. A continuous change in the land utilisation pattern has rendered many farm households redundant and jobless. The policy interventions will have to reflect on these rapid changes that have been occurring in the recent past. Again, the construction sector in the rural context can emerge as the key pathway to employment-rich growth.

The agglomeration economies or external economies are important which result in higher productivity growth in large cities. Both the producers and the workers try to locate in such spaces, as concentration raises the efficient utilisation of resources and reduces the job search cost simultaneously. Agglomeration economies and construction share a strong positive relationship. With productivity growth, demand for physical assets grows. So, a great deal of employment potential can be realised. Parts of the productivity gains are transferred to the workers in terms of higher real wages which are known to the potential rural migrants. Hence, the population movement process from the rural space gets directed to the large cities. The caste-kinship bonds and other informal networks which facilitate the job market information flow and help the potential migrants shift to the urban space involve a strong inclination in favour of the large cities in spite of land and housing scarcity. The policy initiatives will have to recognise this fact that though the rural to all-urban migration rates are moderate, the rural to large city migration is rapid and rational from the point of view of the migrants. Hence, an exclusionary urban policy is not the right method of reducing the city burden; rather, creating amenities and empowering the poor to take advantage of the agglomeration economies should be a cost-reducing way of generating growth by creating employment. Several of the low-income households in cities create a great deal of value addition and in return earn much less than the contribution they make. Infrastructure support and other income support measures are indeed justified.

As the economy emerges gradually from the prolonged lockdown, it is time to have a new vision for the urban informal sector. What safety nets can be provided to the workers to create sustainable sources of livelihood is an important question. In activities where people take self-initiative credit assistance can be provided. More informative support for product innovation and marketing can be extended through online services. The lack of legal security of space for the operation of the informal sector enterprises is a major problem which has resulted in poor marketing of goods and profitability. The rent seeking behaviour of the government officials and the local government employees can be discouraged if the land tenure issue is considered effectively. The introduction of training for the informal sector workers will contribute to skill formation and facilitate upward mobility. Recognition of the skill acquired through informal channels may expand the scope of gainful employment. The linkages between the formal and the informal enterprises will have to be strengthened and the roles of the intermediaries need to change drastically. Though the recent reforms of labour codes say goodbye to the inspection raj in the organised sector, how the exploitative roles of the business and labour contractors can be changed is a critical question. Workers in retail trade and services within the informal sector are not the same as the production workers. Their requirements in terms of assistance, space, and support will have to be considered separately for a society to emerge with the initiation of shared prosperity. On the whole, both the rural non-farm and urban informal sectors will require special attention from the policy makers, and for the post-pandemic and post-lockdown period, sincere efforts will have to be pursued to develop these sectors with great potentiality for the creation of sustainable livelihood. However, the construction activities are instrumental to many of these new dreams and thoughts. In fact, the employment potential of the construction sector must be kept in view while developing the employment strategies.

Subcontracting issues are quite serious in the Indian context. Most of the workers receive work consignments through contractors and subcontractors. The wages are manipulated significantly, reducing the attractiveness of this sector. Many workers prefer to remain unemployed or engage themselves in petty activities instead of joining the construction sector because of large-scale exploitation.

Conclusion

After the COVID-19 pandemic and the subsequent lockdown hit economic activities drastically, the concern for job creation has become even greater as the phenomenon of sluggish employment growth already took the centre stage even during the pre-COVID times. The long-run employment elasticity has been low: mechanisation and the poor human capital are some of the reasons. Technological advancement which contributes to total factor productivity growth is welcome. However, a mere increase in labour productivity prompted by capital accumulation is not the right indicator of progress because it does not ensure a rise in the total factor productivity growth. Improvements in total factor productivity growth can lead to enhanced investments which may contribute to employment creation. Even if the application of advanced technology is expected to reduce labour required per unit of output, the expansion in economic activities from the rise in total factor productivity may compensate for the employment loss and instead add to new opportunities. Rapid growth in the construction sector can mitigate the employment challenges and create vast sources of livelihood in both rural and urban areas.

During the post-lockdown phase, the government is trying to stimulate effective demand so that normalcy returns soon and the economy is able to experience a reasonable rate of growth. However, given the major losses in livelihood and the slump conditions that the economy has encountered, it is difficult to revive effective demand instantaneously. An alternate way would be to provide encouragement to the producers to augment supplies so that, with a rise in production, factor income will increase and demand will rise subsequently. After all, the purchasing power of the consumers has a major impact on GDP. Any reduction in employment can have an adverse effect on output so much so that there can be a steady deceleration in effective demand. When most of the countries are struggling to revive their economies, it would be a far-fetched approach to rely on export demand to pick up and sustain the growth of the economy. Export demand has a number of constraints; unless the level of competitiveness is extremely high, it is unlikely that the exports can sustain the long-run growth. Hence, the classical conceptualisation of a close association between growth and employment is instrumental to the long-run steady state of the economy. As a first step to providing a boost to the economy, the potentiality of the construction sector will have to be utilised and employment opportunities in this sector can come as a great saviour. Government-sponsored construction activities need to be pursued on a large scale so as to create wage income, which, in turn, may provide an impetus to the normal functioning of the economy.

The second wave of COVID hit the economy and the lockdown of 2021 caused massive employment loss. In low-income countries, even under normal circumstances, a large number of households are vulnerable to precarity and the likelihood of loss of livelihood. Their capacity to withstand such employment loss is highly limited, as they do not have an asset base or the flexibility to switch occupations. The strategy of livelihood diversification requires an enormous amount of guidance from both government and non-government agencies, which may have had the requisite experience. While the distribution of food and provision of health support are indeed the short-run rescue measures at the time of crisis, massive planning will be required to create employment in both rural and urban areas. The urban employment guarantee programmes will be relevant for the urban poor and members of low-income households who have been residing in the urban areas for a very long time with little access to the rural areas. In this context, the designing and revamping of the construction sector can play a major role in giving a realistic shape to the dream of having an urban employment guarantee programme in place. Through the revival of the construction activities, particularly in the public sector, the employment guarantee programme can be implemented successfully in both rural and urban areas.

Notes

1 https://www.google.com/search?q=sustaible+urbanisation&rlz=1C1GGRV_enIN896IN896&oq=sustaible+urbanisation&aqs=chrome..69i57j0i13i457j0i13l6.7931j0j15&sourceid=chrome&ie=UTF-8

2 https://www.statista.com/statistics/801831/india-annual-real-gva-growth-in-construction-industry/

References

Dandekar, V. M. and N. Rath (1971), *Urban Poverty in India*. Pune: Indian School of Political Economy.

Eichengreen, B. and P. Gupta (2013), The two waves of service-sector growth, *Oxford Economic Papers*, 65(2013), pp. 96–123.

Ghose, A. (Undated), Services-led growth and employment in India.

Gordon, J. and P. Gupta (2003), Understanding India's services revolution, *International Monetary Fund*, November, 12.

Harriss, J. and M. P. Todaro (1970), Migration, unemployment and development: a two-sector analysis, *The American Economic Review*, 60(1), pp. 126–142.

Kundu, A. and P. C. Mohanan (2009), Employment and Inequality Outcome in India, OECD Conference Paris, OECD Paris.

Lall S. V., S. Harris and Z. Shalizi (2006), *Rural-Urban Migraton in Developing Countries: A Survey of Theoretical Predictions and Empirical Finding*. Development Research Group, The World Bank.

Luke, N. and K. Munshi (2006), New roles for marriage in urban Africa: kinship networks and the labour market in Kenya, *The Review of Economics and Statistics*, 88(2), pp. 264–282.

Mitra, A. (2013), *Insights into Inclusive Growth, Employment and Wellbeing in India*. Delhi: Springer.

Mitra, A. (2020), *Urban Headway and Upward Mobility*. New Delhi: Cambridge University Press.

Mitra, A. and J. P. Nagar (2018), City size, deprivation and other indicators of development: evidence from India, *World Development*, 106, pp. 273–283.

Ramaswamy, K. V. and T. Agrawal (2012), Services-led growth, employment and job quality: A study of manufacturing and service-sector in Urban India, Indira Gandhi Institute of Development Research, Mumbai March, http://www.igidr.ac.in/pdf/publication/WP-2012-007.pdf

Singh, G. P. (2017), Wage differences and inequality within construction sector, *Labour and Development*, ISSN 0973-0419, 24 (2), pp. 98–120.

5

DIFFERING APPROACHES TO EMBEDDING LOW ENERGY CONSTRUCTION AND CLIMATE LITERACY INTO VOCATIONAL EDUCATION AND TRAINING

*CIRT Team: *Linda Clarke*[1], *Melahat Sahin-Dikmen*[1], *Christopher Winch*[2], *Vivian Price*[3], *John Calvert*[4], *Pier-Luc Bilodeau*[5], *and Evelyn Dionne*[5]

[1]University of Westminster, London; [2]Kings College London; [3]California State University, Dominguez Hills, USA; [4]Simon Fraser University, Vancouver, Canada; [5]Université Laval, Quebec, Canada

Introduction

The aim of this chapter is to identify the ingredients needed to embed climate literacy successfully into vocational education and training (VET) programmes for construction. The chapter draws on examples from North America and Europe, representing a wide range of approaches to low energy construction (LEC). Programmes for developing the LEC workforce range from long-term efforts to embed climate and energy literacy into comprehensive VET programmes for all construction occupations to short courses imparting the specific skills required to carry out individual tasks. The respective VET systems involve different coalitions of stakeholders, including national, regional and local governments, employers, unions and VET institutions, and range from those in which the public sector, social partners (unions and employers) and the state education system play a key role to those largely relying on private sector, employer-driven initiatives. Whatever the constellation has implications for what is needed to equip trainees and workers to understand and address the urgency of climate change and apply the knowledge, know-how and competences required. Whilst the chapter, therefore, focusses on VET for LEC, it seeks to place initiatives to embed climate and energy literacy within the context of their respective VET systems.

Globally, around 28 percent of carbon emissions from buildings are attributed to the operation phase (i.e., energy needed to heat, cool and power buildings) and 11 percent to the construction phase (i.e., materials and construction process/embodied carbon) (WBC, 2019). The European Union (EU) aims by 2030 to reduce emissions by 32.5 percent and increase the share of renewable energy and energy efficiency by 32 percent compared to

 DOI: 10.1201/9781003262671-5

1990 levels (EC, 2019). In Canada, the targets are 40–45 percent reductions by 2030 and net-zero by 2050 (Canada, 2021a,b), whilst the United States, which rejoined the Paris Agreement in 2021, aims for 50–52 percent by 2030 and net-zero by 2050 (US, 2021).

The chapter draws on a Canadian-based project, Building it Green, that seeks to embed climate literacy into the building trades and to identify good practice examples. The research involves interviews with stakeholders in Canada, the United States and different European countries, including Belgium, Britain, Germany, Ireland and Sweden. Each case is examined in relation to the involvement of different stakeholders, including unions, employers and the state, the VET model involved, and whether equity and social justice are incorporated in efforts to reduce carbon emissions and energy consumption. Above all, our concern is with the approach taken towards labour, whether Taylorist, whereby each activity is broken down into small steps so as to reduce skill and training requirements, or aiming to empower through developing the potential of each worker by means of comprehensive VET programmes, encompassing a broad range of knowledge, skills and competences (Taylor, 1911). However, each case is also very different, as emphasis is placed on particular aspects and good practices are highlighted irrespective of their overall importance to the VET system. Indeed, it is often questionable whether there is a 'system' rather than disparate and even sometimes desperate attempts to develop the construction workforce for LEC. The chapter is not therefore strictly a comparison of different approaches to VET but a depiction of various scenarios, illustrating a range of possibilities.

We anticipate that, in covering such a large number of different approaches across two continents, the chapter is of value to other countries in the Global North with similar systems. It is, however, beyond our scope to embrace the Global South, which can have a different set of priorities and issues, with many countries being also the victims of carbon-intensive policies and practices enacted in the Global North. The chapter begins with a framework for understanding the different VET systems covered, drawing on Hall and Soskice's (2001) distinction between liberal and coordinated market economies (LMEs and CMEs), before sketching the different components needed for effective VET for LEC. There follows a discussion of the CMEs considered, beginning with the Swedish education-based system, followed by Belgium, also college-based but with a more significant social partner component, and finally the German 'dual' system, regulated by the social partners and orchestrated by the state. The next section concerns the LMEs, from the Canadian system based on formal trade apprenticeships and with close union involvement, to the United States, where, in the absence of significant state involvement, VET for LEC relies on active promotion by unions and employers. In contrast to the close involvement of unions in Canada and the United States, the Republic of Ireland and the United Kingdom are employer-based, with the state playing a more proactive and unifying role in Ireland than the United Kingdom. The chapter concludes by considering the pros and cons of the different approaches to VET for LEC.

Framing VET systems

In considering many countries, a framework is needed for understanding the conceptual and institutional differences between VET systems. Comparative VET research has sought to characterise these in a variety of ways, including Marsden's (1999) distinction between 'training' and 'production' approaches. A 'training approach' is institutionally regulated, related to individuals' abilities and certified qualifications, usually collectively and industrially organised and long-term, by equipping them over a working life to operate in specific occupations and sectors. This contrasts with a 'production' approach, where 'skills' are work-based, with training largely dependent on individual employers and on-the-job learning. The distinction resembles Rauner's (2007) between VET systems educating for an occupation and those concerned with the 'employability' of individuals, as with micro-credentialing where workers are trained to perform a limited task.

For our purposes, faced with a range of countries, an appropriate starting point is Hall and Soskice (2001)'s Varieties of Capitalism (VoC) distinction between LMEs and CMEs, which considers groups of countries as representative of particular capitalist economies and hence VET systems (e.g., Hall and Soskice, 2001). In CMEs, most closely associated with European countries such as Germany, formal institutions play a central role in governing the economy and regulating relations with stakeholders, with VET systems characterised by relatively high levels of social partnership between employers and unions. In contrast, coordination in LMEs, associated with Anglo-Saxon countries, occurs primarily through market mechanisms and is hence more oriented towards shareholders, which can mean the relative marginalisation of VET.

The VoC approach has been criticised for its apolitical, institutionalist and firm-centred nature, viewing labour as passive factor of production and ignoring potential antagonisms in capital-labour relations in the production sphere (Ebenau et al., 2015). Above all, it relies on national characterisations, thereby blurring the considerable disparities that exist within countries. Building closely on the approach, however, Bosch and Charest (2008) showed continued divergence of VET systems globally, resulting from historical differences in industrial relations, welfare systems and product markets. Despite its limitations, therefore, the CME/LME distinction of VoC provides a rough means to divide our countries into two groups, consisting of, on the one hand, Belgium, Germany and Sweden, and, on the other Britain, Canada, Ireland and the United States.

VET research on the construction sector has further elaborated the institutional and conceptual contrasts between the national VET systems (Clarke et al., 2021a). The study of VET for bricklaying in eight European countries, for instance, applied four comparators – education, governance, labour market and competence – and identified qualitative differences between approaches (Clarke et al., 2013). At one extreme is an 'occupational' approach typical of Belgium and Germany, resting on a statutory framework, social partnership, recognised qualifications, comprehensive, broad and recognised VET programmes, multi-dimensional competence, occupational capacity and knowledge, general and civic education, permeability and educational standards related to curriculum content. At the other extreme is a 'skill-based' approach, typical of the British system, resting on weak statutory framework and stakeholder involvement and characterised as employer-based, with poor labour market currency, fragmented narrow skill sets, a functionalist-behaviourist conception of competence built on task descriptors, minimal underpinning knowledge, lack of permeability and learning outcomes as performance criteria related to defined workplace tasks. Our cases are therefore best conceived on a spectrum, from the occupational approaches of CMEs, whether heavily dependent on the state and educational institutions or more reliant on social partnership, to the skill-based approaches of LMEs, whether involving employers and unions or entirely employer-based.

Such contrasting approaches reflect labour market differences, including the structure and organisation of firms, employment status of construction workers, degree of supervision required and industrial relations (Brockmann et al., 2010). A critical issue is lack of equity, including gender diversity and the exploitation of migrant workers, attributable to barriers in VET and to employment and working conditions (Clarke and Sahin-Dikmen, 2021). Meeting the challenge of a green transition in construction opens the possibility to include women and other excluded groups and to change the status of migrant workers. Technologically up-to-date, well-resourced and high-level VET, leading to qualifications valued in the sector and equipping trainees for LEC, could help achieve a just transition to an environmentally and socially sustainable industry.

LEC VET components

LEC requires a highly qualified workforce and a broad scope of abilities incorporated in different construction occupations because it is fundamentally different from conventional construction. The building envelope is defined as a single thermal unit with renewable technologies made up of elements that come together through the social interaction of different occupations, including bricklaying, carpentry, plastering, floor laying, insulation, electrical engineering and plumbing (Clarke et al., 2021b). LEC calls for upgrading existing VET systems to incorporate deeper *knowledge* of energy efficiency, higher technical and precision *skills*, and a holistic approach to the building process (Build up Skills, 2012). A broad set of transversal abilities is needed, involving *competences* not addressed in most VET systems, such as effective communication, project management, problem solving and autonomous working (Clarke et al., 2017; Relly et al., 2022). Cross-occupational coordination on site requires enhanced inter-disciplinary understanding and substantial and varied practical experience, particularly for eliminating thermal bridges in buildings, involving actions at the interfaces of different occupations, such as between the work of electricians and insulators. However, because of the 'skills' conceptualisation of workplace know-how, which goes together with the fragmentation associated with extensive subcontracting, especially in LEC countries, workers are expected to fulfil specified tasks without much reference to those in other occupations or regard to the integrity of the project as a whole.

The Build-up Skills (BUS) programme initiated by the European Commission in 2010 and covering 30 European countries revealed that VET systems are adequately equipped to integrate LEC competences into existing programmes in only a small number of European countries, such as Belgium and Germany where significant progress has already been made in mainstreaming LEC competences (Clarke et al., 2019). There is persistent evidence that energy performance requirements specified are not met in practice because of incorrect and poor-quality installation, effectively jeopardising emissions savings targets, providing further evidence of the need for improved VET quality (Sunikka-Blank and Galvin, 2012; Zero Carbon Hub, 2014). Such shortcomings might be addressed through regulation and procurement policies but are instead frequently left unrectified for cost reasons as contractors, especially in LME countries, tend to compete on price rather than quality.

It is not just a question of developing the knowledge and know-how necessary to carry out a particular occupation to stringent zero carbon construction standards, but also important for trainees to become climate literate so that they know why they are doing what they are doing and how this relates to climate change. As one union trainer in Canada summarised:

> We do need to make sure that we bring it back to why we're making this energy transition and understand that there is severe impact to how we've been doing things

> for these decades upon decades and we're feeling those consequences now. Our time is running short of when we can really make change to avoid some of these disasters that are coming, so we need to make sure that the climate is part of the dialogue. That we're encouraging solar, not just because we want to put our members to work ... we want to make people's lives more resilient ... to make sure that the environment is there for future generations.

Yet interviews with trainers reveal a variety of different approaches to teaching climate material, depending on a specific occupation's direct involvement in low carbon construction practices and the interests of individual trainers. Thus, teaching how to install solar panels, wind turbines and electric charging stations facilitates electricians in making the link to climate change, just as for plumbers does teaching how to install energy conserving heat pumps and advanced HVAC (heating, ventilation and air-conditioning) systems.

Coordinated market economies

CMEs are typically leading countries in Europe that belong to the EU, including Sweden, Belgium and Germany. As such, the transition to 'green' buildings is guided by EU legislation, including the European Performance of the Buildings Directive (EPBD, 2021/2018/2010), the Renewable Energy Directive (2021/2009) and the Energy Efficiency Directive of 2021. The legislation mandating the transformation of buildings is designed to deliver the EU's growth, energy and climate change strategies, articulated in the 2030 Climate Target Plan (2020) and the European Green New Deal (2019) as well as the Renovation Wave for Europe (2020) set in motion with the objective of reaching net zero in 2050. All member states are required to transpose EU legislation into national law so that these initiatives serve as important drivers. It is in this context that the VET systems associated with the CMEs discussed below are being developed.

Education/school-based systems

Of the countries covered in this chapter, both Sweden and Belgium can be classed as having predominantly school-based systems, in the sense that the responsibility for VET rests predominantly with the state education authority. This does not, however, mean that there is no work-based element or that the social partners play no role.

Sweden

Sweden, along with other Scandinavian countries, has a long history of energy efficient building, insulation being common practice since the 1970s. In addition to EU legislation, the Swedish Government's Climate Policy Framework (2017) commits to implementing the Paris Agreement, reaching net zero emissions by 2045. As a school-based construction VET system, the country represents one end of the spectrum from education/state-based VET within CMEs to employer-based systems within LMEs. The Swedish National Agency for Education decides on the VET and qualifications framework and issues guidelines for teachers responsible for developing construction curricula and syllabi, whilst the role of the social partners is advisory. For the trainee, the first three years are spent in a school or college, followed by two years as apprentice in a firm. The result is a greater emphasis on the knowledge component in the curriculum than the skills and tasks required in the workplace (Grytnes et al., 2018).

The VET curriculum was reformed in 2010–2012, when the Swedish National Education Agency issued general guidelines relating to energy efficiency and sustainability. Whilst curricula, for instance, for carpenters, electricians and plumbers cover technical aspects of LEC, the three 'school-based' years for those on the vocational pathway also follow the national secondary school curriculum, designed to ensure students develop general competencies, in addition to construction specific knowledge, understanding and skills. This includes subjects such as the Swedish language, mathematics, history and social studies as well as exposing students to social and political issues including climate change. The national construction curriculum gives students opportunities to develop: understanding of the industry's role in society and in regard to sustainable development; knowledge of what rational, safe and environmentally sustainable development means for the sector and its responsibilities; knowledge of common professions and work processes; and the abilities to cooperate and communicate with others and use professional language appropriately as well as to use sustainable resources, such as material handling, storage, minimisation of waste and sorting of construction waste.

In these respects, the curriculum is underpinned by climate literacy, offering a broader and deeper understanding of climate change and its relationship with building construction. It also encompasses components required for LEC, including a holistic understanding of buildings and transversal abilities such as communication, cross-occupational coordination and interdisciplinarity. However, a common refrain from interviewees was that guidelines are not sufficiently detailed or occupation-specific and that knowledge related to energy efficiency is insufficiently contextualised and related to climate change and to the United Nations (UN) Sustainable Development Goals (SDGs).

Following the reform of 2011, problems have been the lack of training for teachers and the focus only on pedagogical, rather than technical or occupational-specific, issues. As a result, the Swedish Construction Federation in collaboration with the social partner-led training boards developed a short continuing education programme in which a third of VET teachers participated (Tullstedt and Douhan, 2013). The remaining two-thirds remains a key challenge as enhanced energy efficiency competence in the workforce and having qualifications are not requirements to work in construction, except for electricians.

Belgium

As in Sweden, the construction VET system in Belgium is largely college-based, with only one quarter of the three-year education period spent with an employer. However, this is then followed by a two-year, paid apprenticeship, working for an employer and essential for achieving certification as a fully qualified person. In contrast to the Swedish system, VET is governed through social partnership whereby education boards for each construction occupation, involving representatives of employer associations, unions and teachers, work together in an advisory capacity to the National Education Agency. Like Sweden, the national curriculum for vocational secondary schools provides general guidelines that inform the development of occupation-specific course specifications.

Belgium has a national structure for handling construction VET through the long-established social partnership body *Constructiv*, which does not include electrical and plumbing. *Constructiv* is financed by a levy and involves both employers and unions in drawing up and updating occupational profiles, producing indicative syllabuses in the form of handbooks, and arranging onsite continuing VET. It is, however, the training institutions that are ultimately responsible for constructing programmes within the parameters of the occupational profiles. Whilst most of initial VET takes place in colleges, students receive extensive on-site experience and *Constructiv* finances the training of mentors working on-site and workplace-based training that meets certain standards. This also helps ensure that the existing workforce is kept up to date with new technologies, practices and regulations, though continuing VET is largely demand-driven so dependent on employer initiative.

All occupational profiles, for example the *Couvreur-Étancheur* (roofer-installer), contain some common elements and have a common format that details Knowledge (*Connaissance*), Know how (*Savoir faire*) and Attitude (*Savoir être*), with knowledge applied to know-how (theory into practice) but with an appropriate attitude of commitment, consideration, attention to detail, etc. Workers mastering the knowledge, know-how and attitudes set out in the occupational profiles and taught via a curriculum are able to act independently and in teams across broad interfaces.

VET for LEC is based on the principle that LEC-related competencies are incorporated or mainstreamed into existing occupational profiles and curricula of each occupation. Both Belgium and Germany are identified in the European construction social partner project, *Inclusive Vocational Education and Training for Low Energy Construction* (or VET4LEC – Clarke et al., 2019), as relatively well prepared. The Belgian VET system meets the five key requirements identified for successful LEC: communication, coordination, problem-solving, project management and precision. For example, the know-how requirement to 'keep labels and markings of materials used' in relation to 'quality awareness' for the Belgian occupation of *Couvreur-Étancheur* assumes the worker knows how to trace products and justify the work carried out. Furthermore, the know-how is exercised by working with care, precision, patience, economically, autonomously, with professional conscience and even aesthetically. Workers thus need to recognise the practical importance of the knowledge acquired and use their discretion and judgement, including in 'clarifying when others carry out poor quality work' (Clarke et al., 2020a).

Though deeper understanding of climate change and its relationship with construction associated with climate literacy is not embedded, nevertheless Belgium's standards-based approach to VET is paramount for meeting EU NZEB (nearly zero energy building) quality standards. There have been no major subsequent reviews of occupational profiles, as *Constructiv* focusses on combating serious labour shortages and improving workforce diversity, which as in Germany is very poor. Though in both countries, for instance, women make up around 9 percent of the construction workers, this is primarily women in administrative and office positions; the number actively working on construction sites is much smaller, and of those in VET the average percentage of female trainees is 1–2 percent (Women Can Build, 2020).

Social-partner-based VET in Germany

Germany's construction VET system differs from Belgium's in its 'dual' nature, being regulated by the social partners with the state responsible for setting the legislative framework and supervision and unions and employers associations involved formally in training and education bodies at all levels. The system covers over 20 construction occupations, whereby trainees apply to a company and levy-funded training is spread roughly equally between three locations: the company (practical), training centre and vocational school. The three-year VET programme is stepped, whereby trainees begin in the first year with a broad introduction to all the different construction occupations, then specialise in the second year into finishing, building or civil engineering, and only concentrate on a particular occupation in the final year (Clarke et al., 2019). This has the advantage of providing an overview of the work of different construction occupations and their interactions, so conceiving the building envelope as a single unit, though building services belong to a different sector from building, and hence come under different social partners.

LEC elements are as in Belgium mainstreamed into VET programmes of existing construction occupations, and national curricula for each construction occupation (including building services) incorporate these and provide detailed syllabi through pedagogic materials. For example, the textbook for the plasterer (*Stukkateur*) includes the purpose of insulation, internal climate control, costs of heating and energy use, environmental protection and thermal bridging as well as explanations of the nature of climate change, so necessary for developing climate literacy (Handwerk und Technik, 2014: 172–9). VET programmes are constantly reviewed and adjusted, taking account of technological changes, economics, the legal framework and social conditions (Clarke et al., 2020b). Social partnership structures ensure the representation of all relevant perspectives and inclusion of critical elements, overseeing curricula and publishing detailed pedagogic materials covering both practical and theoretical elements of VET for LEC.

Considerations

Of the above CME countries, Belgium and Germany stand out as more advanced in embedding LEC elements into curricula, whilst Sweden and Germany also incorporate the wider context of climate change. For Belgium and Germany, BUS recommended only specific changes, including improving theory-practice integration and teacher training in Belgium and strengthening systems thinking and interdisciplinarity in Germany (EC, 2014). A high proportion of the existing workforce is skilled and holds a recognised qualification, signifying workers possess basic knowledge and competence to master new concepts and techniques. Both VET systems are resourced and up to date, combining school based and practical learning through a substantial off-site, workshop-based component as well as work placements. Their broad-based occupational capacity provides a suitable framework for developing knowledge and understanding of energy efficiency, a holistic view of construction to enhance occupational coordination, and transversal abilities (Clarke et al., 2013; Winch, 2014).

The broad understanding of agency imparted thus responds to the demands of an LEC labour process that workers operate independently, apply expertise acquired appropriately, problem solve as necessary and take responsibility for meeting specified standards and quality. Levy-grant arrangements facilitate co-ordinated development and responses to new developments, such as insulation in Belgium and 'certified renewable energy specialists' in Germany. In both countries, the construction labour market is regulated and less fragmented than many others, providing an infrastructure for the work-based element of VET difficult to achieve elsewhere, so conforming to Marsden's (1999) 'training' approach. However, industrial relations are very different as unions are weaker in Germany, with a unionisation rate of 18 percent compared to 50 percent in Belgium and 68 percent in Sweden. Nevertheless, the works councils in Germany, designed as are the unions to protect the rights and interests of employees, must be consulted by the employer on VET issues and, since 2021, with the Act to modernise works councils (*Betriebsrätemodernisierungsgesetz*) can call the conciliation board if no mutual agreement can be reached.

Liberal market economies

VET systems in the LME countries considered – Britain, Canada, Ireland and the United States – differ significantly from each other and from those in CMEs, in the way they are organised, the terminology applied and their policy frameworks. For instance, in contrast to reference to 'trainees', 'occupations' and 'skilled workers' in CMEs, Canada and the United States refer to 'apprentices', 'trades' and 'journey workers' respectively, so underlining Clarke et al.'s (2013) distinction between occupational and skill-based approaches. These LME systems, therefore, whilst varied, represent skill-based systems, focussed more on labour output than developing labour potential or power.

Union and employer-based VET

North American VET for construction differs from that in Britain and Ireland in the closer involvement of unions, whilst the British and Irish systems are essentially employer-based.

Canada

Like the EU, Canada has a policy framework designed to meet its ambitious climate objectives including through: the *Pan Canadian Framework for Green Growth and Climate Change* (Canada, 2016), co-ordinating provincial and federal climate policies; *Bill C-12* (Canada, 2021a) establishing a roadmap to achieve net zero by 2050; *Commissioner of the Environment* (Canada, 2021c), on lessons learnt; and the *Canada Green Building Strategy* (Canada, 2022) with options to accelerate climate initiatives. The federal government has allocated significant funds to promote union-led training programmes through the *Union Training and Innovation Program* (UTIP) and increased regulatory tools such as building, energy, plumbing, fire and other construction codes to push the construction industry, incrementally, towards net zero (Canada Green Building Council 2020). While national model codes provide the templates, the Constitution allocates responsibility for VET largely to the provinces, which can adapt them to specific circumstances and control update timing (Lockhart et al., 2020; Pride, 2020). Code enforcement is a provincial responsibility, often delegated to municipalities, though inspections focus primarily on public safety, which takes precedence over such environmental concerns as monitoring higher LEC standards.

Canada Green Building Council (2020) companies have positioned themselves as leaders in LEC, having built model projects and promoted LEED (leadership in energy and environmental design) gold and platinum buildings, though LEED points can be awarded for features unrelated to reducing energy or carbon emissions. Despite the advances, mainstream industry remains overwhelmingly wedded to conventional building practices and opposes government regulation, maintaining the market can address and industry shape climate measures (Haley and Gade, 2020; Haley 2021; Haley and Kantamneni, 2021).

VET system[1]

Canada's approach to VET in construction is based on formal apprenticeship in recognised construction 'trades', with provincial governments maintaining qualification standards and working with the Federal Government to harmonise these across the country. Apprenticeship lasts between three to five years, depending on the trade, in which – in contrast to CME countries – 80 percent is spent employed under certified journey workers with the remainder classroom-based (Canadian Apprenticeship Forum, 2018; Mate, 2020). Unions support their apprentices' employment by recommending them to employers with whom they have collective agreements, which normally establish ratios of apprentices to qualified trades' workers, thereby guaranteeing them jobs. Both Federal and provincial governments also support pre-apprenticeship programmes, particularly targeting those traditionally excluded from construction. Federal Government coordinates the national Red Seal Program, which oversees curriculum content through consultation with provinces, employers, colleges and unions, and administers national examinations. Red Seal standards for each trade are amended periodically to reflect industry developments. Provinces also award Certificates of Qualification (CQs), reflecting their different climates, geography, economies and construction industries. About two thirds of graduating apprentices acquiring provincial CQs also obtain Red Seal certifications, confirming them as qualified 'journey workers' and thus enabling them to work in any province, facilitating national mobility.

Beyond the CQ, there is no 'higher level' trades' certification, although various upgrade micro-credentials are offered by manufacturers, public colleges and private training organisations, such as the Canada Green Building Council, in specific areas such as Passive House construction, varying from a few hours to several weeks. Provincial authorities designate trades as either 'compulsory', whereby work is legally restricted to journey workers with a provincial CQ, or voluntary, whereby anyone can do the work. Unions favour compulsory trades to promote member employment and limit employers' ability to hire less skilled workers at lower rates. Government support for these standards thus facilitates unionisation as well as guaranteeing worker competency. In the unionised sector, unions fund trade-specific training facilities jointly with employers through collective agreements, enabling them to operate their own facilities and giving them a major role in apprenticeship training (Calvert and Tallon, 2017). Otherwise, provincial governments provide training through public colleges.

Apprentices bear the bulk of training costs themselves through their paid employment, though provinces partly subsidise the in-class components, while the federal government provides income support for apprentices' classroom time through Canada's employment insurance system, scholarships and tax credits. Nevertheless, loss of earnings during classroom training is a financial problem. To be certified, apprentices must document having worked the required hours for each year of their apprenticeship, but, as construction is precarious, finding sufficient working hours to graduate can be challenging and registered apprentices must also leave their job to return to school for the required two months each year. Approximately half fail to complete, though unionised apprentices are more successful as collective agreements require employers to hire apprentices and unions take responsibility for finding employment (Coe, 2013; Calvert, 2014).

As in CME countries, women and indigenous workers remain substantially under-represented in the construction trades despite efforts governments and unions to expand training opportunities through targeted projects (Construction Council of British Columbia,

2018). New registrations of women in traditionally male dominated Red Seal trades vary from 3 percent in Manitoba to 6 percent in Alberta and British Columbia. Whilst approximately 6 percent of apprentices are indigenous, completion rates remain low and indigenous workers are significantly underrepresented in many large resource projects in their traditional territories (Canadian Apprenticeship Forum, 2018, 2019; Arrowsmith, 2019; Buildforce Canada, 2018).

Including climate literacy in VET

Red Seal Standards focus primarily on specific tasks, not on climate science and more generic problem-solving capacity, whilst reference in provincial curricula to how construction is impacted by climate change or what the industry or the trade can do to mitigate or adapt is absent, though some mention sustainability, environment, LEED and other low carbon building systems. Instructors must teach the skills specified in the Red Seal Standards and complain there is insufficient time to incorporate new climate material. Modifying the Red Seal to include climate change is not easy; standards for each trade are revised approximately every five years and multi-stakeholder committees must agree on amendments. Additional classroom time is not generally supported by employers; provinces are reluctant to pay for extra classroom time; and, for the apprentices, more classroom time means less paid time on the job. Some instructors consider that the skills currently taught in their apprenticeship programmes can be applied to new climate-related challenges, whilst others recognise the potential role of their trade in addressing climate change and consider systems thinking, teamwork, understanding buildings as integrated projects and shared responsibility for outcomes should be in the curriculum. As expressed by one trades' trainer, the system teaches apprentices the 'how' of construction work, not the 'why' of achieving climate objectives. The curriculum lacks emphasis on worker agency and the positive role of the industry and its construction workforce in meeting climate goals.

The siloed training system also presents a barrier to developing a climate literate workforce as each trade has its own curriculum and there is limited cross trade interaction though the need for trades to know more about others on sites is widely accepted. Construction unions normally represent one trade or group of related trades and negotiate to maintain jurisdiction over work they regard as theirs, upholding clear trade demarcations. These distinctions are reflected in the 'green book', while in Quebec they are part of the by-laws of the province and enforced by the labour relations board.

The role of unions in VET for LEC

The unionisation rate for construction in Canada in 2021 was 31 percent, though varying between provinces, being significantly higher in Quebec, at 40 percent, with collective agreement coverage at 57 percent. Despite Red Seal constraints, unions play an important role in VET for LEC, though there are differences between trades and across the country over how to raise climate change awareness and implement climate literacy training. Examples range from unions including climate material in their curricula, incorporating energy efficiency and green skills into apprenticeship training programme, developing climate literacy modules for instruction, remodelling training facilities to prepare apprentices to work with green technology, and actively organising to leverage public policies to promote employment in the green economy. Unions support public funding of new 'green' construction and major retrofits and advocate community benefits agreements (Calvert and Tallon, 2016; Bridge and Gilbert, 2017). For example, the Canadian Building Trades Unions (CBTU), representing 600,000 workers in 14 construction unions, successfully applied to the Federal Government for funding for a 5-year project focused on addressing emerging climate issues and involving consultation with affiliates on the climate literacy content of a future curriculum, to be tested in training programmes and potentially resulting in changes to the Red Seal curriculum (Calvert, 2022).

As affiliates of the US international unions, Canadian construction unions also cooperate on training, even modelling programmes on their US counterparts. For example, UA Canada, representing plumbers, pipefitters and allied trades, uses US training modules from the Urban Green Council's Green Professional or GPRO programme.

The United States

Whilst in many respects similar to Canada, the construction VET system in the United States has structural particularities that affect how it addresses climate change (Belman and Ormiston, 2021). There is, for example, far less government funding of apprenticeship, placing more responsibility on individual workers, their unions and employers to train the workforce.

VET system

One critical difference from the Canadian model is the lack of national testing and certification, magnifying the challenge of broad-based integration of subject matter such as climate literacy across the United States. Most trades have centralised curriculum development, such as electricians, plumbers and pipefitters and sheet metal unions, with centres providing core curriculum and modules that local union organisations, known as locals, around the country can request depending on their regional economies, capacity and preference for developing new areas of work (Woods, 2012).

The primary role of the North American Building Trades Unions (NABTU) is to work with local unions in the 14 affiliate trades in the United States and Canada along with state and local building trades councils (Lerman and Rauner, 2012). The unionisation rate in the United States, whilst varying across the country, is considerably lower than in Canada, at about 12.6 percent of the construction workforce. In terms of VET, NABTU's focus is to support and promote their apprenticeships and pre-apprenticeships and partner with stakeholders to diversify the trades and protect labour conditions. To replace retiring workers and increase diversity, they organise the Apprenticeship Readiness Program using a Multi-Core Curriculum, which requires 120 hours of instruction and began in 2007. NABTU maintains an extensive portfolio of course material including a core set of classes and the highly respected GPRO Climate Change Fundamentals, as well as courses NABTU develops or that have been donated by various locals, ranging from detailed training in wind turbine and solar installation to labour history, anti-racist, sexual harassment and anti-homophobic training. The local Building Trades Council and its educational partners (e.g. high schools, colleges, transitional programmes for the formerly incarcerated, women's programmes) determine what training to offer.

As in Canada, VET programmes that are jointly run by unions and employers are funded with member contributions matched by employers, and based on negotiated clauses in collective agreements (Glover and Bilingsoy, 2005). In practice, this gives unions the ability to operate their own facilities and play a strong role in curricula development. In some states, joint apprenticeships work together with community colleges, and trades classes may be given credit towards matriculation; in some, unions have also successfully pushed legislation to require that trade workers pass a state licensing test and then periodically take a continuing education course. In many unions, upgrades and certifying courses are also offered, providing possible entry points for a climate literacy class for those who have journeyed out.

Embedding climate literacy

US national climate policy has been stymied, especially due to the power of the fossil fuel industry to underwrite political candidates at federal level. Opposition to addressing climate change in Congress and the Supreme Court leaves city and state led initiatives to generate policies that motivate employers and training programmes to plan for a zero or low emission economy. For example, US Ironworkers have developed excellent climate manuals, though these are not widely used. Electrical workers have incorporated solar, wind farm and electric vehicle charging content into their curriculum and one local also includes extensive material on climate science. Mechanical insulators teach energy audits and GHG savings from insulation. An instructor interviewed argued that training should include climate science because apprentices need to think not only about their jobs, but about contributing to a greater good.

While most construction trades are eager for work in the renewable arena, they also seek to retain existing employment. At a trade union's train-the-trainer course on clean energy, for example, the director explained that climate change education and training in green skills did not represent lack of support for members continuing working in the oil and gas industry. The course used the GPRO fundamentals curriculum, which explicitly addresses global warming and the causal link to fossil fuel and is followed by GPRO trade-specific sessions. Training directors and instructors from around the country attending this multi-day session revealed that teaching about climate change is controversial in many locals but becoming more common with the growth of low carbon jobs.

Equity issues

The building trades struggle with inclusiveness despite efforts for change. The 1937 National Apprenticeship Act (NAA) grants the Federal Government certain powers, such as permitting the US Department of Labor (DOL) to issue regulations protecting the health, safety and general welfare of apprentices, to which equity was later added. Since the 1970s, the DOL has responsibility to monitor diversity among apprentices by race and gender, and government intervention has made a difference in diversifying the trades though they remain male dominated (Luke et al., 2017). While concentrated in manual trades, Latinos have gained ground in many construction unions and constitute 32.6 percent of the construction workforce, while African Americans constitute only 6.3 percent of the trades. The proportion of women in the overall construction industry is 11 percent, though those working on the tools is approximately 3 percent (Gallagher, 2022).

Some building trade unions recognise that legislation and funding for LEC jobs calls for plans to recruit, train and employ women and those from low-income groups. Biden's

National Building Performance Standards Coalition, for example, overtly connects electrification and equity with working with low-income communities. Unions in various states have formed coalitions with environmental justice communities to develop programmes for youth and unemployed adults as part of their green jobs legislative package, for example the Illinois Climate and Equitable Jobs Act. Unions competing with non-union companies often appeal to local governments for project labour agreements, so that public money goes to union or signatory contractors, arguing these offer a career path for workers. Unions have also utilised 'community benefits plans' to show they are serious about recruiting underrepresented groups for publicly funded jobs though this, as well as preparing members for a low carbon economy, remains a challenge.

Employer-based VET in Ireland and UK

Unlike the United States, Canada and the CME countries, unions play a more marginal role in VET in the LME countries in Europe considered, Ireland in the EU and Britain outside. The overall unionisation rate in Britain and Ireland, at about 24 percent, is nevertheless higher than in Germany, though considerably lower than in Sweden and Belgium and for the construction industry. Unlike Ireland, in Britain unions are explicitly marginalised from the VET system, as are employer associations, as it is based on individual employers, though at local level and for building services involvement can be significant, particularly in local authorities and on large projects.

Republic of Ireland

In Ireland, in contrast to Britain, the government plays a proactive role in driving forward the national climate strategy, aligning Irish national plans to EU policies for decarbonising building construction, and including stakeholders (e.g. employers, training providers and union representatives) in developing LEC training. Meeting new building standards is a key measure in the Government of Ireland's Climate Action Plan (2021a), which specifies the obligation to upskill contractors and other industry players in deep retrofit, NZEB and new technology installations. Taking a 'just transition' approach, it commits to: improving the fabric and energy efficiency of existing buildings; rolling out zero-carbon heating solutions, predominantly heat pumps and district heating networks; planning for the full phase out of fossil fuels in buildings by 2050; progressive strengthening of building standards for all buildings types; and promoting the use of lower carbon materials and behavioural change in occupants' energy use.

A National Retrofit Plan is part of the Climate Action Plan, and the Public Sector Energy Efficiency Programme foresees an enhanced role in setting standards through good practice examples. Support for expanding LEC training centres is part of the government strategy to upskill the construction workforce. The Action Plan for Apprenticeships (2021–2025) aims to review and upgrade apprenticeships across all industries to respond to skill needs, create a more inclusive system and double apprentice numbers in the next ten years (Government of Ireland 2021b). As a result, workshop and classroom-based elements of the VET system have been considerable improved, with, for instance, LEC centres set up in Waterford for new build and retrofit, with very-well equipped workshops, including a 'mock' house to demonstrate air tightness and insulation examples.

LEC curricula and training

As part of the EU's Build-up Skills programme, the QualiBuild project (2014–2016) addressed challenges identified through two pilot training schemes: The Foundation Energy Skills Programme (FES) and Train the Trainers. A key feature of QualiBuild was to involve a range of stakeholders, supported by a steering group from industry and education, and to set up an online register of LEC-trained construction workers. The FES course was adapted to different trades through ten separate short courses, designed to support the upskilling of experienced workers but suitable for adaptation for apprentices or initial VET. The courses are 'assured' by City & Guilds, which works with governments, organisations and education centres to validate work-based learning programmes, and are at the equivalent of the apprenticeship completion level, whilst the courses for qualified construction workers are higher. Syllabi were agreed and the first course delivered in 2018, and the first trade-specific course in 2019, with a steady increase in attendance, including for LEC fundamentals, electrical, retrofit, plumbing, ventilation and carpentry. A notable feature of this process has been collaboration, including with employer organisations and unions, training bodies, and private companies with expertise in energy-efficient construction, and the active support of the Irish government.

The United Kingdom

Though there exists an apprenticeship framework, construction VET in Britain is hampered by the undervaluing of Further Education (FE) Colleges responsible for classroom and workshop elements and lack of regulation and of a training infrastructure for the work-based element given extensive self-employment, subcontracting and dominance of micro-firms (Relly et al., 2022). Consequently, most construction trainees are full-time in colleges, in anticipation of obtaining a work placement. For instance, in a London FE College covering all construction occupations, there were 500 full-time and 200 part-time students and only 150 apprentices. Young people can enter an FE college full-time, acquire a construction qualification after two years and in principle go on to an apprenticeship with an employer, so echoing the Swedish system.

Individual employers responsible for reviewing the plumbing or electrical apprenticeship frameworks are themselves members of employers' associations and of the Joint Industry Boards covering the building services of electrical, plumbing and heating and ventilating and consisting of employers and unions. As a result, unions and employers' associations in these areas have a say, albeit indirectly. Electrical contractors should be registered on the Microgeneration Certification Scheme, requiring work such as heat pump installation be done correctly, and the Electrical Contractors Association (ECA) encourages members to use PAS35 as a framework for installation for local authority and housing association work, which ensures coordination of work carried out. However, electrical and plumbing work is beset by private training providers and companies purporting to train electricians to, for instance, install solar panels in a few weeks. The union demands a competent person be a judicial entity, but only in Scotland is a minimum level of competency specified.

The Electrical Joint Industry Board (JIB) has introduced LEC elements into the domestic electrician apprenticeship framework, which focuses on new technologies and includes on-site generation, such as small wind energy storage, heat pumps, load control, and smart technology to reduce energy consumption and maximise efficiency, control systems etc.

Of the pathways to becoming an intermediate craft plumber, few sign up to environmental technologies and gas remains popular, though the plumbing qualification framework is set for review through an employer-led process in which educationalists are not involved. A qualification handbook is given to training providers, such as FE Colleges, based on an assessment plan containing plumbing and electrical apprenticeship standards.

VET for LEC initiatives

Attention is increasingly given to the VET required for retrofit programmes, largely focussed on technical issues concerning insulation and the installation of heat pumps, though including transversal skills such as communication as well as the roles of Retrofit Coordinator and Retrofit Assessor. The statutory levy-based Construction Industry Training Board (CITB) is developing national occupational standards for insulation and LEC skill requirements. However, incorporating climate literacy into the curricula of different construction occupations is at a low level, due largely to lack of government regulation and recognition of national occupational standards, insistence on an employer-led system, and marginalisation of unions and employers' associations.

Many initiatives by different industry organisations and regions represent attempts to overcome these restrictions. For instance, a Welsh FE college visited has a flourishing apprenticeship programme, knowledgeable staff and extensive college-based construction programmes, with well-equipped workshops, including a mock house to demonstrate air tightness, solar panels and heat pumps. Procurement policy is used to drive LEC through for example the government agency National Resources Wales and the Welsh Government's own housing programme. Colleges across Britain rely on initiatives from local and metropolitan authorities, for instance, the Greater London Authority's Mayor's Construction Academy, which coordinates and promotes 'green skills' through hubs of key stakeholders, including employers, universities and local authorities but rarely unions. A key problem, however, is the facilities, though FE Colleges envisage that more stringent procurement requirements, including for supply chain contractors to adhere to PAS35 standards, will help create demand.

In Scotland, City Building Glasgow provides an inspiring alternative approach, revolving round an alliance between the local authority – Glasgow City Council, unions, and a housing association (Clarke and Sahin-Dikmen, 2019). All 2,200 staff are directly employed on the construction and maintenance of public works such as social housing, care homes and schools. The accredited apprenticeship scheme delivered at City Building's own Training Centre is a comprehensive, four-year programme, with a diverse intake of about 60 apprentices a year and high completion rates. The programme covers LEC (e.g. insulation, installation of renewable technologies) and provides all-round care to apprentices, including substantial on-site practice, support plans and post-training employment opportunities, with 80 percent staying on as employees. City Building's LEC schemes contribute to the Scottish Government's ambitions to reduce carbon emissions and tackle fuel poverty, with social housing schemes delivering two-thirds reduction in energy costs. One project involves a district heating network installation, utilising air source heat pump to 350 properties, and part funded by British and Scottish governments. Collaboration with the Council's in-house architects, a comprehensive training programme, direct employment of the workforce, monitoring subcontractors through a framework agreement and setting employment and quality standards provide a favourable set up for meeting energy efficiency standards. The high unionisation rate and active union engagement with management through the Joint Trade Union Council underpin this strong social ethos.

All in all, much effort is being put into developing and training a workforce for green construction across Britain, particularly by FE Colleges and local authorities. Everywhere, colleges are improving workshop facilities, such as model houses and renewable energy installations. Most noteworthy are events organised to promote green construction to women, combining greening the industry with improving its inclusivity. Many local authorities have declared a climate emergency and set carbon emission reduction targets to address climate change. Many are carrying out assessments of the green construction employment and training required in their areas and drawing up retrofit strategies. The sector is, however, beset with structural problems, partly connected with the domination by micro-businesses and self-employment, leading to difficulties in meeting low energy design specifications, providing training and integrating interdependent processes on site. With half the workforce classed as 'self-employed' and over 95 percent of firms classed as small, the construction labour process remains extremely fragmented and hardly provides an infrastructure for the work-based training needed for young people to transition from FE colleges into the labour market.

Conclusions

LEC means a transformation of VET systems to encompass deeper knowledge of energy efficiency, higher technical and precision skills and, above all, a holistic approach so that the building envelope is conceived as a single thermal unit and the social interaction of different occupations is understood. The high-quality construction labour process required involves teamwork and cross-occupational coordination, which imply interdisciplinarity, as well as transversal abilities such as communication, project management, precision, problem solving and coordination. But over and above this, climate literacy is needed to give meaning to the knowledge, skills and competences acquired so that trainees and workers are empowered and can appreciate why they are doing what they are doing and recognise their contribution to creating a safer and more equitable society. Climate literacy is tied to social equity and climate justice, comprising affirmation of the social contribution and responsibility construction workers, their unions, and the industry have to reduce emissions and the influence they have in determining policy direction.

In addressing these LEC requirements, each of the VET systems outlined has its own strength and weakness. With the CMEs, whilst the construction curriculum of the Swedish school-based system, with its emphasis on knowledge and general competences, is underpinned to an extent by climate literacy and the inclusion of transversal abilities, it is insufficiently detailed or occupation-specific. The curriculum of the other largely school-based system, Belgium, succeeds in mainstreaming LEC elements, breaking down broad occupational profiles into knowledge, know-how and attitudes and developing transversal abilities, so facilitating trainees to work independently and in teams across broad interfaces. Yet, climate literacy is not directly embraced. The other CME, Germany, has the advantage of a stepped programme of gradual specialisation, helping trainees to understand the whole building envelope, as well as covering climate change relating to different occupations, but is weakened by low unionisation and dependence on individual employers to take on trainees.

In many respects the construction VET systems of LMEs, more constrained by market conditions, are at a disadvantage in their ability to meet LEC requirements and incorporate climate literacy. They have nevertheless the advantage of being less constrained in developing new initiatives, particularly at local level and by the unions, as evident in the countries considered. The regulatory and joint funding arrangements for the Canadian apprenticeship system, for instance, which allows unions to have a say in construction VET and even to run their own training facilities, has potential to develop climate literacy in the curriculum and strengthen worker agency. It is, however, limited in doing so by lack of time allocated to classroom and workshop-based training and by siloed trade divisions, leading to demarcation disputes and impeding the development of a holistic, industry-wide approach. A similar situation applies regarding the US VET for LEC system, whose strength lies in the active promotion by unions of effective equity measures and courses such as GPRO, which explicitly address global warming and its causes, but which is thwarted by lack of standards for the trades and low union density.

VET for LEC initiatives in North American LMEs are rooted in the unions, but this is far less the case in LMEs such as Ireland and Britain, where unions play a marginal role in developing curricula and in the VET system. In Ireland, the state is pivotal in supporting LEC and the training courses needed for the existing workforce through the Training Boards and Institutes of Technology, though curricula lack the emphasis on climate literacy and the impact of climate change on different trades evident in the US GPRO courses and efforts to embed climate literacy in apprentice training lack urgency. The success achieved in turning a pilot LEC training course into a (nearly) national programme tailored for different occupations owes much to a process that mimics aspects of co-ordination and stakeholder partnership found in CMEs. In Britain, despite the employer-based VET system, unions play a role in promoting LEC elements in building services curricula, and politically accountable local authority direct labour building departments together with the FE colleges provide an important alternative model. In other respects, VET suffers from lack of regulation and curricula are narrow and largely restricted to developing technical skills for LEC rather than the underpinning knowledge required or climate literacy.

Whilst each system has its pros and cons, what is evident is that the 'skill'-based LEC VET systems need to move towards the 'occupation'-based CME approaches if LEC requirements are to be addressed in curricula. Whether CME or LME, only rarely do curricula address climate literacy, though this represents an important means to empower construction workers, providing both a motivation for building a zero-carbon economy and a threat to the status quo. Above all, achieving equity remains a critical issue in construction and for construction unions if they are to be a positive force in transitioning to a green economy. Indeed, valuing labour is key to valuing the environment and combatting climate change.

Notes

* CIRT or the Climate and Industry Research Team is responsible for the research underpinning the Canadian Building Trade Unions' Building it Green project, funded by the Canadian government.

1 The content of this section mostly refers to English Canada; labour market policies in Quebec covering the construction industry are closer to those of CMEs (Charest, 2003).

References

Arrowsmith, E. (2019). *Promoting Careers in the Skilled Trades to Indigenous Youth in Canada.* Ottawa: Canadian Apprenticeship Forum. www.caf-fca,org

Belman, D. and Ormiston, R. (2021). Creating a Sustainable Industry and Workforce in the US Construction Industry. In Belman, D., Druker, J. and White, G. (eds), *Work and Labour Relations in the Construction Industry*. New York: Routledge.

Bosch, G. and Charest, J. (eds). (2008). *Vocational Training in the 21st Century: A Comparative Perspective on Systems and Innovations in Ten Countries*. London: Routledge.

Bridge, T. and Gilbert, R. (2017). *Jobs for Tomorrow – Canada's Building Trades and Net Zero Emissions*. Vancouver: Columbia Institute.

Brockmann, M., Clarke, L. and Winch, C. (eds). (2010). *Bricklaying Is More than Flemish Bond: Bricklaying Qualifications in Europe*. Brussels/London: CLR.

Buildforce Canada (2018). Representation of Indigenous Canadians and Women in Canada's Construction and Maintenance Workforce. July. https://www.buildforce.ca/system/files/documents/Indigenous_Canadians_Women_in_Canadas_Construction_Workforce.pdf

Build up Skills (2012). *Vocational education and training for building sector workers in the fields of energy efficiency and renewable energy*, German report by Peter Weiss, Richard Rehbold and Elisa Majewski, Intelligent Energy Europe, September.

Canada (2016). *Pan-Canadian Framework on Clean Growth and Climate Change*. https://www.canada.ca/en/services/environment/weather/climatechange/pan-canadian-framework/climate-change-plan.html

Calvert, J. (2022). Labour and Climate Change. In Peters, J. and Wells, D. (eds), *Canadian Labour Policy and Politics*. Vancouver: UBC Press.

Calvert, J. and Tallon, C. (2017). *Promoting Climate Literacy in British Columbia's Apprenticeship System: Evaluating One Union's Efforts to Overcome Attitudinal Barriers to Low Carbon Construction*. Toronto: York University ACW Working Paper 201

Calvert, J. and Tallon, C. (2016). *The Union as Climate Advocate: The BC Insulators' Campaign to Green the Building Industry in BC*. Toronto: York University

Calvert, J. (2014). *Overcoming Systemic Barriers to 'Greening' the Construction Industry: The Important Role of Building Workers in Implementing Climate Objectives at the Workplace*. Alternative Routes. https://www.alternateroutes.ca/index.php/ar/article/view/20596

Canada (2021a). Bill C-12: *An Act respecting transparency and accountability in Canada's efforts to achieve net-zero greenhouse gas emissions by the year 2050*. Chapter 22. June 29, 2021.

Canada (2021b). *Achieving a Sustainable Future: A Draft Federal Sustainable Development Strategy for 2022 to 2026*. Gatineau: Environment and Climate Change Canada.

Canada. (2021c). *Report 5: Lessons Learned from Canada's Record on Climate Change*. Commissioner of the Environment and Sustainable Development for the Parliament of Canada, Ottawa: Office of the Auditor General of Canada.

Canada. (2022). *The Canada Green Building Strategy*. Natural Resources Canada Discussion Paper. July.

Canada Green Building Council. (2020). *Canada's Green Building Engine: Market Impact and Opportunities in a Critical Decade*. https://delphi.ca/wp-content/uploads/2021/01/canadas-green-building-engine.pdf

Canadian Apprenticeship Forum. (2018). *Making Apprenticeship a National Skills Priority*. Ottawa. August.

Canadian Apprenticeship Forum. (2019). *National Strategy for Supporting Women in the Trades*. Ottawa.

Charest, J. (2003). Labor Market Regulation and Labor Relations in the Construction Industry: the Special Case of Quebec within the Canadian Context. In Bosch, G. and Philips, P. (eds), *Building Chaos: An International Comparison of Deregulation in the Construction Industry*, London: Routledge, 95–113.

Clarke, L. and Sahin-Dikmen, M. (2021). Why Radical Transformation Is Necessary for Gender Equality and a Zero Carbon European Construction Sector. In Magnusdottir, G. L. and Kronsell, A. (eds), *Gender, Intersectionality and Climate Institutions in Industrialized States*, Routledge.

Clarke, L., Westerhuis, A. and Winch, C. (2021a). Comparative VET European Research since the 1980s: Accommodating Changes VET Systems and Labour Markets, *Journal of Vocational Education and Training*, 73/2, 295–315

Clarke, L., Duran-Palma, F. and Sahin-Dikmen, M. (2021b). Towards Nearly Zero Energy Building in Europe: Challenges of Vocational Education. In Filho, W.L., Azul, A.M., Brandli, L.L., Salvia, A.L. and Wall, T. (eds), The *Encyclopaedia of the UN Sustainable Development Goals. Industry, Innovation and Infrastructure*. Springer

Clarke, L., Sahin-Dikmen, M. and Winch, C. (2020a). Transforming Vocational Education and Training for Nearly Zero-Energy Building, *Buildings and Cities*, 1(1), 650–661.
Clarke L., Sahin-Dikmen M. and Winch C. (2020b). Overcoming Diverse Approaches to Vocational Education and Training to Combat Climate Change? The Case of Low Energy Construction in Europe, *Oxford Review of Education*, 46/5: 619–636
Clarke, L., Gleeson, C., Sahin-Dikmen, M., Winch, C. and Duran-Palma (2019). *Inclusive Vocational Education and Training for Low Energy Construction: VET4LEC*, a) Final Report; b) Country Summaries, European Federation of Building and Woodworkers and European Construction Industry Federation, Brussels
Clarke, L. and Sahin-Dikmen, M. (2019). City Building (Glasgow): inspirational model of low energy construction and direct labour, *Scottish Left Review*, 113, September/October
Clarke, L., Gleeson, C. and Winch, C. (2017). What Kind of Expertise Is Needed for Low Energy Construction?, *Construction Management and Economics*, 35/3, 78–89.
Clarke L., Winch, C. and Brockmann, M. (2013). Trade-Based Skills versus Occupational Capacity: the Example of Bricklaying in Europe, *Work, Employment and Society*, 27/6, 932–951.
Coe, P. (2013). Apprenticeship Program Requirements and Apprenticeship Completion Rates in Canada, *Journal of Vocational Education and Training*, 65/4, 575–605.
Construction Council of British Columbia. (2018). *A Community Benefits Agreement*. BC Infrastructure Benefits Inc. and Allied Infrastructure and Related, July.
Ebenau, M., Bruff, I. and May, C. (eds). (2015). *New Directions in Comparative Capitalisms Research: Critical and Global Perspectives*. London: Palgrave.
European Commission (EC) (2019). *Clean Energy for All Europeans*. Brussels: Directorate General for Energy.
European Commission (EC) (2014). *Build-up Skills: EU Overview Report*, Brussels: Staff Working Document, Intelligent Energy Europe.
Gallagher, C.M. (April 2022). *The Construction Industry: Characteristics of the Employed, 2003–2022*, Bureau of Labor Statistics, U.S. Department of Labor
Glover, R. W. and Bilginsoy, C. (2005). Registered apprenticeship training in the US construction industry. *Education + Training*, 47, 337–349.
Government of Ireland (2021a). *Climate Action Plan: Securing Our Future*. Department of the Environment, Climate and Communication.
Government of Ireland (2021b). *Action Plan for Apprenticeship 2021–2025*, Department of Further and Higher Education
Grytnes, R., Grill, M., Pousette, A., Törner, M. and Nielsen, K.J. (2018). Apprentice or Student? The Structures of Construction Industry Vocational Education and Training in Denmark and Sweden and their Possible Consequences for Safety Learning, *Vocations and Learning*, 11, 65–87
Haley, B. and Gaede, J. (2020). *Canada Needs an Ambitious Energy-Retrofit Plan for Buildings*. Ottawa: Policy Options. February 11:
Haley, B. (2021). *Federal Mandate Letters Signal Changes to Building Codes, Deep Retrofits and Electric Vehicles*. Efficiency Canada. December. 17.
Haley, B. and Kantamneni, A. (2021). *Efficiency for All: A Review of Provincial, Territorial Low-income Energy Efficiency Programs with Lessons for Federal Policy in Canada*. Ottawa: Efficiency Canada
Hall, P. and Soskice, D. (2001). *Varieties of Capitalism: The Institutional Foundations of Comparative Advantage*. Oxford: Oxford University Press.
Handwerk und Technik (2014). *Lernfeld Bautechnik Stuckateur Fachstufen*, Hamburg: Handwerk und Technik GmBH
Lerman, R.I. and Rauner, F. (2012). Apprenticeship in the United States. In Barabasch, A. and Rauner, F. (eds), *Work and Education in America*, 175–193. Dordrecht: Springer.
Lockhart, K. and Haley, B. (2020). *Strengthening Canada's Building Code Process to Achieve Net-Zero Emissions*. Efficiency Canada/Carleton University. October.
Luke, N., Zabin, C., Velasco, D. and Collier, R. (2017). *Diversity in California's Clean Energy Workforce: Access to Jobs for Disadvantaged Workers in Renewable Energy Construction*. University of California, Berkeley Labor Center report.
Marsden, D. 1999. *Theory of Employment Systems: Microfoundations of Societal Diversity*. Oxford: Oxford University Press

Mate, G. (2020). *A Critical Analysis of Apprenticeship Programs in British Columbia*. PhD thesis, Faculty of Graduate and Postdoctoral Affairs, Carleton University
Pride, A. (2020). *Tiered Energy Codes: Best Practices for Code Compliance*. Efficiency Canada/ Carleton University. September.
Rauner, F. 2007). Vocational Education and Training – A European Perspective. In Brown, A., Kirpal, S. and Rauner, F. (eds), *Identities at Work*. Dordrecht: Springer
Relly, S.J., Killip, G., Robson, J., Emms, K., Klassen, M. and Laczik, A. (2022) *Greening Construction – A complex challenge for jobs, skills and training*. Edge Foundation
Sunikka-Blank, M. and Galvin, R. (2012). Introducing the prebound effect: the gap between performance and actual energy consumption, *Building Research & Information*, 40(3), 260–273
Swedish Government (2017). Climate Policy Framework, Ministry of Environment and Energy
Taylor, F.W. (1911). *The Principles of Scientific Management*. New York: Harper and Brothers.
Tullstedt, L. and Douhan, A. (2013). *Build Up Skills Sweden Roadmap: Building Skills for Energy Efficient Buildings*. The Swedish Construction Federation
US (2021). *The Long-Term Strategy of the United States*. The White House, 2021,
Winch, C. (2014). Education and Broad Concepts of Agency, *Educational Philosophy and Theory*, 46/6
Woods, J.G. (2012). An analysis of apprentices in the US construction trades: an overview of their training and development with recommendations for policy makers. *Education and Training* 54/5, 401–418.
Women can Build (2020). *Towards an equal construction industry*, European Union Erasmus Report, October https://www.womencanbuild.eu/wp-content/uploads/2021/01/WCB_IO2_Joint-Report_immersive_experience-for-women.pdf
World Building Council (WBC) (2019). *Bringing Embodied Carbon Upfront*, London.
Zero Carbon Hub. (2014). Closing the gap between design and as-built performance. End of Term Report

6

GREEN JOBS AND CLIMATE JUSTICE IN THE BUILT ENVIRONMENT

Lessons from American cities[1]

Edmundo Werna[1], *Mônica A. Haddad*[2], *Erin Ritter*[3], *and Anne Wurtenberger*[4]

[1]London South Bank University, UK; [2]Iowa State University, USA; [3]Grinnell College, USA; [4]Iowa State University, USA

Introduction

On his first month in office, US President Joseph R. Biden re-joined the Paris Agreement and signed a series of environmental executive orders, which will be "empowering American workers and businesses to lead a clean energy revolution that achieves a carbon pollution-free power sector by 2035" (The White House, 2021). While these are pieces of good news, more needs to be done. The current socio-economic structure of society is characterised by a large income gap between the richest and the poorest. If this gap is overlooked during the transition to a low-carbon economy, then climate change mitigation through the built environment may not improve livelihoods for vulnerable populations but replicate the unfair status quo with lower greenhouse gas (GHG) emissions.

Within this context, one of the pillars of climate change mitigation should be climate justice. Climate change with justice means making "a 'just transition' to a post-carbon economy and [providing] assistance to vulnerable communities" (Schlosberg & Collins, 2014, p. 366). This 'just transition' should focus on issues such as ethnicity, income, and gender in both the Global North and South (McCauley & Heffron, 2018). In the United States, this is crucial because the socio-economic structure is characterised by inequalities that affect the non-white population to a greater extent. The concept of 'just transition' was first put forward by labour unions in the 1980s "in response to new regulations to prevent water and air pollution" (Healy & Barry 2017, p. 454). It evolved in an international collaboration to promote "green jobs" as necessary elements of a just transition (p. 474), and today is part of the wording of the Paris Agreement: "the imperative of a just transition of the workforce and the creation of decent work and quality jobs in accordance with nationally defined development priorities" (UNFCCC, 2015: 21). At the UN Climate Change Conference 21, also called Convention of the Parties (COP) 21 in Paris, on 12 December 2015, Parties to the United Nations Framework Convention on Climate Change (UNFCCC) reached a landmark

DOI: 10.1201/9781003262671-6

agreement to combat climate change and to accelerate and intensify the actions and investments needed for a sustainable low carbon future.

Cities can, and should, play a central role in climate change mitigation (Fitzgerald, 2020). In 2020, 75 percent of all global GHG emissions were derived from urban areas (Cities Alliance, 2021). It is estimated that around 59 percent of GHG emissions come from the built environment (University of California, 2017). Pressure within US climate politics for more working-class jobs in clean energy and socially just climate policies should stem from a place-based political-economic approach and be connected to environmental justice (Knuth, 2010). Climate change related plans have been designed and implemented in some American cities (Deetjen et al., 2018), and other cities have begun to add to their public agendas isolated actions to decrease greenhouse gas emissions (Rice, 2010). However, the complex urban arrangement which encompasses several actors, local politics and diverse leadership styles entails a challenge to address urban climate change with fairness and further climate justice. According to Bulkeley et al. (2014), questions about who stands to gain and lose through mitigation of climate change should be raised. Spatial inequalities in cities lead one to question existing standards for assessing fair distribution of resources.

The aim of this chapter is to provide a review of the literature encompassing how US local governments are engaging in climate change mitigation and climate justice, provision of green jobs that improve the built environment, and undertaking workforce development and training for vulnerable residents. Upon reviewing previous studies, it was found that these themes were dispersed in the literature, making it difficult to utilise the knowledge as a basis for policy making or to further research in the combined themes. The chapter starts by introducing the method. Next, it provides the analysis and its results. To conclude, the findings are used to provide recommendations for future research and for local action – with an emphasis on community training programmes.

Method

The method adopted to undertake the review of the literature was based on a combination of keywords and snowball sampling. The keywords, used via Google Scholar to identify references, included the thematic boundaries of the research: US local governments and climate change mitigation, transition in the built environment, climate justice, green jobs, workforce development and training for vulnerable residents.

The list of relevant studies, only written in English, was scrutinised and expanded by the research team, the members of which have academic experience on the themes of the chapter, together with access to documentation provided by the International Labour Office, in particular, its Green Jobs Programme. Only a few technical reports were included; the majority of the works were peer reviewed.

The above methods were combined with literature snowball sampling: once a study was included in the literature review, then, using Google Scholar, main articles that cited that study were also examined. Snowball sampling has been traditionally and widely used in academic research and continues to be a popular approach (see, for example, Biernacki & Waldford, 1981; Parker et al. 2019; Thiollent, 2000).

This combination provided the basis for the review and also delineated its boundaries. Such boundaries were further confirmed by applying the concept of saturation: at a particular point of the review, new authors echoed previous ones, leading to a loop in the literature, without any innovation (Baker & Edwards, 2012; Guest et al., 2006; Marshall et al., 2013).

Findings

The section starts with the broad concepts of social justice and climate justice. The first subsection concludes by linking such concepts to the green economy, which is the subject of the next subsection. The analysis of the green economy is then elaborated upon, again with emphasis on the issue of justice. This leads to how the green economy generates (green) jobs. Such jobs can be a conduit for providing climate justice to vulnerable populations. The final subsection focuses on training for green jobs. All subsections are specifically related to urban areas and to jobs related to the built environment.

Social justice and climate justice

Justice, equality and fairness are the foundations for all aspects of change (Bulkeley et al., 2013). Climate justice, "the mobilization of justice with respect to climate policy," aims to use those foundations to respond to the unequal impacts of climate change through focusing on local and community experiences (Schlosberg and Collins, 2014). Climate justice emerged from the broader concept of environmental justice. The two remain intertwined (Schlosberg and Collins, 2014). Also, the concept of environmental justice is inextricably linked to the racial justice which was the foundation of the movements of the 1980s, themselves building upon earlier civil rights movements, in an attempt to put human rights at the centre of the environmental debate (Schlosberg and Collins, 2014). In North America, climate initiatives that focus on justice are typically led by non-government organisations (NGOs) (Bulkeley et al., 2013). While more specific frameworks exist, such as "access and allocation," the notion of climate justice resonates powerfully within both academic and policy debates, suggesting a push for change (Grecksch and Klöck, 2020).

This review included papers that connect climate justice, social justice and environmental justice in urban areas and the central role the built environment plays when addressing climate change. EJLF (2020) and McKendry (2016) note that locations in the United States characterised by high GHG emissions coincided with low-income communities, who are preponderantly Afro-Americans. Thus, it is important to ensure that the spatial distribution of actions resulting from low-carbon strategies implemented in the built environment are not biased towards the wealthy; otherwise, this will reinforce the status quo and not promote climate justice (McKendry, 2016).

It is clear from the literature that the right to enjoy the benefits of climate mitigation in cities is not equally distributed. Ethnicity, class and gender influence how injustices shape urban areas when climate change mitigation strategies are implemented (Granberg and Glover, 2021; McClure et al., 2017; Chu et al., 2021; Dunn, 2010; Knuth, 2010; Schlosberg, 2012; Shi et al., 2016; Soanes et al., 2021). Using the case of solar energy, Mulvaney (2013) notes a disparity. On the one hand, it produces clean energy and promotes green jobs. On the other hand, some of the stages in the manufacture of panels can involve toxic materials, bringing environmental and health risks to communities.

Many studies advocate for a more progressive shift in planning and policy development to fight these injustices and promote climate justice. According to Bulkeley et al. (2014), the shift should be from a primary goods distribution approach to a capabilities approach, based on the work of Sen (1999). He argues that a government should be measured against the concrete capabilities of its citizens; that is, he tried to make the loose concept of 'rights' tangible (capabilities). In his seminal book, *Development as Freedom*, Sen (1999) argues that development is much broader than a simple focus on growth of GDP or income per-capita. It is about advancing the real freedoms that people enjoy.

Examples of such shifts are: involving disconnected youth in the green economy (Grobe et al., 2011; Martinson and Stanczyk, 2010); having key participants in the processes who have a keen interest in disadvantaged communities (Shi et al., 2016); and designing policies that incorporate the specifics needed for a green economy (Arcand, 2020). Ensuring that marginalised populations have access to living-wage green jobs, and strengthening alliances between community climate organisations are also strategies that could be used to promote climate justice (Robinson 2020).

To make this picture even more complex, issues of climate mitigation and social justice do not have one universal solution (McKendry, 2016). For instance, Knuth (2010), when examining mitigation in Philadelphia's suburbs, proposes a place-based strategy with more socially just climate policies. For local solutions to work, there is a need to improve governmental policies and action which supports adaptation planning, and to bridge the gap between academia and actors in climate adaptation initiatives in communities (Shi et al., 2016). Additionally, Soanes et al. (2021) indicate that when local actors know about predictable funding and how to access these funds, communities can effectively influence adaptation that can lead to improvements in skill and innovation in businesses, tailored for application at the local level. Moreover, more information is needed on what solutions work and the ones that do not work, when connecting climate action and social inclusion (Robins et al., 2019).

Bringing these ingredients together: responsibility for action, right to benefits, progressive shifts and one size does not fit all, the green economy can play a central role to recover and preserve the natural capital in urban built environments, while improving the quality of life of citizens with sustainable social welfare. Nature, which provides the elements which are crucial for (urban) life, such as clean water and good air quality, is having its capacities being put under severe pressure due to human agency through a variety of environmental degradation processes. The transition to a green economy is a pressing need for the health of the planet (ILO 2018; Robinson, 2020).

The green economy and climate justice

A green economy prioritises the environment through low-carbon solutions and mindful use of resources (Heshmati, 2014; Sulich & Zema, 2018; Robinson, 2020). Economic wealth and the protection of the environment should not be the only indicators (Sulich & Zema, 2018; Heshmati, 2014; McKendry & Janos, 2015). The green economy should also incorporate the social dimension of development (Heshmati, 2014; Robinson, 2020, Gupta et al., 2019). That is where climate justice comes into place.

There are criticisms of the concept of the green economy. According to UNEP (2023), the green economy is basically an economy with low carbon, and resource efficiency. In a green economy, growth is driven by public and private investment in economic activities which reduce carbon emissions and pollution, enhance energy and resource efficiency, and prevent the loss of biodiversity and ecosystem services.

UNEP (2023) also mentions the social inclusion aspect of the green economy. However, despite the social concerns of that agency whose core mandate is the environment, of particular importance to the discussion in this chapter is the argument that the green economy does not necessarily address equity (McKendry & Janos, 2015; Krause, 2018), and it is "less explicit about how social equity and wellbeing can be addressed" (Krause, 2018 p. 513). To show an example, according to Healy and Barry (2017), in California, the green economy has come under scrutiny due to "African American workers largely being left out" (p. 455) of the green job solar boom, as the solar market actually profits from exploiting unskilled workers.

Green policies can alter the level and composition of labour demand, inducing changes in prices that strengthen clean sectors and hurt dirty ones (Kruse et al., 2017). Reductions in labour taxation linked to market-based instruments could improve environmental quality through environmental tax reform, boosting both economic growth and employment. Policies such as carbon taxes are important for cities (Mi et al., 2019). However, American cities are not ahead with regards to carbon taxes, and should implement them as a central element of their policies to reduce emissions, following in the footsteps of the 23 other jurisdictions across the world that have done so (Metcalf, 2019).

Green legislation at the national level in the US centres on economic concerns, whether that of a recession or of driving growth. For example, the 2007 US Green Jobs Act allocated $125 million to the creation of green job training programmes (Elliott & Lindley, 2014). The 2009 American Recovery and Reinvestment Act (ARRA) pledged $500 million with stipulations for creating jobs in the renewable energy sectors (Popp et al., 2020; Elliott & Lindley, 2014). When examining the ARRA, Popp et al. (2020) showed that "the pre-existing level of green skills matters," because green stimulus works best in areas that already have a local base of green economy to support it. The existing capabilities of both individual workers and communities have an impact on the success of such investments (Popp et al., 2020).

Local governments can work with green programmes and policies that directly affect the built environment in order to see significant growth and positive impacts on the economy and environment (EPA, 2020). Detroit and Philadelphia have implemented green infrastructure programmes for stormwater management with notable success (Riedman, 2021; Shokry et al., 2020). Building codes directly affect the production of the built environment and can be adopted by cities to stimulate the growth of carbon-neutral housing stock, aiding climate change mitigation while influencing national housing policy to be greener (Osthorst, 2021). For instance, Los Angeles established the first residential cold roof legislation in the United States (Gilbert et al., 2016). Additionally, strong construction policies aid energy efficiency by committing to renewable energy on all types of buildings and stimulate innovation (Kammen & Engel, 2009). Knowing the influence that building codes have on construction, many green jobs could be created in cities that have green building codes.

To assist city staff working with climate mitigation, analyses should provide practicable policy guidance, filling the gap that "exists between researchers and policymakers on city-level climate change mitigation" (Mi et al., 2019: 587). Localising and integrating policy actions between domains such as urban planning, transportation and energy infrastructure help local development (Osthorst, 2021). Shokry et al. (2020) suggested that cities can promote local workforce development using green resilient infrastructure by allying with community efforts aimed at low-income groups. Also, local green workforce initiatives can be tied to broader labour reforms to promote social justice (Shokry et al., 2020).

Critics of local level policies argue that: (a) the local approach overlooks regional strategies; (b) the dominant adaptation trajectories render bottom-up approaches unable to gain traction because they do not operate at a larger governmental scale; (c) multilevel governance affects local actions as local governments in many countries lack control over key areas central to urban adaptation; (d) strategies need to be "country-specific" and create alliances between social and ecological movements, labour unions, minor communities and in this case, energy sector workers (Shi et al., 2016; Henrique and Tschakert, 2020; Healy and Barry, 2017). While these points should be considered, there is really no conflict between local and supra-local actions. These actions complement, not contradict, one another.

Climate justice, equality and human rights should guide the green (urban) economy. To understand the impacts of the green economy on urban vulnerable populations, equality embedded in the relationship between technology and society should be examined (Heshmati, 2014). Knowing that climate and jobs goals can appear contradictory, if the creation of jobs regardless of their environmental impact is prioritised, an economic approach that centres human rights is needed to reconcile these goals with taking advantage of the benefits of structural change (Stephenson, 2010).

Green jobs for climate justice

Green jobs play an essential role in the transition to a low carbon economy, and are an important indicator of sustainable development (Sulich & Zema, 2018). There are different definitions of green jobs. This study adopts the one which argues that they "contribute to preserving or restoring the quality of the environment while also meeting requirements of decent work, adequate wages, safe conditions, workers' rights, social dialogue, and social protection" (Renner, Sweeney and Kubit, 2008: 3). Bringing the concept of decent work to the green economy is important because green jobs are not always quality jobs (Littig, 2017, p. 323; Osterman and Chimienti, 2012). Decent work is a concept advocated by the ILO. In essence, there is a need to provide workers with adequate pay, job security, good and safe working conditions and career prospects, and to advocate for workers' rights (Renner et al. 2008; Van Empel and Werna, 2010). There are positive signs in the literature for green jobs growth in the United States. Bowen et al. (2018) argue that green processes of production are rapidly becoming a major creator of profits and jobs.

Technology plays a central role in the growth of the green economy and skilled workers are crucial for managing low-carbon technologies. However, this is not a straightforward path to follow. As technological evolution accelerates, the equipment and skills needed for green jobs will likely change, especially in technology-heavy industries such as renewable energy (ILO, 2018). Thus, policy makers should strive to achieve a balance between implementing regulatory and qualification standards, adding a flexibility that acknowledges the reality of changing technologies (ILO, 2018). Moreover, there might be a need for the market and the public sector to work together to address the possible failures related to the provision of low-carbon skills, as shortages of such skills are a reality in many economies (Jagger et al., 2013). To bring in an example from outside the United States, in the United Kingdom, training is suboptimal, making "the adoption of new technologies such as low carbon technologies more difficult" (Jagger et al., 2013, p. 47). Indeed, technological changes and green jobs go hand-in-hand, and technological development should be on the radar of those involved in training the labour force.

Labour unions are important actors in advocating especially for green-collar workers, among others. Coalitions between environmental and labour groups at the local level seem to be a promising way to achieve the common goal of climate justice (Robinson, 2020). These coalitions should be based on trust and collaboration to motivate organisations to make compromises and to create hybrid identities (Robinson, 2020). As noted by Robinson

(2020), labour unions should look past immediate self-interest to try to be greener, not only when it is financially beneficial. Moreover, labour unions sometimes defend fossil fuel jobs against green ones; thus, they might have a "jobs versus the environment/climate" mentality (Healy and Barry, 2017). On the other hand, regardless of how their motivations are perceived, union density is associated with reduced levels of greenhouse gas emissions, showing that the role of labour unions in policymaking can be environment-friendly (Hyde & Vachon, 2019). The engagement of trade unions and their confederations in favour of the environment has evolved significantly. For example, the policies of the Building and Wood Workers International (BWI).

Green jobs related to the production of the built environment are the focus of this study. By improving the urban built environment through the creation of jobs, climate change can be counteracted, and local economic development can see immense growth (Werna, 2013; Deschenes, 2013). This can, in turn, prevent potential negative impacts on people living in certain areas of risk. For example, studies show that African Americans disproportionately experience "deaths during heat waves and from increased air pollution" (Schlosberg and Collins, 2014, p. 362). Green jobs in the creation, operation and maintaining of the urban built environment range from work on infrastructure to buildings. Green infrastructure can lead to improvements in public health (such as the benefits from improving drainage), urban aesthetics, job creation in areas with low-income and high racial and age diversity, and support the local economy with the increased needs for jobs in construction and maintenance (Dunn, 2010).

There are negative impacts. Shokry et al. (2020) argued that as green and resilient infrastructure is commissioned, lower-income residents of colour are omitted from the benefits. Gentrification and green investments are often seen to be occurring in the same areas (Shokry et al., 2020). For example, in Philadelphia, green resilience infrastructure was not implemented in a way that benefits the most socio-ecologically vulnerable residents. Shokry et al. (2020) identified green resilience infrastructure programmes as opportunities for "transformative urban climate justice and reparations efforts."

Despite the arguments that cities are the cause of environmental problems, McKendry and Janos (2015) discussed how many are praising cities as being possible keys to sustainability which reinforces the ideas of working to reduce GHG emissions and promote green urban space, business, and industry. McCoy et al. (2012) argued that if excellence in building is the goal, there needs to be standards embedded in the culture of the built environment.

Cities attempting to create green jobs are often driven by economic motivations, such as the need to address a larger recession, or the promotion of non-green industry (McKendry & Janos, 2015). In the face of a difficult economic situation, local politicians and citizens often argue that any industry should be welcomed, drawing attention away from specifically green industry (McKendry & Janos, 2015). As Dunn (2010) noted, cities are more likely to pursue green projects when funding is available. She argues for more funding from the local, state, and national levels to prioritise green infrastructure to be located in high poverty locations.

From a labour perspective, cities have a "comparative advantage for employment creation [such as] the concentration of supply and demand for a wide variety of services and products in urban areas, that potentially offers great employment opportunities" (Van Empel and Werna, 2010:2). In an era when cities are increasingly moving forward to being carbon-free, green jobs can provide the opportunity to change the urban built environment. For this to be realised, training is essential – especially for low-income workers for the purpose of climate justice.

The importance of green policies for the built environment

As mentioned in the section on the green economy, green policies can alter the level and composition of the demand for labour, inducing changes in prices that strengthen clean sectors and hurt dirty ones (Kruse et al., 2017). The most popular type of green policy is a carbon tax, a price-based mechanism that aims to decrease the dependence of the economy on fossil fuel energy (OECD, 2017). Governments implement taxes on any carbon emission from industries such as the production of energy from natural gas or coal, and also from residential use (OECD, 2017). Carbon tax is a cost-efficient mitigator because the emitters can choose to lower their emissions at a cost below the fixed tax (Mi et al., 2019). Despite all that, carbon taxes are unpopular with the influential industries that they impact, making it difficult to implement them (ILO, 2018). Additionally, some climate activists view carbon taxes as legitimising an unacceptable level of carbon emissions (ILO, 2018).

Carbon taxes are discussed above. To expand the analysis, policies related to carbon taxes are necessary for cities because they are a mechanism for climate change mitigation available at the urban level (Mi et al., 2019). To learn from another country: in British Columbia, Canada, the carbon tax proved to be successful, even though the majority of the public was initially opposed to the tax. In addition to per capita GHG emissions declining by 10 percent, this tax led to the creation of around 10,000 jobs. The increase in employment occurred in clean and service-oriented industries, while the most polluting and trade-exposed sectors experienced a decline in employment. After a few years, the public was mostly in support of the tax (OECD, 2017 p. 7). In the United States, carbon taxes are part of the set of national policies proposed by the Green New Deal, showcasing that they are growing as a legitimate form of climate change mitigation that can be impactful across a large geographical area (ILO, 2018; Bezdek, 2020). As noted in the previous section, American cities are not ahead with regards to carbon taxes despite the benefits they bring, and should consider implementing them.

As noted in a previous section, green legislation at the national level in the US centres on economic concerns. The majority of the successful pieces of legislation funding green jobs have been passed by local governments. To give one example from the national level: when examining the American Recovery and Reinvestment Act Popp et al. (2020) show that "the pre-existing level of green skills matters," because green stimulus works best in areas that already have a local base of the green economy to support it. Therefore, intentional investment in technical education is necessary in order to complement green investments in communities with fewer existing green skilled personnel. In other words, the existing capabilities of both individual workers and communities have an impact on the success of green investments (Popp et al., 2020).

Moreover, local governments can work with different green Programmes and policies that directly affect the built environment in order to see significant growth and positive impacts on the economy and environment (EPA, 2020). To illustrate this, the City of Chicago, IL, has a green policy that requires landscaping be included in building plans to reduce heat and air pollution while promoting adaptation (EPA, 2020). The City of Detroit, MI, is implementing green stormwater infrastructure as a way to mitigate sewer overflows and stormwater pollution that are increasing due to climate change (Riedman, 2021). The cities of Indianapolis, IN, Washington D.C., and New York, NY have invested in green roof programmes, which include actions like putting solar panels, vegetation, and rain barrels on the roofs of urban structures (Bowman et al., 2015).

Building codes directly affect the production of the built environment, and can be adopted by cities to stimulate the growth of a carbon-neutral housing stock, aiding climate change mitigation while influencing national housing policy to be more green (Osthorst, 2021). For example, in 2012, Chula Vista was a statewide pioneer in California when officials revised its codes to adopt firmer cool roof requirements (Gilbert et al., 2016). In 2013, Los Angeles, California, also updated its building codes to require cool roofs for every new and refurbished house, establishing the first residential cold roof legislation in the United States (Gilbert et al., 2016). Since they reflect the experiences of local workers, local building codes will need to be updated as greening expands the skills required in the industry (McCoy, 2012). Additionally, construction policies can aid by committing to renewable energy (Kammen & Engel, 2009). Knowing that with the influence building codes have in construction, many green jobs could be created, cities that have green codes in their legislation need to ensure that complementary green job training is locally available.

For a low-carbon way of living to be realised in urban areas, green policies targeting the built environment would be required. As recommended by Kammen and Engel (2009), green government policy advocating for renewable energy should have continuity, predictability and reliability to succeed. Knowing that the built environment plays a large role in determining the outcomes of sustainability efforts, particularly by influencing greenhouse gas emissions (Angelo & Wachsmuth, 2020), this trio - policy continuity, predictability and reliability - should be extended beyond green energy to other aspects of the built environment such as the use of renewable building materials, and reduction of pollution during the construction process and from buildings while in use.

Green jobs for the built environment

Despite the arguments that cities are the cause of environmental problems, McKendry and Janos (2015) discuss how many are praising them as possible keys to sustainability, which reinforces the ideas of working to reduce greenhouse gas emissions and promote green urban space, business, and industry. McCoy et al. (2012) argue that if excellence in building is the ultimate goal, there needs to be standards embedded in the culture of the built environment.

Green jobs related to the production of the built environment are the focus of this study. By advancing the urban built environment through policies for green jobs creation, climate change can be counteracted, and businesses and overall employment can see significant growth (Werna, 2013). Indeed, green jobs are important as they help to reduce environmental problems related to climate change (Deschenes, 2013), which can, in turn, prevent potential negative impacts on people living in certain areas of risk.

By separating already marginalised residents from its long-term benefits, green infrastructure exacerbates existing socio-spatial inequities (Riedman 2021). Shokry et al. (2020) suggested a dual process that as green resilience infrastructure is commissioned, lower-income residents of color are omitted from the benefits. With green investments occurring in rapidly gentrifying areas, lower-income residents of color continue to be forced out in an unjust process (Shokry et al., 2020). Shokry et al. (2020) found that in Philadelphia, green resilience infrastructure was not being sited or planned in a way that benefits the most socio-ecologically vulnerable residents. Residents of American cities increasingly worry that as newly greened neighbourhoods become more desirable, poorer residents may be driven out through gentrification exacerbated by the greening itself (McKendry & Janos, 2015).

Gentrification and green investments are seen to be occurring in the same areas (Shokry et al., 2020). Shokry et al. (2020) identified green resilience infrastructure Programmes as opportunities for "transformative urban climate justice and reparations efforts." Green initiatives that promote social justice are not solely limited to infrastructure plans (Shokry et al., 2020).

When implementing green projects, such as green stormwater infrastructure, to mitigate impacts of climate change in cities in the United States, volunteer citizen labour is often required for the projects to succeed (Riedman, 2021). However, Riedman (2021) argues that policies usually fail to consider what impact the request of this labour has on the project and the community. Citizens are unfairly requested and unpaid. (Riedman, 2021). Recognising and paying volunteer labour would not only support equitable development of low-income communities, but also, it would create gainful green jobs (Riedman, 2021).

For green structural change to legitimately function as an apparatus against climate change, there needs to be a legal approach that considers the right to work (Stephenson, 2010). Stephenson (2010) identifies work-related security, the right to work, as a fundamental part of employment policy that governments must include in future climate change legislation. Jobs protection, included as part of larger social protections, should be implemented so that climate change policy works for marginalised groups and is not overshadowed by profit drives (Stephenson, 2010). Green jobs training Programmes can be considered in that light. Similarly, Scully-Russ (2018) suggested that the transition to a green economy could make work more equitable and improve the functioning of labour market institutions. Green economies offer social opportunities from the structure, nature and scope of the new labour market which is malleable and open to influence. This emergence of green economies allows purposes, relationships, ethics, structures, policies, and practices to be reevaluated.

As stated by Andrić et al. (2019), the connection between climate change and the built environment is complex. The potential impacts can be grouped into four categories: impacts on building structures, building construction, building material properties, and indoor climate and energy use.

Greening local initiatives

With an increasing number of cities participating in and taking climate mitigation actions, cities play an essential role in the low-carbon transition, that should have its climate justice as its basis (Mi et al., 2019; Osthorst, 2021; Angelo & Wachsmuth, 2020; Harris, Weinzettel, & Levim, 2020). Indeed, Vogel and Henstra (2015) described local governments as "key actors in the formulation and implementation of climate adaptation policies" (p. 117). Additionally, Gilbert et al. (2016) advocated that other institutions such as schools and local organisations should also implement strategies for climate mitigation, and that this should not be left to only city governments.

Non-institutional mechanisms are crucial for coalitions around green jobs to succeed (Robinson 2020). In places without a clear community voice to ensure that the impacts of greening are equitable, do not contribute to gentrification and do not accentuate existing socio-economic differences, the economic benefits of greening should become the dominating element in the discussion of the green economy (McKendry & Janos, 2015). Local organising is valuable in ensuring that climate justice does not take a back seat to economic concerns, particularly in the conversation on the need for jobs (Robinson, 2020).

Supporters of localism in climate mitigation initiatives argue that each city has its own spatial configuration and that affects how people understand environmental problems and solutions (Angelo & Wachsmuth, 2020). The spatial characteristics of a place are a necessary precondition to implement climate mitigation policies, particularly in cities (Curley, 2018; Osthorst, 2021). However, there are different understandings of how cities are constructed on an ideological level, which is based on the dynamics that exist between the urban built environment and nature (Angelo & Wachsmuth, 2020). For instance, a common way in which cities attempt to combat climate change is through moving to sustainable energy sources. The existing physical and spatial relationships within unclean energy will influence how communities intentionally facilitate transitions by making space for sustainable energy (Curley, 2018).

However, that is a challenging road to take. Again, learning from elsewhere: of the ten European cities that Harris et al. (2020) studied, none were on course to achieve the greenhouse gas objectives of the Paris Agreement.

If these objectives are to be met, cities must make intentional efforts to fight climate change (Harris et al., 2020). The works in the literature studied note different strategies to facilitate meeting the objectives. In order to assist city staff working with climate mitigation, local analyses should provide practicable policy implications, filling the gap that "exists between researchers and policymakers on city-level climate change mitigation" (Mi et al., 2019, p. 587). Localising and separating policy actions between domains such as urban planning, transportation, and energy infrastructure could help prioritising development at the local level (Osthorst, 2021). Additionally, Mi et al. (2019) suggest that city-based practices should be globalised through cooperation among cities to exchange knowledge and good practice. Shokry et al. (2020) suggested that cities can promote local workforce development using green resilient infrastructure by allying with low-income and minority community efforts. Additionally, local green workforce development initiatives can be tied to broader labour reforms to promote social justice amongst different urban populations (Shokry et al., 2020).

Climate change adaptation and mitigation should be at the centre of policy and action at the local level. Critics argue that this local approach overlooks both regional and metropolitan strategies and the multi-scalar context in which local planning occurs (Shi et al., 2016). Henrique and Tschakert (2020) argue that the dominant adaptation trajectories render bottom-up approaches unable to gain traction because they do not operate at a larger governmental scale. Multilevel governance, mainly in Europe, Canada and Australia, affects local actions as "local governments in many countries lack control over key areas central to urban adaptation" (p. 134). For instance, "long-term infrastructure upgrades and policies" that have the potential to improve overall equity, often need state or national funding and coordination. Because of lack of control, there are more short-term activities in city adaptation such as "integrate climate considerations into land-use plans" than policies (Shi et al, 2016, p. 134). Healy and Barry (2017) argued in their policy paper that strategies need to have a strategy that creates alliances between social and ecological movements, labour unions, minority communities, and energy sector workers. As there is more disagreement than agreement at the regional level because of political barriers, local policies are more effective.

Attempts by cities to create green jobs are often limited to economic motivations, such as a larger recession, or the promotion of non-green industry (McKendry & Janos, 2015). In the face of a difficult economy, local politicians and citizens often argue that any industry should be welcomed, drawing attention away from specifically green industry (McKendry & Janos, 2015). Cities are more likely to pursue green projects when funding is available

(Dunn, 2010). Dunn argues for more funding from the local, state, and national levels to prioritise green infrastructure to be placed in high poverty locations.

From a labour perspective, cities face "comparative advantage for employment creation [such as] the concentration of supply and demand for a wide variety of services and products in urban areas, that potentially offers great employment opportunities" (van Empel and Werna, 2010 p. 2). In a time that cities are increasingly moving forward to carbon-free areas, green jobs are the types of jobs necessary to change the urban built environment.

Green job training for low-skilled workers

Green jobs training programmes can be an essential part of ensuring that vulnerable populations, specifically unemployed minority young adults, are not left out of the opportunities that come with the greening of cities (Falxa-Raymond et al., 2013). Chandler et al. (2016) argue that employers, the government and providers of job-training programmes should work together. While partnerships between community environmental and social justice organisations have, traditionally, not been sustained, those alliances will prove essential for identifying and overcoming the barriers that prevent people in underserved communities from entering green training programmes and jobs (Chandler et al., 2016). Chandler et al. (2016) also highlighted investments in community college programming as well as connecting underserved workers with manufacturer-run training programmes. Another strategy is to increase the number of partnerships between youth programmes, post-secondary institutions, industry and other workforce entities, to provide a larger mix of resources to boost the green workforce (Grobe et al., 2011, Martinson and Stanczyk, 2010). Scully-Russ (2018) argues that workforce development strategies must emerge from within local relationships and conditions.

The curriculum is an important ingredient of green jobs training, and it needs to be adapted to different needs (Werna, 2013). In California, educatio providers are adding flexibility to the green jobs' curriculum (McCoy et al., 2012). Chandler et al. (2016) specifically identify labour unions as playing a key role in both designing curricula for green job programmes and connecting their graduates to union-track careers.

Labour unions can provide support and assistance to workers to gain access to apprenticeships, and give advice on the Programmes and curriculums. The involvement of employers and unions would lead to the alignment of Programme activities (Grobe et al., 2011).

The literature reviewed includes a number of case studies on green job training Programmes for disadvantaged residents in US cities. Scully-Russ (2013) analysed two federally funded green job Programmes in New England, USA. He also considers the ability of the Center for Sustainability and the Pacific Northwest Solar, Inc. to connect learning opportunities to the disadvantaged residents. Borken (2011) examined the Sustainable South Bronx, Bronx Environmental Stewardship Training Programme, and Baltimore's B'More Green scheme. Falxa-Raymond et al. (2013) studied the MillionTreesNYC. Weigensberg et al. (2012) analysed six Chicago-based Programmes. Located in marginalised urban neighborhoods, these programmes support the local workforces while improving the urban economy with initiatives embedded in environmental justice (Borken, 2011). Weigensberg et al. (2012) suggested that commonalities between programmes be identified and explored to establish standards, develop ways in which individuals can analyse outcomes, and expand what is measured in each programme with its processes and outcomes.

Existing programmes still face challenges, which include "misaligned infrastructure, lack of synchronisation in the labour market, and workforce gaps" (Scully-Russ, 2013). McKendry & Janos (2015) argue that green jobs offered through certain training programmes are limited in number and in levels of pay (McKendry & Janos, 2015). By focusing on the Navajo Nation, Curley (2018) noted how green jobs programmes may serve the individual residents but fail the larger community by still allowing coal companies to enter the community by co-opting the rhetoric of environmentalism. However, existing challenges do not mean that training programmes should be discontinued. Challenges need to be addressed and efficiency increased. All in all, the literature is positive regarding the existing efforts to train low-income people in green jobs, to improve climate justice.

Conclusions

This chapter reviewed the literature on how US local governments are engaging in climate change mitigation and climate justice, green jobs that improve the built environment, workforce development, and training for vulnerable residents. This consideration is important because these topics are treated in a fragmented manner in the literature, which makes it difficult to use the knowledge to catapult integrated research, and suggest appropriate action.

Most of the authors of the works reviewed asserted the importance of the green economy and related jobs in the context of the built environment and that local governments can play a strategic role in implementation. However, it was found in the literature that this may not be a panacea, especially if the process does not consider the social aspects. Environmental conservation may take place without social development – or indeed, it can be used as a smokescreen to cover the lack of social policies. This brings to the fore the importance of climate justice, which the chapter analysed in the context of the built environment. Programmes of green jobs training for vulnerable groups constitute an important set of actions in this regard. As noted in the discussion, existing programmes are not devoid of problems. The way forward is to correct and improve, not to ignore or abolish.

It is important for programmes which have already been implemented to bring practical lessons to the body of knowledge. Some of the authors of the works reviewed also highlight areas in which further research is needed. Starting at the broad level, Angelo and Wachsmuth (2020) suggest that research be undertaken on the interconnectedness of urban policies and social ideologies on nature. Scully-Russ (2013) points towards the implications of the green economy for the labour market. Falxa-Raymond et al. (2013) propose that more comprehensive analyses of green job training be conducted with the goal of identifying how to serve those who fail to graduate or remain unemployed. Future research should also address the ownership of the built environment in the communities that green jobs programmes are being introduced into to ensure sustainability throughout the entire process (Curley, 2018). Weigensberg et al. (2012) suggest that similarities among successful programmes be defined to establish good-practice standards and analyse long-term outcomes of graduates. Borken (2011) suggested that programmes should continue to expand and believed that the best practices should be "acknowledged, along with the additional suggestions" (p. 49) as the programmes develop and show success.

This context opens up an opportunity for understanding how cities' vulnerable groups can participate in development strategies that target the urban built environment in the movement towards curbing carbon pollution. It is important to learn from programmes which have already been implemented, to bring in practical lessons. Hence, the authors of the present chapter consider the possibility of a follow-up with a research study based on a comparative analysis of green jobs training programmes across the United States, with the main goal of critically examining how programmes adopted by cities, or partnered with cities, can improve the livelihood of low-skilled workers by including them in the green workforce, after their participation in local green jobs training.

Notes

1 This chapter was developed from the paper presented by the authors at the World Building Congress 2022 in Melbourne in June 2022, entitled "The more local, the better: green jobs and climate justice in the U.S. urban built fabric". The authors thank the anonymous reviewers of the paper and this chapter, and George Ofori for their comments and suggestions. The usual disclaimers apply.

This chapter is part of a research project that was funded by the Iowa State University Center for Excellence in Arts and Humanities - Research Grants (2021-2022).

References

Andrić, I., Koc, M. and Al-Ghamdi, S. G. (2019) A review of climate change implications for built environment: Impacts, mitigation measures and associated challenges in developed and developing countries. Journal of Cleaner Production, 211, pp. 83–102.

Angelo, H. and Wachsmuth, D. (2020) 'Why does everyone think cities can save the planet?', Urban Studies, 57(11), pp. 2201–2221.

Arcand, B. (2020) 'Ontario's Energy Transition: A Successful Case of a Green Jobs Strategy?', Environmental Policy: An Economic Perspective, pp. 193–212.

Baker, S.E. and Edwards, R. (2012) How many qualitative interviews is enough. National Centre for Research Methods Review Paper. NCRM. University of Southampton.

Bezdek, R. H. (2020) The jobs impact of the USA new green deal. American Journal of Industrial and Business Management, 10(6), pp. 1085–1106.

Biernacki, P. and Waldford, D. (1981) 'Snowball sampling: problems and techniques of chain referral sampling', Sociological Methods & Research, 2. pp. 141–163.

Borken, C.L. (2011) Successful attributes of green-collar job training Programmes in urban environmental justice communities: A case study. PhD thesis. Duke University, USA [Online].

Bowen, A., Kuralbayeva, K. and Tipoe, E.L. (2018) 'Characterising green employment: The impacts of "greening" on workforce composition', Energy Economics, 72, pp. 263–275.

Bowman, J. , Kington, L. , Julian, N. , Grunewald, E. and Greenslade, E. (2015) Green Job Training and Smart Roofs in Columbus, Ohio.

Bulkeley, H., Carmin, J., Broto, V.C., Edwards, G.A., and Fuller, S. (2013) 'Climate justice and global cities: Mapping the emerging discourses', Global Environmental Change, 23(5), pp. 914–925.

Bulkeley, H., Edwards, G.A. and Fuller, S. (2014) 'Contesting climate justice in the city: Examining politics and practice in urban climate change experiments', Global Environmental Change, 25, pp. 31–40.

Chandler, S., Espino, J. and O'dea, J. (2016) 'Delivering opportunity: how electric buses and trucks can create jobs and improve public health in California', Union Of Concerned Scientists, Berkeley, Ca.

Chu, E.K. and Cannon, C.E. (2021) 'Equity, inclusion, and justice as criteria for decision-making on climate adaptation in cities', Current Opinion in Environmental Sustainability, 51, pp. 85–94.

Cities Alliance. (2021) 'Ending Urban Poverty and Promoting the Role of Cities', Available at: https://citiesalliance.org/ (Accessed: 2 Feb. 2021).

Curley, A. (2018) 'A failed green future: Navajo green jobs and energy "transition" in the Navajo Nation', Geoforum, 88, pp. 57–65.

Deetjen, T. A., Conger, J. P., Leibowicz, B. D. and Webber, M. E. (2018) Review of climate action plans in 29 major U.S. cities: Comparing current policies to research recommendations. Sustainable Cities and Society, 41, pp. 711–727.

Deschenes, O. (2013) 'Green jobs', IZA Policy Paper, 62.

Dunn, D. (2010) 'Siting green infrastructure: legal and policy solutions to alleviate urban poverty and promote healthy communities', BC Envtl. Aff. L. Rev, 37(41), pp. 41–66.

Elliott, R.J. and Lindley, J.K. (2014) Green jobs and growth in the United States: Green shoots or false dawn? Department of Economics, University of Birmingham.

EPA Environmental Protection Agency (2020) The Local Action Framework: A Guide to Help Communities Achieve Energy and Environmental Goals. [Online]. Available at: https://www.epa.gov/greeningepa/green-buildings-epa.

EJLF Environmental Justice Leadership Forum (2020) Green Jobs Report: Creating A Green Workforce, Community-Based Solutions for a Diverse Green Jobs Sector. [Online]. Available at: https://weact.org.

Falxa-Raymond, N., Svendsen, E. and Campbell, L.K. (2013) 'From job training to green jobs: A case study of a young adult employment Programme centered on environmental restoration in New York City, USA', Urban Forestry & Urban Greening, 12(3), pp. 287–295.

Fitzgerald, J. (2020) Greenovation: Urban leadership on climate change. Oxford University Press.

Gilbert, H., Mandel, B.H. and Levinson, R. (2016) Keeping California cool: Recent cool community developments. Energy and Buildings, 114, pp. 20–26.

Granberg, M. and Glover, L. (2021) 'The Climate Just City', Sustainability, 13(3), pp. 1201.

Grecksch, K. and Klöck, C. (2020) 'Access and allocation in climate change adaptation', International Environmental Agreements: Politics, Law and Economics, 20(2), pp. 271–286.

Grobe, T., O'Sullivan, K., Prouty, S. and White, S. (2011) A Green Career Pathways Framework: Postsecondary and Employment Success for Low-Income, Disconnected Youth [Online]. Available at: https://www.jff.org/

Guest, G., Bunce, A., & Johnson, L. (2006) How many interviews are enough? An experiment with data saturation and variability. Field methods, 18(1), pp. 59–82.

Gupta, S., and Alatriz, T. (2019) Achieving Climate Justice in the Southside Green Zone: Recommendations for City of Minneapolis Work Plan Action (2020-2025).

Harris, S., Weinzettel, J. and Levin, G. (2020) Implications of low carbon city sustainability strategies for 2050. Sustainability, 12(3), p. 5417.

Healy, N. and Barry, J. (2017) 'Politicizing energy justice and energy system transitions: Fossil fuel divestment and a "just transition"', Energy policy, 108, pp. 451–459.

Henrique, K.P. and Tschakert, P. (2020) 'Pathways to urban transformation: From dispossession to climate justice', Progress in Human Geography, 45(5), pp. 1169–1191.

Heshmati, A. (2014) An empirical survey of the ramifications of a green economy. IZA Discussion Papers. Institute for the Study of Labour (IZA), Bonn

Hyde, A. and Vachon, T.E. (2019) 'Running with or against the treadmill? Labour unions, institutional contexts, and greenhouse gas emissions in a comparative perspective', Environmental Sociology, 5(3), pp. 269–282.

ILO (International Labour Office) (2018) World Employment and Social Outlook 2018: Greening with jobs. International Labour Office: Geneva.

Jagger, N., Foxon, T. and Gouldson, A. (2013) 'Skills constraints and the low carbon transition', Climate policy, 13(1), pp. 43–57.

Kammen, D.M. and Engel, D. (2009) Green jobs and the clean energy economy (Vol. 4). Copenhagen Climate Council Thought Leadership Series, Copenhagen. Retrieved from: http://rael.berkeley.edu/

Knuth, S.E. (2010) 'Addressing place in climate change mitigation: Reducing emissions in a suburban Landscape' Applied Geography, 30(4), pp. 518–531.

Krause, D. (2018) 'Transformative approaches to address climate change and achieve climate justice', In Jafry, T., Mikulewicz, M., and Helwig, K. (eds.) Routledge Handbook of Climate Justice. London: Routledge, pp. 509–520.

Kruse, T., Dellink, R., Chateau, J. and Agrawala, S. (2017) Employment implications of Green Growth: Linking jobs, growth, and green policies. [Online]. Available at: https://www.oecd.org/environment/Employment-Implications-of-Green-Growth-OECD-Report-G7-Environment-Ministers.pdf

Littig, B. (2017) 'Good Green Jobs for Whom?: A feminist critique of the green economy', In MacGregor, S. (ed.), Routledge handbook of gender and environment. London: Routledge, pp. 318–330.

McCauley, D. and Heffron, R. (2018) Just transition: Integrating climate, energy and environmental justice. Energy Policy, 119, pp. 1–7.
Marshall, B., Cardon, P., Poddar, A., & Fontenot, R. (2013) Does sample size matter in qualitative research?: A review of qualitative interviews in IS research. Journal of computer information systems, 54(1), pp. 11–22.
Martinson, K. and Stanczyk, A. (2010) 'Low-Skill Workers' Access to Quality Green Jobs. Perspectives on Low-Income Working Families' Brief 13.Urban Institute (NJ1).
McClure, L.A., LeBlanc, W.G., Fernandez, C.A., Fleming, L.E., Lee, D.J., Moore, K.J. and Caban-Martinez, A.J. (2017) 'Green collar workers: an emerging workforce in the environmental sector', Journal of Occupational and Environmental Medicine, 59(5), p. 440.
McCoy, A.P., O'Brien, P., Novak, V. and Cavell, M. (2012) 'Toward understanding roles for education and training in improving green jobs skills development' International Journal of Construction Education and Research, 8(3), pp. 186–203.
McKendry, C. (2016) 'Cities and the challenge of multiscalar climate justice: Climate governance and social equity in Chicago, Birmingham, and Vancouver' Local Environment, 21(11), pp. 1354–1371.
Metcalf, G.E. (2019) 'On the economics of a carbon tax for the United States', Brookings Papers on Economic Activity, (1), pp. 405–484.
McKendry, C. and Janos, N. (2015) 'Greening the industrial city: equity, environment, and economic growth in Seattle and Chicago', International Environmental Agreements: Politics, Law and Economics, 15(1), pp. 45–60.
Mi, Z., Guan, D., Liu, Z., Liu, J., Viguié, V., Fromer, N. and Wang, Y. (2019) 'Cities: The core of climate change mitigation', Journal of Cleaner Production, 207, pp. 582–589.
Mulvaney, D. (2013) Opening the black box of solar energy technologies: Exploring tensions between innovation and environmental justice. Science as Culture, 22(2), pp. 230–237.
OECD. (2017) Employment implications of Green Growth: Linking jobs, growth, and green policies.
Osterman, P. and Chimienti, E. (2012) 'The politics of job quality: A case study of weatherization', Work and Occupations, 39(4), pp. 409–426.
Osthorst, W. (2021) 'Tensions in Urban Transitions. Conceptualizing Conflicts in Local Climate Policy Arrangements', Sustainability, 13(1), p. 78.
Parker, C., Scott, S. and Geddes, A. (2019) Snowball Sampling. In Atkinson, P., Delamont, S., Cernat, A.; Sakshaug, J. & Williams, R. (editors). Series Research Design for Qualitative Research. SAGE Publications: Newbury Park.
Popp, D., Vona, F., Marin, G. and Chen, Z. (2020) 'The employment impact of green fiscal push: evidence from the American Recovery Act (No. w27321)', National Bureau of Economic Research.
Renner, M., Sweeney, S. and Kubit, J. (2008) Green Jobs: Towards Decent Work in a Sustainable, Low-Carbon World: Report for United Nations Environment Programmeme.
Rice, J. L. (2010) Climate, carbon, and territory: Greenhouse gas mitigation in Seattle, Washington. Annals of the Association of American Geographers, 100(4), pp. 929–937.
Riedman, E. (2021) 'Othermothering in Detroit, MI: understanding race and gender inequalities in green stormwater infrastructure labour', Journal of Environmental Policy & Planning, 23(5), pp. 616–627.
Robins, N. , Gouldson, A. , Irwin, W. and Sudmant, A. (2019) Investing in a just transition in the UK How investors can integrate social impact and place-based financing into climate strategies. Lse. Ac. Uk.
Robinson, J.L. (2020) 'Building a green economy: Advancing climate justice through environmental-labour alliances', Mobilization, 25(2), pp. 245–264.
Sen, A. (1999) Development as freedom. Oxford University Press: New York.
Schlosberg, D. (2012) 'Climate justice and capabilities: A framework for adaptation policy', Ethics & International Affairs, 26(4), pp. 445–461.
Schlosberg, D. and Collins, L.B. (2014) 'From environmental to climate justice: climate change and the discourse of environmental justice', Wiley Interdisciplinary Reviews: Climate Change, 5(3), pp. 359–374.
Scully-Russ, E. (2013) 'The dual promise of green jobs: A qualitative study of federally funded energy training Programmemes in the USA', European Journal of Training and Development, 37(3), pp. 257–272.

Scully-Russ, E. (2018) The Dual Promise of Green Jobs: Sustainability and Economic Equity. In The Palgrave Handbook of Sustainability (pp. 503–521). Palgrave Macmillan, Cham.
Shi, L., Chu, E., Anguelovski, I., Aylett, A., Debats, J., Goh, K., Schenk, T., Seto, K.C., Dodman, D., Roberts, D. and Roberts, J.T. (2016) 'Roadmap towards justice in urban climate adaptation research', Nature Climate Change, 6(2), pp. 131–137.
Shokry, G., Connolly, J.J. and Anguelovski, I. (2020) 'Understanding climate gentrification an shifting landscapes of protection and vulnerability in green resilient Philadelphia', Urban Climate, 31, p. 100539
Soanes, M., Bahadur, A., Shakya, C., Smith, B., Patel, S., del Rio, C.R., Coger, T., Dinshaw, A., Patel, S., Huq, S. and Musa, M. (2021) Principles for locally led adaptation: Climate Resilience. Retrieved from: https://www.wri.org/initiatives/locally-led-adaptation/principles-locally-led-adaptation.
Stephenson, S. (2010) 'Making jobs work: The right to work, jobs and green structural change.' Sustainable development law on climate change.
Sulich, A. and Zema, T. (2018) 'Green jobs, a new measure of public management and sustainable Development', European Journal of Environmental Sciences, 8(1), pp. 69–75.
The White House. FACT SHEET: President Biden Takes Executive Actions to Tackle the Climate Crisis at Home and Abroad, Create Jobs, and Restore Scientific Integrity Across Federal Government (2021). Retrieved from https://www.whitehouse.gov/briefing-room/statements-releases/2021/01/27/fact-sheet-president-biden-takes-executive-actions-to-tackle-the-climate-crisis-at-home-and-abroad-create-jobs-and-restore-scientific-integrity-across-federal-government/ (Accessed 1 Feb. 2021).
Thiollent, M. (2000) Metodologia da pesquisa-ação. Cortez / Autores Associados: Sao Paulo.
UNFCCC (2015) Conference of the Parties (COP), 2015. Adoption of the Paris Agreement. Proposal by the President. In Conference of the Parties (COP), 2015. Adoption of the Paris Agreement. Proposal by the President. p32. doi:FCCC/CP/2015/L.9/Rev.1.
UNEP (United Nations Environment Programme) (2023) The Green Economy. https://www.unep.org/regions/asia-and-pacific/regional-initiatives/supporting-resource-efficiency/green-economy. Acceded 17 April 2023.
University of California. (2017, April 17) Where do greenhouse gas emission come from? University of California Carbon Neutrality Initiative. Retrieved February 02, 2021, from https://www.universityofcalifornia.edu/longform/where-do-greenhouse-gas-emissions-come#:~:text=Electricity%20%26%20Heat%3A%20This%20is%20power,of%20greenhouse%20gas%20emissions%20worldwide
Van Empel, C. and Werna, E. (2010) Labour-oriented participation in municipalities: how decentralized social dialogue can benefit the urban economy and its sectors. Working Paper: Sectoral Activities Dept. ILO: Geneva.
Vogel, B. and Henstra, D. (2015) Studying local climate adaptation: A heuristic research framework for comparative policy analysis. Global Environmental Change, 31, pp. 110–120.
Weigensberg, E., Schlect, C., Laken, F., George, R., Stagner, M., Ballard, P., DeCoursey, J. (2012) Inside the black box: What makes workforce development programmes successful, Chicago: Chaplin Hall at the University of Chicago.
Werna, E. (2013) 'Working in green cities: Improving the urban environment while creating jobs and enhancing working conditions' In Simpson, R., and Zimmerman, M. (eds.) In The Economy of Green Cities: A World Compendium on the Green Urban Economy. New York, NY: Springer, Dordrecht (pp. 57–70).

7

LABOUR CONTRACTING, MIGRATION AND WAGE THEFT IN THE CONSTRUCTION INDUSTRY IN QATAR, CHINA, INDIA, US AND THE EU

Jill Wells

ENGINEERS AGAINST POVERTY, UK

Introduction

Massive restructuring has taken place in the organisation of production and employment in many sectors of the economy over the last 30–40 years. In 2014, David Weil introduced the term 'fissured workplace' to describe fundamental changes in firms' competitive strategies that have reshaped the way that workers are employed in the twenty-first century. He shows how lead firms have pulled back to focus on their core competencies, subcontracting non-core tasks to other businesses while off-loading significant risk and responsibility for their workforce to complicated networks of smaller business units and individual workers (often cast in the role of independent businesses in their own right). These '*lower level businesses operate in more highly competitive markets, creating downward pressure on wages and benefits, murkiness about who bears responsibility for work conditions and increased likelihood that basic labour standards will be violated*' (Weil, 2014, p.8). The fissured workplace offers precarious jobs with deteriorating and insecure wages for workers, many of whom are employed by labour market intermediaries at the bottom and periphery of complex, multi-layered contractual chains.[1]

The construction industry has led the way in the changes described by Weil (2014). The variety of trades and skills involved in this industry means that subcontracting of tasks that require specialist expertise has always been important, but the last two decades of the twentieth century saw a dramatic increase in subcontracting in construction industries around the world. Over the same period, in a search for more 'flexible' labour practices, contractors and subcontractors shed their permanent and directly employed labour force in favour of out-sourcing labour through intermediaries. These processes were documented in some detail in a wide variety of countries in a report by the International Labour Office at the turn of the century (ILO, 2001). This was a seminal document which inspired many others and generated a small literature on the subject, of which the work of Weil (2014) is but one example.[2]

DOI: 10.1201/9781003262671-7

The changes were most dramatic in the industrialised countries where the significant stabilisation of the construction workforce that had been achieved in the 1950s and 1960s began to unravel from the 1970s. But there were also big changes in the developing world with a decline in contractors' directly employed workforce and increase in the proportion of workers employed less formally through subcontractors and intermediaries. For example, in the emerging economy of Brazil the permanent registered employees of general contractors halved from 41% of the construction workforce in 1981 to 20.9% by 1999, while unregistered and self-employed workers rose from 56.7% to 74.6% (ILO, 2001, p. 18). At the same time there was an enormous expansion in the number of informal employers, most of whom are assumed to be self-employed workers or labour contractors, disparagingly referred to as *gatos* (the Portuguese word for cat).

The processes identified by the ILO in 2001 have accelerated into the twenty-first century, leading the construction industry in many countries to be characterised by 'hollowed out' firms at the head of long chains of subcontracted businesses. While most (sub)contractors will have a small core workforce directly employed, a large and increasing segment of the low-skilled workforce is being supplied through intermediaries (agencies/brokers) for use by (sub)contractors on a temporary basis, as and when needed. This has been the traditional way of recruiting the bulk of the construction workforce in many developing countries, but such arrangements (often described as 'Worker Dispatch') have increased dramatically in both industrial and emerging economies and are seen as an important part of fissurisation (Nakdubo and Araki, 2017).

The process of intermediaries supplying labour takes different forms, with variations in the division of responsibility for supervision of the work between the contractor (the user of the labour) and the supplier of the labour. But it generally involves a triangular employment relationship based on two quite different contracts. Workers are engaged by contractors through a commercial contract with a labour supplier who in turn has a contract of employment with the workers. The absence of a direct contractual employment link between the contractor and the workers blurs the lines of accountability and allows contractors to distance themselves from the risks associated with the employment and management of labour. The risks, which include interruption of the work due to inclement weather etc., are passed down to the labour contractors in the form of no work and no pay. They may also be passed directly to the workers through illegal misclassification as 'independent contractors' whereby they forego all rights and benefits as employees.

Shifting labour costs and liabilities to small businesses, labour intermediaries or independent contractors, which has been accompanied by a steep decline in trade union membership, has significantly increased the vulnerability of workers. In addition to the insecurity that comes from short term contracts and increased risk of not having work, other aspects of vulnerability suggested by Weil (2014) include increased exposure to health and safety hazards in the workplace, exclusion from a range of work-based benefits (sickness, health care, unemployment, etc.) and various types of wage violation.

In construction, concern has traditionally focused on improving the health and safety of workers in a notoriously hazardous industry where loss of life may be the most significant risk. Wage violations are far less widely monitored and reported. However, that may now be changing in the context of mounting evidence of incidents of late, partial or non-payment of wages to construction workers in many parts of the world.

Exposure to the risk of wage violations appears to be greatest in low-wage construction jobs occupied by unskilled or low-skilled workers. *Searching for 'construction labour' in*

Google, we find that: 'in the construction industry, labour is the term usually given to what is normally termed "manual labour" or "site labour". They may be doing a variety of jobs including, digging, cleaning, catering, carrying and lifting. Typically, such workers have little training or qualifications and may be employees of a contractor or sub-contractor, or temporary workers hired for part or all of the project'.

These jobs are often filled by migrants. A review of the literature documenting the incorporation of international migrants into construction labour markets concluded, on the basis of evidence presented, that wage theft and wage withholding are systemic characteristics of many local construction industries but especially experienced by migrants employed at the bottom of subcontracting chains and those without legal status or access to trade unions, who may have additional vulnerabilities due to immigration status and/or language (Buckley et al., 2017).

Wage violations among migrant workers in Qatar

This was the situation found by Engineers Against Poverty (EAP) researching in Qatar during a period of intense construction activity in the lead up to hosting the World Cup in 2022. Some 90% of all construction workers in Qatar are international migrants from low-wage economies in South Asia (Nepal, India, Sri Lanka, Bangladesh, Pakistan). These workers often have to pay costs and fees to recruiting agents for the promise of a job in Qatar for which they may have to borrow money, so many arrive already in debt. Once they start work in the host country the late payment of wages is the single most serious issue they face. Almost all (93%) of the complaints handled by the Ministry of Labour in 2010 were about delayed salaries and two-thirds of complainants were construction workers (Wells, 2014). In 2012 the Ministry received 6000 complaints, with salaries and denial of benefits the main issue. Further evidence of late and non-payment of wages can be found in reports by Amnesty International (2013) and Human Rights Watch (2012) at around the same time. *The Business and Human Rights Resource Centre* recorded 17 public allegations of labour rights abuse by construction companies in the Gulf between 2017 and 2018, 71% of which concerned late or non-payment of workers' wages (BHRRC, 2016).

On the basis of evidence provided through in-depth interviews with international contractors operating in Qatar, clients and senior government advisers, the immediate cause of late payment was traced to changes in the organisation of employment similar to those described by Weil (2014). In a search for greater flexibility in the employment of labour, here as elsewhere in the world, large construction companies that used to employ workers directly, have devolved significant risk and responsibility for their workforce to networks of smaller businesses, through extensive subcontracting and further outsourcing of labour through intermediaries. Attaining flexibility in labour supply is difficult in the countries of the Gulf where the sponsorship system does not allow workers to move easily between employers and where employment contracts are generally signed for two years. But the past few decades had seen the development of manpower companies which specialise in providing temporary labour for the building trade. Acting contractually as the employer they are able to side-step the restrictions imposed by *kafala* laws[3] which tie foreign workers to a particular job, and move workers among contractors as needed (Buckley, 2012). A survey of the workforce composition of major construction companies in Qatar and UAE conducted by the Business and Human Rights Resource Centre (BHRRC) in 2016 found up to 50% of workers to be employed by intermediaries supplying labour to (sub)contractors (BHRRC,

2016). The same proportion (50%) was working on projects building stadia for the World Cup in 2022, under the direction of the Supreme Committee for Delivery and Legacy.

The vast majority of these intermediaries supplying labour were very small firms from labour sending countries, many with fewer than 10 employees and some former migrant workers themselves. With few assets and limited liquidity, they are unable to pay wages until they have received payment from the subcontractor to whom they supplied workers, who may also still be waiting for payment from those above him in the subcontracting chain (Wells, 2018). Despite denials by government advisers, clients do not pay on time, often due to disputes over quality and/or progress of the work. At the same time, the common acceptance in the region of the practice of companies only paying subcontractors after payment has been received from the contractor in the tier above, which is known in the region as '*back to back*' and internationally as *pay when paid,*[4] means that contractors in the upper tiers of subcontracting chains can legitimately hold back payment to boost their own cash flow, starving the lower tiers of funds. Hence the immediate cause of late payment of wages was traced to slow movement of interim payments for construction work down a subcontracting chain, which was lengthening with sub-subcontracting among labour suppliers.

However, lack of funds to pay wages when they are due can also be used as an excuse, and may not be the only reason for late payment of wages. It is also likely that the immediate employers cannot pay because there is insufficient money in the contract for labour supply, in which case late payment will turn into non-payment. The predominant form of procurement of construction projects is international competitive bidding with contracts awarded to the lowest priced bidder and competition for contracts in the GCC construction industry is intense (DLA Piper/MEED, 2018). Competition to participate in projects as subcontractors or as labour suppliers at the bottom of the supply chain is likely to be even more intense, as predicted by Weil (2014), so reducing prices in order to win contracts becomes a priority. In this context cheating workers of their wages would seem to be inevitable.

The wage protection system

When workers have not been paid the first step in efforts to recover their money must be to have proof. A few contractors interviewed in Qatar in 2013 were paying wages to their directly employed workers through electronic bank transfer into workers' bank accounts, but the majority paid in cash, maintaining that workers were paid too little to open a bank account. The preliminary recommendation in the report by Engineers Against Poverty (EAP) in January 2014 was that payment of wages through electronic bank transfer should be mandatory as this would provide workers with a record of payment and the evidence needed to prove when they have not been paid (Wells, 2014).

In April 2014 an independent review of the legislative and enforcement framework of Qatar's labour laws, which had been commissioned by the State of Qatar in October 2013, was published by leading international law firm, DLA Piper (2014). The authors strongly supported the recommendation that paying wages through electronic bank transfer should be mandatory and went further in proposing that the government monitor the process to expedite the detection of non-payment. The authors also expressed concern about undue delay in the payment of project funds which could impact on the payment of wages to migrant workers through subcontractors, or be used as an excuse for late payment. A reduction in the payment period in construction contracts from 90 to 60 days was recommended with sanctions on late payment throughout the chain of contracting. In addition to financial penalties, the proposed sanctions include the right of subcontractors to stop work and labour suppliers to withdraw labour if they have not been paid.

The endorsement of DLA Piper gave real impetus to the introduction of an electronic Wage Protection System (WPS). In June 2014 the Governor of the Qatar Central Bank (QCB) issued a circular which required all banks to accept employee WPS accounts, regardless of their income level, and some 17 banks were subsequently approved for participating in the WPS. The legal framework for the WPS came into effect in February 2015 with an amendment to Article 66 of the Labour Law (14) of 2004 which made clear that workers are entitled to wages either monthly or fortnightly and employers are not deemed to have paid their workers unless this is done through the WPS.

Immediate employers (the small intermediaries supplying labour) are central to the operation of the system (ILO, 2019). Each month the employer is required to submit to his bank a Salary Information File (SIF) which is an excel spread sheet carrying the details of remuneration for each employee. There are various checks on the file and the data and if accepted it undergoes a financial audit to verify the employer's bank account and whether there are sufficient funds in the account to cover the payment of salaries. If passed, the system automatically distributes salaries from the employer's bank to the bank accounts of the employees who are notified by SMS of the amounts paid.

The same information is automatically and simultaneously transmitted to the QCB and the WPS unit (WPSU) under the Labour Inspection Department (LID) at the Ministry of Administrative Development, Labour and Social Affairs (MADLSA) for monitoring and follow-up on payment discrepancies and any non-payment of wages (ibid). At the heart of the WPSU monitoring system are the '*checkers*' and '*blockers*', key personnel at the forefront of detection of WPS non-compliance and subsequent action. The *checkers* respond to possible violations by seeking additional information from employers, while the *blockers* are labour inspectors with authority to impose sanctions in the form of limitations on the issue or renewal of work permits.

In November 2017, the ILO entered into a three-year technical cooperation agreement with the State of Qatar and a project office was established in Doha in 2018 to support the government's labour reform agenda. An assessment of the WPS commissioned by the ILO (2019) found 50,000 enterprises registered with the WPS covering 1.3 million workers. Almost three quarters of the enterprises in the system at the time (39,000 enterprises) were being penalised with blocks on applications for work permits or the renewal of current work permits. Fourteen recommendations were set out to extend and improve the system and most were swiftly implemented.

The annual progress report (ILO, 2020) showed considerable strengthening of the WPS with coverage extending to 96% of eligible workers and 94% of eligible enterprises and penalties for wage related violations increasing to prison sentences of 12 months. Also in 2020 the most problematic and restrictive elements of the *Kafala* sponsorship system were dismantled by the removal of the requirement for workers to obtain exit permits to leave the country and/or no-objection certificates (NOC) to change employers.

In 2021 Law number 17 of 2020 came into force establishing a basic non-discriminatory minimum wage of 1,000 Ryals per month with minimum allowances for food and accommodation of 300 and 500 Ryals, respectively, and the SIF was improved with a more detailed breakdown of the wage (ILO, 2021). Also in 2021 the government established an online platform for workers to submit complaints anonymously including complaints on behalf of multiple workers. Between October 2020 and October 2021 MADLSA received 24,650 complainants both online and in person with the top three complaints concerning non-paid wages, end of service benefits and/or annual leave not being granted or paid.

In early 2020 the promised 'wage guarantee fund' (to be known as the '*workers support and insurance fund*') had become operational. Funded by 60% of the fees paid for work permits and their renewal, the fund aims to support workers and, in particular, is tasked with '*Paying the workers' benefits, which are settled by the labour dispute settlement committees, and subsequently reclaiming those amounts from the employer*' (Law number 17 of 2018, Article 2). By March 2022 the Fund had disbursed a total sum of QAR 358 million (almost US$ 100 million) to 35,000 workers (ILO, 2022). However, there is evidence that not all workers whose case was approved by the dispute settlement committees have received payment. It is also not known how much of the money paid out has been collected from employers. If there are real difficulties in getting immediate employers to pay up – because they will not or cannot pay – there may be difficulties in re-stocking the fund, as well as serious moral and political implications of allowing employers to get away with not paying.

To date, the government of Qatar has refused to take action to ban '*pay when paid*' or to adopt other measures that would improve the flow of project funds down the subcontracting chain, such as recommended by DLA Piper in 2014. It has also failed to introduce a '*joint liability' scheme to ensure that all workers on a project are paid on time and in full through the WPS,* as recommended in the assessment report (ILO, 2019, recommendation 4) and in a report for the ILO Regional Office for Arab States (ILO, 2018). Extending liability for wages up the subcontracting chain to the users of the labour, up to the principal contractor or the client, would provide additional companies to call upon when wages have not been paid. It is a policy that dates back to the twentieth century in Brazil and eight countries in Europe and is the direction in which policy has been moving in many other parts of the world, including China and the United States (Wells and da Graca Prado, 2019).

The Supreme Committee for Delivery and Legacy (SC) is one major client that has shown the way in accepting responsibility for the payment of wages to workers on its projects when they have not been paid. Through its careful selection of contractors, together with auditing and pre-approval of subcontractors and labour suppliers, the SC has in fact been undertaking the kind of due diligence that main contractors would be more likely to undertake if they were held jointly liable for wages, as shown in a recently published report (Wells, 2022).

Why do they keep coming?

After an eight-month suspension of recruitment of migrants to Qatar in 2020 due to the COVID-19 pandemic, recruitment resumed in November 2020. Despite significantly higher costs of migration because of the cost of tests and the need to quarantine, South Asian workers were still determined to travel to Qatar. In January 2021 Bangladeshi labourers who had worked in Qatar protested in front of the Ministry of Foreign Affairs in Dhaka, demanding that their government facilitate their return to the Gulf.[5] This raises the question, *why do they keep coming*?

Ethnographic researcher Andrew Gardner, whose research has focused on labour migration to the Gulf for many years, notes that the extraordinary challenges that many labour migrants face co-exist with the fact that a vast flow of remittances still stream from the Gulf Cooperation Countries (GCC) to the countries the migrants came from. These two facts are often presented together suggesting that labour migrants yield a certain proportion of their rights and render themselves vulnerable to exploitation in order to secure the economic benefits provided by work in the wealthy Gulf states. However, this explanation obscures the fact that the exploitation that foreign migrants often encounter is unequally borne by the poorest members of this transnational population – the unskilled and semi-skilled labourers (Gardner, 2012).

In his own ethnographic field work, following a diverse group of ten low-income labour migrants through a year and a half of their lives in Qatar, Gardner reports frequently encountering men and women for whom a stay in the Gulf States has been a financial catastrophe. The economic incentives are still real and are the primary attraction for many migrants. But workers may migrate for a variety of other reasons, including entanglements with the law. Moreover, it is usually a family that makes the decision and the young men who migrate for work in the Gulf are very often the emissary of a household livelihood strategy. The decision about which member of the family will migrate is decided at the level of the extended family, hence some young men may migrate simply because the family insists on it (Gardner, 2012).

Deception is central to their experiences, particularly over contracts (salary, hours, type of work, etc.). The contractually promised wage levels in the Gulf States remain competitive in South Asian terms and generally surpass salary levels in the poorest nations from which migrants come. But they often arrive to find a job very different from the one they were promised and many receive no pay for many months. It is difficult to locate responsibility for the level of deception but labour recruiting agents are often working with poor information. The system itself produces further disinformation and this is perpetuated by labour migrants themselves who go to great lengths to present false images of what life is like to family members back at home. When they have not been paid they may even borrow to send money home, in order to keep up appearances. Migrants may also become complicit in the contractual deception that commonly occurs as *chain migration* (whereby migrants are used to recruit others) remains a vital conduit for tens of thousands of migrants to the Arabian Peninsula (ibid 2012).

Gardner suggests conceptualising the migration system in the Gulf – including the money lenders in south Asian villages, the labour brokers in sending countries, the manpower agencies in the Gulf state and the citizens who sponsor foreign labour, as well as the managers and supervisors who serve as these sponsors' proxies, as a ***migration industry*** that is not only geared to increasing profit from the labour of foreign migrants, but is also capable of deriving profit from the migration process itself.

However, it is important to understand that this should not be taken as evidence of how transnational migrants are simply exploited by Qataris in Qatar. The exceptional wealth of Qatar means that most citizens are able to secure public sector employment and few Qataris are actually employed in positions that directly supervise and manage low-income migrants. It is typical of Qatar that the individuals directly involved in the governance of contingent foreign labourers are themselves migrants (Gardner et al., 2013).

Wage theft and wage withholding in China and India

Stories of late payment, non-payment and deliberate withholding of wages to construction workers led us to look at other countries and regions. In the two emerging economies of

India and China, which together account for more than 100 million construction workers (almost half the estimated number of construction workers in the world as a whole) contractors and subcontractors have also largely off-loaded the business of employing workers to intermediaries. Workers in both countries are overwhelmingly internal circular migrants from the countryside, employed for limited periods by labour contractors who have emerged form the process. They are routinely not paid until their immediate employer receives payment from the (sub)contractor to whom workers are supplied (the users of the worker) and this is usually at the end of a project or the end of the year - an extreme case of *pay when paid.* With-holding payment until completion of a year-long contract ties the workers to a single employer and creates a high risk of working for a year without pay. In both countries workers are all too often not paid in full or not paid at all (Breman et al., 2009; Pun and Lu, 2010; Parry, 2014; Lerche et al., 2017; Mimi Zou, 2017; Pang, 2019).

These findings were reinforced in a research project into *labour conditions and the working poor in India and China* funded by the UK government. Surveys and interviews were carried out with construction workers in Delhi and Shanghai between 2011 and 2013 by a team of researchers from the Centre for Development Policy and Research at the School of Oriental and African Studies (SOAS), University of London. Two thirds of the workers in Delhi were found to have been given a small advance payment with the remaining wages paid only at the end of the 8–10 month working year (Development Viewpoint 80). In Shanghai, only 9% received their wages on a monthly basis, 25% received them at the end of the construction project and over half received wages at the end of the year (Development Viewpoints 77, 78).

Developments in China

Labour subcontracting had a long history in both countries until late 1958 when construction work in China began to be organised under state owned or collective enterprises. However, in 1980 a World Bank project, Lubuge Hydropower in Yunnan Province, challenged socialist practices in the construction industry by adopting international competitive bidding for its work, leading Deng Xiaoping to point out that construction could be a profit making industry. Deng-era reforms brought an end to socialist labour practices in the construction industry and the labour subcontracting system re-emerged (Pun and Lu, 2010). The stage was set when in 1984 public regulations were issued preventing general contractors from directly employing their field workforce and employing labour contractors instead.[6] These regulations accelerated change in the management of the construction industry and the composition of its workforce, leading to the emergence of a multi-tier labour subcontracting system.

Since the 1980s the construction sector in China has undergone further dramatic changes, reorienting itself to the needs of a more market-oriented economy. Freed from state planning State Owned Enterprises (SOEs) began constructing homes and commercial buildings that could be sold in evolving domestic property markets (Development Viewpoint 77, 78). Since the 1990s there has been a remarkable construction boom driven mainly by real estate development in rapidly expanding cities and towns. These developments have been stimulated by the state and local authorities and appear to have been built on extracting the maximum work effort from a large peasant migrant workforce with virtually no rights.

In their classic study of migrant construction workers in China, Pun Ngai and Lu Huilin (2010) show how the rapid development of the construction industry over the previous thirty years had enabled a highly exploitative labour contracting system to emerge. Workers interviewed by the authors in Beijing and Hebei between 2007 and 2009 had been promised a daily pay rate ranging from 50 to 120 Yuan, depending on the type of job and the skill required. But all were eventually paid at a substantially lower rate and some were at risk of not receiving any pay at all. Instead of the payment of weekly or monthly wages construction workers are usually paid a living allowance (*shenghuo fei*) of between 10 to 20% of their promised monthly wage, barely enough to cover food and other daily expenses. Many subcontractors had to use their own money to provide this living allowance and some workers received nothing because their subcontractor claimed to have no money. The authors concluded that '*Significant extraction of labour value in the production process was made possible when wages were replaced by living allowances and when subcontractors justified this practice by saying that there were no funds for salaries coming from their contractor*'(Pun and Lu, 2010).

There is little doubt that the industry in China has developed a complex and deliberately opaque hierarchical structure, constructed to absolve the higher echelons of any direct responsibility for labour conditions in the sector (Development Viewpoint 77). At the top of the pyramid are usually SOEs which are the property developers. Below them are the contractors (privately or collectively owned companies) and at the bottom of the hierarchy are the labour subcontractors. From the sample of workers interviewed on a construction site in Shanghai by the UK team from SOAS, 80% were employed by labour contractors or even more indirectly and illicitly by what Irene Pang (2019) calls '*petty labour contractors*'.

Pang (2019) explains that 'labour service contractors' in China can be categorised into different groups. Those supplying labour to specialised contractors (electrical or mechanical construction) and smaller boutique contractors (responsible for finishing work and aesthetics) usually maintain permanent crews of skilled workers as they must meet higher licensing standards. However, labour services contractors for the bulk of civil and structural construction are not so constrained and often contract out illicitly to what she calls '*petty labour contractors*'. In her research in Beijing it was not uncommon for five or six tiers of petty labour contractors to be engaged as links in the subcontracting chain.

In the former case, when workers are recruited as skilled workers and employees of specialised labour service contractors, the relationships between the workers and their employers are unambiguously recognised legally as labour relations and they are protected under the general framework of labour protection as well as sector specific provisions. But when unskilled workers are employed indirectly through *petty labour subcontractors* the relationship becomes ambiguous as the petty labour contractors do not have legal standing as employing entities, hence no real basis from which to claim their rights when wages are delayed or never paid. Without a labour contract it is also impossible to join the only legal trade union (ACFTU). Only 22% of the workers interviewed in Shanghai by the SOAS team had a labour contract and most were skilled workers. 41% of the total pointed to the lack of a labour contract as their main problem (Development Viewpoint 78).

Workers are also differentiated between skilled and unskilled in terms of their right to be in the urban areas. The vast majority are migrants from the countryside who enter the industry through informal and indirect regimes of recruitment and with no permanent right of residence in urban areas. Only a tiny top layer of migrant workers have access to urban *hukou* (Nguyen, 2013). Around one-fifth of construction workers interviewed in Shanghai had residence rights in urban areas and they were mostly skilled or semi-skilled. These workers tend to be younger and better educated (Development Viewpoint 78). Unskilled workers were older and had been in the industry longer, with 58% having worked in the sector for more than 10 years. They were hired for just less than a year from rural households in neighbouring provinces (48%) or from further away as Sichuan (18%).

Pang (2019) writes further about subcontracting, which is endemic in the construction sectors of both China and India and has introduced multiple risks, the burden of which is borne disproportionately by workers and low-level petty labour subcontractors. First, it functions through each tier extracting value from the tier below so that the more tiers of subcontracting the greater the costs accrued which are ultimately extracted from the workers' wages. Second, since subcontracting transactions are poorly /seldom documented it is difficult for workers to reconstruct the subcontracting chain to recover payments if others along the chain run off with project funds. Third, since project payments are calculated per unit volume along the subcontracting chain down to the lowest tier, but workers are usually paid at per diem rates, delays to progress of the work could create further liquidity problems for petty labour contractors in paying workers.

The wage arrears campaign

The issue of late or non-payment of wages became a major cause of labour unrest in China in the 1990s. Studies carried out in the early 2000s estimated that half of construction workers have experienced payment default, but others have considered the figure to be close to 70%, with cases of salaries still being withheld for over a year (Walk Free Foundation, Global Slavery Index, 2018). To tackle the problem the Chinese Government started a Wage Arrears Campaign back in in 2003. A circular issued at the time emphasised the urgent need to solve the issue of migrant worker payment in the construction industry. This was followed by a series of policies and laws to facilitate the timely and full payment of wages.

According to a recently published report by an expert on wage arrears in Zhejiang Province, three legal documents are regarded as critical landmarks in the governance of wage arrears (Kun Huang, 2022). The first is an amendment, adding one paragraph to Article 276 of the criminal law of the Peoples' Republic of China in 2011 which has made it a criminal offence to evade paying workers' wages. The second is the '*Opinion on Comprehensive Management of Wage Arrears of Migrant Workers*' issued by the State Council in 2016 which highlights the need to identify the responsibilities of all parties involved in the supply chain and that the main contractor should take overall responsibility for the payment of migrant workers' wages (Mayer-Brown, 2016). The third and most important document is the '*Regulation on Ensuring Wage Payment to Migrant Workers*' which was signed by Chinese Premier Li Keqiang on December 30, 2019 and came into effect on May 1, 2020. The 2019 Regulations for the first time set a clear requirement that the general contractor has the primary responsibility for wage payments. Article 28 of Regulation 2019 requires that the general contractor or the subcontractor signs a labour contract with the migrant worker and registers all workers in his/her real name.

With the legal framework in place, Kun Huang (2022) explains how central and local government agencies have attempted to improve the efficiency of the labour inspection system in dealing with wage arrears to workers, including through the use of information technology. Three key processes are involved:

- a digital system for the real name registration of migrant workers;
- a separate financial account for workers' wages (separating labour costs from other costs)

- a wage payment system through which the general contractor is required to pay wages to workers directly instead of through subcontractors

The Ministry of Housing and Urban-Rural Development (MOHURD) commissioned the China Construction Industry Association to develop a national digital real name registration system for construction workers. In 2017 the National Construction Workers Management and Service Information System (NCWM&SISystem) was officially launched and the MOHURD subsequently urged local administrations to speed up the construction of the system in their regions. As of August 2021 all 31 provinces, autonomous regions and municipalities directly under the central government in mainland China were connected to the NCWM&SR System, with 21.5 million construction workers registered. However, the number of workers registered to work varies greatly from more than 2 million in five provinces (including Zheijiang) down to less than 200,000 in Xinjiang. Quinghai, Tibet, Yunnan and Heilongjiang Province.

Zhejiang province has developed and established the Enterprise Wage Payment Online Supervision System (EWPOSS) based on the real name system, using block-chain technology. General contractors play a central role, registering wage account information, allocating labour costs required and entrusting banks to pay wages to the worker's bank account on a monthly basis for the very first time. The labour inspectorate of Zhejiang believes that the use of information technology has significantly improved the effectiveness of labour inspection in the construction sector as EWPOSS makes it possible to monitor the whole process in real time, from workers engagement at the construction site to the monthly wage payment. If a project fails to pay workers at the stipulated time the system will automatically alarm the labour inspectorate to intervene.

What is different in India?

For many years, labour contractors (variously known as *labour intermediary*, *gang leader, jobber, mukkadam, kangani, sardar, arkati, maistri*, etc.) have been central to India's organised construction industry. They are responsible for hiring labour, controlling and supervising the workers during the contract period, taking them away when the job is finished and bringing them back when there is a new job (Vaid, 1999, cited in ILO, 2001). The loyalty of the labour is ensured through the payment of an advance (*peshgi*) from the contractor via the labour intermediary and this provides a bond between the contractor, intermediary and worker. The intermediary is the guarantor of the contractor's money and the workers' employment and a continuous link between the two. For this service he gets a commission from the contractor and a cut from the wages of the workers (ILO, 2001).

Papers published in Breman et al. (2009) confirm that the labour contractor is a key mediator in India's economy and is generally associated with triangular employment relationships whereby the legal employer is separated from the person for whom the work is carried out. However, as the construction industry has grown rapidly in India and become more concentrated at the top, the chain of contracting of work and workers has changed – becoming longer and more complicated and involving many different players.

At the same time, the demand for labour generated by investments in large scale urban construction projects has given rise to new forms of migration. In a major study of migration of construction workers to five cities in south Asia, researchers Sunil Kumar and Melissa Fernandez (2015) highlight the growth of what they call *transient contract migration* to refer to migration to join a workforce in the city, as opposed to traditional forms of migration that were predominantly in search of work.[7] They maintain that the use of such labour in large scale urban construction is certain to increase as it provides a flow of migrant workers whose rural livelihoods are increasingly untenable and whose alternative income earning opportunities are elusive. The practice of housing these workers in 'gated' labour camps with variable freedom to move in and out makes the labour force invisible and hard to reach by state of non-state actors seeking to address deprivation, but also has implications for the possibility of workers organising collectively.

The Delhi fieldwork included as part of the research project into *labour conditions and the working poor in India and China* would seem to provide an example of *transient contract migration* with its focus on labour conditions in the organised medium and large-scale segments of the construction sector in the greater Delhi area (Development Viewpoint 80). Eighty-four percent of the workers interviewed on the projects were recruited and employed by labour contractors and 99% of them were migrants from other regions (Bihar, Uttar Pradesh and West Bengal and other Indian states with widespread poverty). Many were recruited by large scale labour contractors located in their own region and delivered *en masse* to firms subcontracted to carry out construction work in Delhi. Others were recruited by labour contractors based in Delhi, some of whom were supplying hundreds or thousands of workers across multiple work sites. Although they may recruit Delhi-based workers these contractors also despatch agents to other regions to recruit the bulk of their workforce. All workers were housed in temporary sheds in fenced and guarded worksites or labour camps with very basic conditions.

The vast majority of migrant workers recruited in this way are unskilled and stay on the job usually only four to nine months (ibid). Most of those interviewed were young, with little formal education. They work for an average of 10 hours per day for an average daily wage for that is only about two-thirds of the official minimum wage. Some were recruited for even shorter periods, such as two months and paid even lower wages. Many workers complained about low pay, no payment for overtime, no paid day off work and irregular payment of wages, with the majority given only a small advance payment with the remaining wages paid only at the end of the working year.

While the labour contractors might be formally registered they keep their workers unregistered, thus denying them a formal employment relationship. 94% of the sample had no formal employment contract and hence no labour rights in Delhi. They are in effect second-class citizens for up to 8–10 months a year. At the construction site, their attendance and payments continue to be the responsibility of the agents of the labour contractor. The construction firms argued that they had no duties towards the workers since their employers were the labour contractors (ibid).

There are clearly many similarities between the situation for construction workers in the urban areas of India and China, but there is one key difference: in India many workers (almost 25% of the sample interviewed in Delhi) are female and they have moved to the city and the construction site with their families including many young children. After agriculture, construction is the largest employer of labour in India and up to one-third, of construction workers are women. All women working on construction sites are regarded as unskilled and working for an average monthly wage one-third lower than that of unskilled male workers.

To understand more about the role of women in the construction industry we can draw on the account by Van der Loop (1992) and on a more recent study in Bhilai (Chhattisgarh) the site of a major publicly owned steel industry, by anthropologist Jonathon Parry (2014). The regular Bhilai Steel Plant (BSP) workforce is permanently and formally employed (what Parry calls '*the local aristocracy of labour*'). But over the years permanent employees have been systematically replaced with cheap contract labour, paid at a daily rate 1/7 or 1/8th that of the lowest paid regular workers and without any fringe benefits or sick or holiday pay. They move readily and frequently among contract work in the plant and casual labour on construction sites with no security of work or over what, or even whether, they will be paid.

While the permanent workforce is exclusively male, around one-third of contract labourers are women. The overwhelming majority are local Chhattisgarhis as this is one of the poorest states and a major source of labour migrants to other towns and cities. Women are usually employed as *Rejas,* unskilled general labour, the female equivalent of a coolie. Coolies and Rejas share many tasks but the women do most of the carrying, especially of bricks which they carry six or eight at a time on their heads. Coolies mix and shovel sand and cement and lift the bricks onto the Rejas heads, but no self-respecting coolie would carry them on his own. Over the past two decades coolies have consistently received wages at least 20 % higher than Rejas and masons two and half to three times as much. These differentials are largely unquestioned. In 2004 a typical contractor was paying Rejas 40 rupees, coolies 50 rupees and masons 100–150 rupees. Hence, a household headed by a coolie or Reja would fall well below the official poverty line and even if both husbands and wives are working 80–90% of household income is likely to go on food. The situation is very similar elsewhere (Parry, 2014).

In Bhilai the practice of bonding labour by offering advances, which is common in construction elsewhere, is absent. Contractors suspect that workers will abscond or stop working if they get an advance, so instead of bonding by advances, labour is bonded by payment in arrears. As elsewhere, contractors use the fact that they themselves have not yet been paid by the main contractor or the owner of the building (which is often true) as an excuse for not paying workers all that is owed, retaining a percentage – perhaps one day's pay per week. Workers are reluctant to walk out because they cannot afford to forego the money and the further behind the contractor falls in payment the more the worker is bound (ibid).

In their conclusions from the ESRC-DFID research project on *Labour standards and the working poor in China and India,* Lerche et al. (2017) highlight the 'Triple Absence of Labour Rights' in both countries: the absence of (i) recognised employers and labour rights (ii) the right to organise (ii) rights other than those directly related to labour relations. However there are differences and they go beyond the fact that women play a major role in construction in India. Workers in Shanghai were ready to take collective action for their rights if need be, while the Delhi workers were not (ibid). In Shanghai pay and conditions were slightly better and only there were there signs that the market was tightening both for unskilled workers (due to ageing and few joining) and skilled workers who were younger, better educated and often had succeeded in getting contractual rights even though they were still agency workers.

However, it may be concluded that the biggest difference is in the role of the state. State leveraged and state implemented improvements are only of real importance in Shanghai. In India, pro-labour initiatives are not on the agenda as the state gains legitimacy from delivering growth and only secondarily from poverty reduction (ibid). After agriculture, construction is the largest employer of labour in India, sustaining the livelihood of 16% of the population. But, as Parry writes: '*Although bonded labour is illegal, even on government-funded construction projects, it is common. By the 1980s there were 25 separate laws that supposedly regulated the terms and conditions of workers in the construction industry in India:* ***All are routinely ignored*'** (Parry, 2014, page 1252).

Wage violations in the construction industry in industrialised countries

Wage violations are also common in many of the older industrialised countries where the construction industry has seen significant restructuring in recent decades, often accompanied by the entry of migrants into the construction workforce. The ILO investigated the role of international migration for work in the construction industry in the late 1990s and found increasing migration of construction workers from low to high wage economies particularly in East Asia (Japan, Korea) and in Europe (ILO, 1995). It was concluded that a combination of health and safety risks plus low and insecure wages was driving workers with a choice of employment (which is mainly those in the richer countries) to shun low-wage work in construction, with the development of labour shortages opening the door to the entry of migrant workers.

A deliberate search for cheaper labour may also be a factor driving the emergence and development of international markets for construction labour. Excess supply of labour in these markets was held in the 1990s to be adversely affecting the terms and conditions of work for construction workers and predicted to eventually impact construction workers around the world (Wells, 1996). The change to outsourcing has certainly made employment in construction even less attractive while facilitating the incorporation of migrants into the construction workforce, as shown by Buckley et al. (2017). The proportion of foreign workers in the construction workforce in some high wage economies (either legally or undocumented) is now significant, accompanied by an increasing number of reports of wage violations.

Wage theft in the United States

One such example is the United States where minimum protections for workers established under the Fair Labour Standards Act are routinely violated under the general concept of 'wage theft', defined as violation of US wage and hour laws. A survey of workers in low-wage jobs in Chicago, Los Angeles and New York in 2008 found wage violations to be common in many sectors of the economy including the private residential construction sector of the industry (Bernhardt et al., 2008). Research into employment practices in residential construction in Massachusetts (USA) documents how workers, the majority undocumented migrants from central and south America, are routinely cheated out of their wages by contractors who pay late, do not compensate for overtime and sometimes do not pay at all (Juravich et al., 2015).

Further evidence of wage theft among contractors in construction can be found in detailed studies in Austin, Texas (Torres et al., 2013) and post hurricane reconstruction in New Orleans (Waren, 2013). The Austin-based Workers defense Project (2013, 2017) conducted two studies and found half of the construction workforce in Texas was foreign born and in six southern states one-third were undocumented. Over 40% of the surveyed group in Texas reported routinely experiencing non-payment of overtime and in many cases non-payment of any wages (Erlich, 2021).

These incidents of wage theft among undocumented migrant construction workers have taken place in parallel with (and could not have occurred without) the quite dramatic changes in the organisation of production and employment in the construction industry over the past decades. The formalised world of labour relations, whereby building trades unions functioned as '*the equivalent of the human resource department for an entire industry*'

(Erlich, 2021, p.4) broke open during the 1980s with the emergence of 'open shop' contractors and the replacement of general contractors by fee-based construction managers which shifted risks to subcontractors. In the last three decades of the twentieth century general contractors' share of the directly employed workforce fell from 35% to 24% while the share of subcontractors rose from 48% to 63% (Bosch and Philips, 2003). These changes were followed by the growth of contingent labour and widescale misclassification of workers as 'independent contractors' (Erlich, 2021, Weil, 2005).

By misclassifying workers as independent contractors employers are able to save up to 30% of the tax and insurance cost of employing workers which is a significant saving in a very competitive industry where the majority of owners select their contractors based on price alone. Hence there is a strong incentive for all employers to misclassify. Union contractors are unable to take advantage of misclassification schemes as collective bargaining agreements require that their workforce be treated as employees, but non-union contractors are not so constrained.

However, the development and rapid growth of misclassification has also been stimulated by legislation: first, by Section 530 of the Revenue Act of 1978 which gave a green light to misclassification on the basis that a significant number of competitors were already treating their workers as non-employees: second, by the Immigration Reform and Control Act (IRCA) of 1986 that introduced penalties against employers 'knowingly' hiring undocumented workers as employees, leading many subcontractors to start hiring them as 'independent contractors' instead. The practice is strongest in those trades where output can be measured, such as dry-wall construction and it first became significant with the entry of French-Canadian drywall installers from across the border in Quebec.

Erlich (2021) shows how, in the wake of IRCA, misclassification and immigration became inextricably mixed. While IRCA was introduced in an attempt to stem the flow of migrants crossing the Mexican border into the United States, all that it achieved was to deny those who were undocumented the right to be classed as employees, thereby exposing them to a work-life characterised by poverty level payment and wage theft. On a 2015 project in Houston 45 of a 60-member crew were undocumented and working for a labour broker, earning $14 an hour with no overtime pay, whereas the 15 documented workers were paid $22 an hour for identical work. However, the costs to the worker of misclassification go beyond the difference in hourly pay to the loss of all legal rights to minimum wage or overtime payments, workers compensation coverage in case of illness, unemployment benefit in case of lay-off, as well as leaving workers solely responsible for any health and retirement benefits.

Rebuilding the Gulf coast after Hurricane Katrina with undocumented workers classified as independent contractors enhanced the use of the labour broker system. However, it is no longer confined to the southern states. In the early 2000s the 'Coyotes' (a colloquial term dating back to the nineteenth century) who had moved large groups of people across the border for a fee, graduated to become labour brokers moving with undocumented immigrants into the mid-west and the northeast. In this way misclassification and brokerage, coupled with the practice of payroll fraud and wage theft, spread from the south to the rest of the country.

The rapid growth of the immigrant workforce, with undocumented migrants estimated at 15% of construction workers in the country as a whole by 2014 (Passall and Cohn, 2016), eventually led employers to not bother with the paper-intensive[8] legal approach to misclassification. Assuming that most migrant workers were undocumented, they just moved to payment in cash. By 2010, 45% of the trades workers in New York city were undocumented migrants. Payroll fraud was claimed to be particularly acute in the city's affordable housing industry where two thirds of the workforce was either working as independent contractors or paid 'off the books' (Fiscal Policy Institute, 2007).

This posed insurmountable obstacles to measuring, let along regulating, employer behaviour. With a decline in funding at the US Department of Labour's Wage and Hour Division (WHD), the energy for innovation in enforcement devolved to the individual states, and even municipal levels. The California Division of Labor Standards Enforcement (DLSA) is one of several state enforcement agencies that has accepted the concept of 'co-enforcement' whereby state departments of labour increasingly rely on unions, work centres, community organisations and 'high road employers' (those who resent having to compete against others benefitting from illegal and unethical employment arrangements) as sources of information about the industries the agencies monitor.

California has also paved the way in addressing the issue of joint employment by passing a bill that holds general contractors liable for their subcontractor's wage theft violations. Assembly Bill No. 1701 came into force in January 2018 making up-chain contractors responsible for wage theft by their subcontractors on all construction projects in the state. In October 2018 a similar law came into force in Maryland (Senate Bill 853) with the exposure of the main contractor up to three times the amount of the delayed wage and this was followed by Oregon with House Bill 41548. Massachusetts has also passed a Bill that would hold general contractors and their subcontractors liable for wages owed due to theft by allowing the Attorney General to seek civil suits for wage theft and issue a stop work order until wage violations are settled. It is yet to be seen whether other states may follow these examples, which only apply to the private sector, and whether public sector contracts will be included in the same liability rule.

In addition to providing a further source of funds from which to compensate workers for unpaid wages, it is expected that extending liability for wages to third parties higher up the subcontracting chain will also cause contractors to develop a better control over payroll records of their subcontractors and to take other steps to avoid wage theft. These steps could be strengthened by making liability 'duty-based' so that, penalties are reduced if checks are carried out. This is important as, from an enforcement perspective, it is preferable to deter rather than punish illegal behaviour (Erlich, 2021, Rogers, 2010).

In a very powerful argument for linking liability to goods or services produced in violation of the Fair Labour Standards Act, Brishen Rogers (2010) argues that holding firms liable for their suppliers' violations could be an effective and relatively low cost way of securing compliance with the wage and hour Act. It would also be fair where wage theft is a foreseeable result of such firms' cost-cutting exercises. On page 46/47 he notes that '*In many instances, wage and hour violations (do) seem eminently foreseeable results of firms' sourcing and contracting practices* ... and ... *at times avoiding wage and hour liability, is firms' primary motivation to subcontract unskilled labour*'. He goes on to argue that such violations are'*acutely foreseeable if a firm plays one contractor or supplier off against another to lower prices ... or ... enters into a contract that does not include sufficient funding for minimum wages to be paid*' (Rogers, 2010, p. 46/47).

Theft of wages and benefits in the European Union

Withholding wages and other benefits owed to construction workers is also a serious problem in the European Union (EU) where subcontracting of both specialised and labour-intensive tasks has been growing rapidly since the early 1990s, raising concerns about possible erosion of workers' rights at the lower end of a subcontracting chain. At the same time, labour market intermediaries in the form of Temporary Work Agencies (TWA) and Employment Placement Agencies (EPA) have grown rapidly, acting as links in the subcontracting chain, and making it easier for contractors to dispose of a possible liability for wages, taxes and social security contributions for an otherwise inevitable own workforce (Heinen et al., 2017).

The situation is complicated by the fact that the EU is a common market with freedom of movement of labour among the member countries as well as freedom to provide services across national boundaries. Workers employed by a subcontractor in country A who has won a contract in country B may be sent to work with their employer in country B and are known as 'posted' workers. The EU enacted the Posting of Workers Directive (PWD) (96/71/EC) which aims to protect the rights of posted workers to enjoy the same terms and conditions as the local workers in the country to which they are posted, as well as to prevent fraud and abuse in the payment of social contributions and taxes.

Following the incorporation into the EU of lower wage economies in Eastern Europe in 2004, 2008 and 2013, genuine subcontracting of construction services across the borders between EU countries did increase, with a growth in legitimate posting of workers. But the concept of 'Posting' has also been massively abused by the rapid growth of intermediaries (EPAs and TWAs) setting up in a member country with the sole purpose of supplying labour across national borders. Intermediaries providing temporary workers were illegal in many EU countries until recently but EPAs and TWAs now operate in all EU member states. Once established in a member country, corporate legal entities can provide services across the whole of the EU, starting a massive externalisation and cross border recruitment of labour.

Comparative national research across EU countries in 2010 on the functioning of the Posting rules within the framework of the free provision of services revealed that problems appear as soon as cross-border labour-only (sub)contracting is presented as the provision of services (Cremers, 2011, 2016). Groups of workers have been recruited via agencies, gang-masters and 'letter-box' companies, advertising and informal networking (platforms) as 'posting' has become one of the channels for the cross border recruitment of cheap labour with downward pressure on wages in the host countries, as well as loss of benefits from social security abuses. There is evidence that 'posted' workers have been left without payment of wages and instances where they were unable to enforce their wage claims against the employer because the company had disappeared, or never really existed.

Difficulties in implementing the PWD led policy makers to search for effective tools to ensure that workers are paid the wages and benefits that they should be entitled to, prompting a debate on the issue of liability in subcontracting chains. Extending liability for the actions of subcontractors is considered to have a preventive and deterrent effect by giving a strong incentive to contractors to choose subcontractors more carefully and to verify that they comply in full with their obligations under host country rules. A series of commissioned studies followed which have sought to assess whether extending liability for wages, plus taxes and social fund contributions, could be an effective measure to protect workers' rights and improve the effectiveness of the PWD. The first study which focused only on construction and is known as the 'Dublin study' (Houwersijl and Peters, 2008) was commissioned by Eurofound and this was followed by a second study by two of the same authors (Jorens, Peters and Houwerzjil, 2012).

The Enforcement Directive (2014/67/EU) which was eventually passed in 2014 and came into effect in 2016 does establish (Article 12) a liability scheme in subcontracting for the construction industry. Liability is extended in respect of outstanding net remuneration at the minimum rate of pay of the host state, as well as contributions to common funds. However the liability in the Enforcement Directive is limited to direct (joint and several) liability, one link up in the chain, which is easily overcome by inserting a letter box company or other form of bogus subcontractor that declares bankruptcy if held liable. The preventive or deterrent effect is also limited. A system of extending liability to the entire subcontracting chain, including the main contractor and even the investor/client, is expected to be a more effective way of protecting the workers in the chain and preventing wage abuse. A third study (Heinen et al., 2017) was commissioned at the request of the European Parliament during the process of revising the posting directive in 2018 to assess whether further action may be needed.

The concept of extending liability further along the subcontracting chain as a way of enforcing the PWD is highly controversial within the EU, given the natural differences in the interests of labour sending countries in Eastern Europe and the labour receiving countries in the West, eight of which have introduced their own national liability schemes over the years. More stringent liability up to 'full chain' (to embrace the client/owner) has been demanded by some trade unions and is the preferred option of the EU, which has argued that such a system would pose a mechanism of self-regulation between private actors and be less restrictive and more proportionate than alternatives such as pure state intervention by inspections and sanctions.[9] However, extending liability has proved too ambitious as some states (including Hungary and the United Kingdom) are reluctant to implement any kind of liability scheme into their national legal systems (Heinen et al., 2017; Voss, 2016).

In their conclusion, Heinen et al. (2017) expressed serious doubts as to whether the Commission's favoured proposal can be generally accepted among the eastern member states on the grounds that these states were very active in watering down the more extensive measures that had been included in the first draft of the Enforcement Directive. Also, the revision of the Posting Directive took 27 months to reach agreement with a clear split between labour sending and labour receiving countries (EU.EURACTIV.fr, 2017). The views of the United Kingdom (an outlier in this respect) can be seen in the government response to the consultation on implementing the Enforcement Directive in 2016.[10]

The situation in the EU has continued to deteriorate with the development of *undeclared work and social fraud* in EU construction. No official definition of 'undeclared work' exists in the EU but it is understood to mean any lawful activities that are not declared to public authorities, the main reason for not declaring activities being to evade payment of taxes, social security contributions and/or minimum wages. Eurobarometer 2019[11] showed 19% of all undeclared work activities take place in the construction industry, with most in home repair and renovation activities. The European platform tackling undeclared work has since been integrated into the European Labour Authority (ELA) which is tasked with coordinating and supporting the enforcement of EU laws on labour mobility and ensuring that all workers receive the minimum wages due to them.

A joint statement by the European social partners in the construction sector (EFBWW and FIEC) at an online conference on 24 September 2020[12] noted that the 'shadow economy' is increasing in construction, with the increase in international migration of workers, including workers now coming from outside the EU, recruited through platforms. Other key concerns raised by the social partners include misclassification of workers as 'independent contractors' which in the European context is known as 'bogus self–employment'.

Conclusion

This chapter has provided some insights into the ways in which construction workers are recruited and employed in a wide variety of settings. We have chosen to focus on the low-wage construction labouring jobs occupied by unskilled or low-skilled workers. These workers are usually employed by intermediaries (labour contractors/agents/brokers) supplying labour on a temporary basis for use by (sub)contractors as and when needed. Many workers are migrants from the countryside, while others may be international migrants from lower wage economies. In all of the countries examined they are likely to experience late, under-payment or non-payment of the wages owed to them.

Some may argue that low-skilled construction workers are exploited because they are migrants. In the case of the United States, Erlich (2021) has provided evidence that many are suffering wage abuse simply because they lack the documentation to prove that they have a right to reside and work in the United States. This clearly puts them in a weak bargaining position with potential employers, the labour contractors/brokers, who are the main employers of unskilled labour since contractors took the business decision to outsource their labour supply. The decision on the part of their employers to also misclassify them as 'independent contractors', thereby denying them the potential benefits that accrue to employees, is also a direct result of their undocumented status.

However, this is not the case in China where construction workers are nationals of the country and urgently needed for work in the cities to support China's remarkable construction boom. Neither is it the case in Qatar, which is totally dependent on foreign labour, with the vast majority welcomed to work legally with visas generally issued for two years. It is difficult to argue that wage abuse in Qatar or in China is due to the fact that workers are migrants (either internal or international).

A more likely explanation for wage violations must stem from the business decision on the part of the contractors to outsource their labour requirements to intermediaries (agents, brokers) which opens up the supply of labour, and hence the wages of workers, to competition and encourages cheating by less than honest brokers. Labour suppliers are also generally less substantial businesses with fewer assets than the contractors they are replacing as employers. There is evidence from both Qatar and China that they often do not have funds to pay wages unless and until they have been paid by the users of the labour they are supplying.

Governments in all of the countries examined (with the exception of India) do see wage abuse as a challenge that cannot be left to the business community to address. While the chapter reveals similarities in the ways that workers are cheated of their wages it also suggests common approaches to addressing the problem. The following actions are put forward for consideration:

- i Wage protection systems that monitor and detect non-payment of wages in real time, as well as providing a record of wages paid (Qatar and China)
- ii Development of separate, ring-fenced project accounts for wages (as in China/Korea)
- iii Improving the payment of project funds down the subcontracting chain (e.g., through adjudication schemes) with the right to withdraw labour until paid

iv Shifting the liability for wages up the subcontracting chain to the users of the labour and making the main contractor/client liable (China, California, Europe, EU)

While none of these is a panacea on its own, if combined and accompanied by an active labour inspectorate there may be a chance of deterring and possibly even preventing wage abuse before it happens.

Notes

1 David Weil (2014) The fissured workplace: Why work became so bad for so many and what can be done to improve it, Cambridge MA, Harvard Press, 2014.
2 See, for example, Guy Standing, 2017; Jan Breman and Marcel van der Linden, 2014.
3 *Kafala* is the Arabic word for the sponsorship system that locks the foreign worker to a particular job and s/he cannot legally obtain other employment in Qatar or leave the country without the sponsor's permission. The *Kafala* exists at the junction between law and custom, reinforced by legal contracts, typically for two years, which are signed by most migrants. The *kafala* sponsorship system (and the labour brokerage system it gave rise to) have been the focal point of a global human rights based critique over the past two decades and Qatar has been exploring change in the system in response (Gardner et al., 2013).
4 The practice of *pay when paid* has grown in significance with the increase in subcontracting and the length of subcontracting chains. It has been outlawed in the UK and several other countries although it is still included as a legitimate practice in FIDIC subcontracting contract.
5 Cited in 'Still struggling: Migrant workers in Qatar during the pandemic' by Zahra Khan, NYU Stern page 9.
6 According to Pun and Lu (2010) the regulations can be found in two documents. The first document stated that 'The state owned construction and installation enterprises shall reduce the number of fixed workers gradually, In future they shall not, in principle, recruit any fixed workers except skilled operatives necessary to keep the enterprise technically operational'. The second regulation, entitled 'Separation of management form field operations' stated that general contractors or contracting companies should not directly employ their blue collar workforce: Rather they were to employ labour subcontractors who were to be responsible for recruiting the workforce. The original source of these statements comes from a document entitled *Xin Zhongguo jianzhu ye wushi nian* (The fifty years of new China's construction industry) published by a study group formed by the construction ministry in 2000. This was part of a deliberate policy to redress the rural/urban imbalance in living standards by moving workers from rural areas to the cities, 38.5% of whom found work in the construction industry (Kaixun Sha and Zhenjian Jiang, 2003).
7 The project was funded by the UK Department for International Development, South Asia Research hub in New Delhi, research was undertaken from March 2014 to October 2015 in five cities in South Asia: Kabul (Afghanistan) Dhaka (Bangladesh) Chennai (India) Katmandu (Nepal) and Lahore (Pakistan). The overarching research question was *How do investments in large-scale urban construction and the demand for labour generated, give rise to varied forms of migration.* https://www.gov.uk/research-for-development-outputs/briefing-note.
8 To remain within the law, workers can be classed as 'independent contractors' and be paid without deductions for tax etc. but are obliged to complete a 1099 form at the end of the year setting out their payments for income, unemployment, social security and Medicare tax, plus provision for workers' compensation and insurance policies. etc. The losses to the state when these payments are not received is huge.
9 Proposal for a Directive of the European Parliament and of the Council on the Enforcement of Directive 96/71/EC concerning the posting of workers in the framework of the provision of services COM/2012/0131 final, page 22 (cited in Heinen et al., 2017).
10 https://www.gov.uk/government/consultations/call-for-evidence-eu-proposal-for-a-posting-of-workers-enforcement-directive

11 https://ec.europa.eu/commfrontoffice/publicopnion/index.cfm/survey/getsurveydetail/instruments/special/surveyky/2250

12 file:///C:/Users/Jill%20Wells/Downloads/2020-09-24%20-%20FIEC-EFBWW%20Joint%20statement%20UDW%20SIGNED.pdf

References

Amnesty International, *The dark side of migration: Spotlight on Qatar's construction sector ahead of the World Cup*, 2013.

Bosch, Gerhard and Peter Philips (2003) *Building chaos: An international comparison of deregulation in the construction industry*, Taylor and Francis.

Bernhardt, Annette, Ruth Milkman, Nik Theodore, et al. (2008) *Broken laws, unprotected workers: Violations of employment and labor laws in America's cities*, UCLA.

Breman, Jan, Isabelle Guerin, and Aseem Prakash (2009) India's unfree workforce: Of bondage old and new.

Breman, Jan and Marcel van der Linden (2014) *Informalizing the economy: The return of the social question at a global level*, Development and Change, 45(5), 920–940.

Buckley, M. (2012) *From Kerala to Dubai and back again: Construction migrants and the global economic crisis*, Geoforum, 43(2), 250–259.

Buckley, M., A. Zendel, J. Biggar, L. Frederiksen, and J. Wells (2017) *Migrant work and employment in the construction sector*, ILO, Geneva.

Business and Human Rights Resource Centre (BHRRC), *On shaky ground: Migrant workers' rights in Qatar and UAE construction*, 2016.

Cremers, J (2011) *In search of cheap labour in Europe: Working and living conditions of posted workers*, CLR studies No.6 (CLR/EFBWW/International Books).

Cremers J. (2016) *Economic freedoms and labour standards in the European union*, Etui, 22(2), 149–162.

Development Viewpoint, Number 77 (2014) *The deterioration of labour conditions in China's construction sector*, Centre for Development Policy and Research, School of Oriental and African Studies (SOAS), University of London: ESRC-DFID Research Project, "Labour Conditions and the Working Poor in China and India".

Development Viewpoint, Number 78 (2014) *Survey results document exploitative labour conditions in China's construction sector*, Centre for Development Policy and Research, School of Oriental and African Studies (SOAS), University of London: ESRC-DFID Research Project, "Labour Conditions and the Working Poor in China and India".

Development Viewpoint, Number 80 (2014) *Documenting the lack of labour rights in India's construction sector*, Centre for Development Policy and Research, School of Oriental and African Studies (SOAS), University of London: ESRC-DFID Research Project, "Labour Conditions and the Working Poor in China and India".

DLA Piper/MEED (2018) *Time for change: Construction in the GCC reaches a tipping point.*

DLA Piper (2014) *Migrant labour in the construction sector in the State of Qatar.*

Erlich, Mark (2021) *Misclassification in construction: The original Gig Economy*. International Labour Review, 74(5), 1202–1230.

EU.EURACTIV.fr reports. *European parliament votes in favour of the revision on posted workers*, 2017.

Fiscal Policy Institute, *The underground economy in New York City affordable housing industry*, April 17, 2007 (cited in Erlich 2021 p.16).

Gardner, Andrew, Silvia Pessoa, Abdoulaye Diop, Kaltham Al-Ghabim, Kien Le Trung, and Laura Harkness (June 2013) A portrait of low-income migrants in contemporary Qatar, *Journal of Arabian Studies*, 3(1), 1–17.

Gardner, Andrew M. (2012) *Why do they keep coming?: Labor migrants in the Persian Gulf States* in Mehran Kamrava and Zahra Babar (editors) *Migrant labor in the Persian Gulf*, Hurst and Company, London.

Heinen A, A. Muller, and B. Kessler B. (2017) Liability in subcontracting chains: National rules and the need for a European framework, European Parliament, 2017, available at: www.europarl.europa.eu/RegData/etudes/STUD/2017/596798/IPOL_STU(2017)596798_EN.pdf

Houwersijl, M and S. Peters (2008) *Liability in subcontracting processes in the European construction sector*, European Foundation for the Improvement of Living and Working Conditions, 2008, available at: www.eurofound.europa.eu/sites/default/files/ef_publication/field_ef_document/ef0894en.pdf

Huang, Kun (2022) *Technological solutions to guaranteed wage payments of construction workers in China*, International Labour Organisation (ILO) Working Paper 48, February 2022.

Human Rights Watch (2012) *Building a better world cup: Protecting migrant workers in Qatar ahead of FIFA 2022*.

International Labour Organisation (ILO) (1995) *Social and labour issues concerning migrant workers in the construction industry*, ILO, Geneva.

International Labour Organisation (2001) *The construction industry in the twenty first century: Its image, employment prospects and skill requirements*, International Labour Organisation, Geneva.

International Labour Organisation (2018) *Exploratory study of good practices in the protection of construction workers in the Middle East*. White Paper, February 2018.

International Labour Organisation (2019) *Assessment of the wage protection system in Qatar*, ILO Project Office for the State of Qatar, Doha, June 2019.

International Labour Organisation (2020) *Progress report on the technical cooperation programme agreed between the Government of Qatar and the ILO*, ILO Governing Body, 2020.

International Labour Organisation (2021) *Progress report on the technical cooperation programme agreed between the Government of Qatar and the ILO*, Governing Body, Pillar Two, Enforcement of Labour Law and Access to Justice, 2021.

International Labour Organisation (2022) Overview of Qatar's labour reforms.

Jorens, Y, S. Peters, and M. Houwerzjil (2012) *Study on the protection of workers' rights in subcontracting processes in the European Union*, Project DG EMPL/ B2-VC/2011/0015, University of Amsterdam, 2012, available at: www.dare.uva.nl/search?identifier=80520784-0f36-4537-ae15-3986a0e38330

Juravich, Tom, Essie Ablavsky, and Jake Williams (2015) *The epidemic of wage theft in residential construction in Massachusetts*. University of Massachusetts Amherst Labor Center, working paper series, May 2015.

Kumar, Sunil and Fernandez, Melissa (2015) *Urbanisation-construction-migration nexus/ 5 Cities/ South Asia*. Briefing Note, LSE Enterprise, London, UK.

Lerche, Jens, Allessandra Mezzadri, Dae-Oup Chang, Pun Ngai, Lu Huilin, Liu Aiyu, and Ravi Srivastava (February 2017) *The triple absence of labour rights: Triangular labour relations and informalisation in the construction and garment sectors in Delhi and Shanghai*, Working Paper 32/17, Published by the Centre for Development Policy and Research, SOAS, University of London.

Mayer Brown, Asia Employment Law Quarterly Review 2015–2016, Issue14, 2016.

Nakdubo, Hiroya and Araki, Takashi (2017) *The notion of the employer in the era of the fissured workplace*. Bulletin of Comparative Labour Relations, edited by Roger Blanpain and Frank Hendrick. Belgium. Kluwer Law International.

Nguyen, Thao (2013) Governing through *Shequ*/Community: The Shanghai example, *International Journal of China Studies*, 4(2), 213–231.

Ngai, Pun and Lu Huilin (2010) *A culture of violence: The labour subcontracting system and collective action by construction workers in post-socialist China*, The China Journal, 64, 150–151..

Pang, Irene (2019) *The legal construction of precarity: Lessons from the construction sectors in Beijing and Delhi*, Critical Sociology, 45(4-5), 549–564.

Parry, Jonathan (2014) *Sex, bricks and mortar: Constructing class in a central Indian steel town*, Modern Asian Studies, 48(5), 1242–1275, London, Cambridge University Press.

Passall, Jeffrey and D'Vera Cohn (November 3, 2016) *Occupations of unauthorized immigrant workforce*, Pew Research Center.

Rogers, Brishen (2010) *Toward third-party liability for wage theft*, Berkeley Journal of Employment and Labour Law, 31(1).

Sha, Kaixun and Zhenjian Jiang (2003) *Improving rural labourers' status in China's construction industry*, Building Research and Information, 31(6), November-December 2003.

Standing, Guy (2017) *The corruption of capitalism: Why rentiers thrive and work does not pay*, London, Bitback.

Torres, Rebecca et al. (2013) *Building Austin, Building Justice: Immigrant construction workers, precarious labor regimes and social citizenship*. Geoforum, 45(2013), 145–155.

Vaid, K.N. (1999) *Contract labour in the construction industry in India*, in D.P.A. Naidu (Ed.) Contract labour in South Asia, Geneva ILO, Bureau for Workers Activities.

Van der Loop, Theo (1992) *Industrial dynamics and fragmented labour markets: Construction firms and labourers in India*, Netherlands Geographical Studies, University of Amsterdam.

Voss, Eckhard (2016) *Posting of workers directive – current situation and challenges*, Study for the EMPL Committee of the European Parliament, 2016.

Walk Free Foundation, Global Slavery Index 2018.

Waren, Warren (2013) Wage theft among Latino day laborers in Port-Katrina new Orleans: comparing contractors with other employers, *International Migration and Integration* 15(2014), 737–751.

Weil, David (2014) *The fissured workplace: Why work became so bad for so many and what can be done to improve it*, Harvard University Press, 2014.

Weil, David (2005) *The contemporary industrial relations system in construction: Analysis, observations and speculations*, Labor History, 46(4). November 2005, 447–471.

Wells, Jill (1996) *Labour migration and international construction*, Habitat International, 20(2) (DLA Piper 2014), 295–306.

Wells, Jill (2014) *Improving employment standards in construction in Qatar*, Final Report, October, Engineers Against Poverty, 2014.

Wells, Jill (2018) *Exploratory study of good policies in the protection of construction workers in the Middle East*. White Paper, International Labour Organisation, Regional Office for Arab States.

Wells, Jill and Maria da Graca Prado (2019) *Protecting the wages of migrant construction workers: Part Three: What can be learned from systems of wage protection in China, EU, US and Latin America, October 2019*.

Wells, Jill (2022) *Out of pocket: A ten year review of paying Qatar's construction workers*, Engineers Against Poverty, November 2022.

Zou, Mimi (2017) *Regulating the fissured workplace: The Notion of the 'Employer' in Chinese Labour Law*, Bulletin of Comparative Labour Relations, 95(2017), 183–203.

8
THE PRECARIAT OF THE BUILT ENVIRONMENT
Decent work and the myth of Sisyphus

Andrés Mella

PRIVATE CONSULTANT

Introduction

The construction sector and the built environment provide jobs and incomes to more than 330 million workers in developed and developing countries around the world, working on projects from large stadiums in Qatar to small refurbishments in Sweden and bustling residential developments in Kenya. It provides remunerated work for migrants, women, and youth facing challenges to access the labour market. Nonetheless, precarious employment and working conditions are becoming a reality for many workers across the world. In 2011, Guy Standing developed a provocative and influential new term: the *precariat* (Standing, 2011). The term combines the concepts of "precarious" work and the proletariat in societal structures.

The present chapter presents evidence on the increasing forms of precarious work in the built environment sector. Who is part of the built environment precariat? What are the employment and working conditions of the precariat in the built environment? How are trends shaping the future of work likely to affect the precariat in the built environment? What can policymakers do to improve working conditions in the sector? The present chapter aims to answer these questions by looking into precarious built environment work around the globe, focusing on its most prevalent forms. The chapter builds on two previous publications as well as discussing and reporting on additional research covering recent studies on precarious working practices. The publications are a report, *Good Practices and Challenges in Promoting Decent Work in Construction and Infrastructure Projects* published by the ILO (2015) and a report commissioned by the former UK Department for International Development through the Infrastructure and Cities for Economic Development facility (ICED, 2018).

The rest of the chapter considers answers to the above-mentioned questions in four distinct sections. In section A, the concepts of the precariat, decent work, and the myth of Sisyphus are presented. This is followed by section B, which lays out major aspects of precarious work and types of workers of the built environment precariat. Then, section C discusses the future of work and the impact of its key determinants and developments on the built environment precariat. In section D, the policy options are explored. The chapter concludes by presenting the main findings and setting out some recommendations.

DOI: 10.1201/9781003262671-8

The precariat, decent work, and the myth of Sisyphus in the built environment

The International Labour Organization (ILO) defines decent work as "productive work for women and men in conditions of freedom, equity, security and human dignity" (ILO, 2010). The concept is based on four elements: employment creation, social protection, rights at work, and social dialogue. Decent work is not categorical in its interpretation but progressive and aspirational. Rather than describing a job as decent or not decent, its multidimensional nature entails that a job is more decent than another depending on the specific dimension. The concept is useful in pointing towards a direction, and its progressive achievement. While the concept is helpful in defining a policy agenda, there is a stark contrast between its vision and reality.

Far from improving job quality, the experience of more and more workers resembles that of the Sisyphus myth. In this Greek myth, Sisyphus is condemned for eternity to roll a rock up to the top of a mountain, only to have the rock roll back down to the bottom every time he reaches the top. Should reaching the top of the mountain represent achieving decent work with stable working conditions, many workers invest in qualifications and obtain high competency levels, to never achieve it.

The myth of Sisyphus, with precarious working conditions, is becoming the reality for more and more workers in all kinds of trades. The phenomenon is so widespread that the workers caught in it have become a social class. In 2011, Guy Standing developed in a provocative and influential book, *The precariat: The new dangerous class*, the term "precariat".

Precarious employment includes employment conditions with wage labour or fixed-term employment, many times without written contracts. The terms are intentionally flexible to cater to company needs and changing business cycles. In addition to the uncertainty of employment, the workers may face the additional burden of other poor working conditions, such as exposure to accidents and diseases, as well as delayed payments.

The reference to the **proletariat** in the term evokes a social class of the nature and importance the proletariat had in other times. Although Standing does not list the categories of workers, it is clear that the precariat is not as homogeneous as those who fell into the category denoted by the classical term proletariat. Furthermore, a worker in this new class does not have a clear-cut identity nor a sense of belonging, stability, and channels of representation (such as trade unions or political parties).

The precariat term builds on two concepts, the poor employment and working conditions and a heterogeneous group of workers that experience such conditions. The chapter sheds light on the reality of the built environment proletariat. There is no intention to deny that the proletariat exists in all sectors of economic activity, but the intention is to provide evidence of the applicability and relevance of the concept for describing the reality of workers in the built environment sector. The next section describes prevalent forms of precarious employment and working conditions in the built environment sector and presents a characterisation of examples of types of workers in the built environment precariat.

Major aspects of precarious work and types of workers of the built environment precariat

Precarious work is becoming the reality for more and more workers in maintenance work, road construction, bridge development, and commercial housing development. Some of the major elements of precarious work in the built environment include poor employment and working conditions. Working conditions cover a broad range of topics and issues, from working time (hours of work, rest periods, and work schedules) to remuneration, as well as the physical conditions and mental demands that exist in the workplace linked to occupational health and safety. Employment conditions form part of the working conditions and refer to the contractual relationship between the employer and employee, its type (part-time or full-time), and length (fixed-term or indefinite). Employment conditions are considered distinctly here among the elements of the working conditions because of their salient importance in explaining precarious work.

Poor employment conditions

The construction industry's project-based nature is the reason behind the proliferation of non-standard forms of employment, including casual and fixed-term arrangements. When a project ends, companies may not have another one to offer all in their workforce. The technical and specialist nature of some trades and the range of occupations involved in a single construction project mean that companies hire additional temporary workers. Construction firms usually focus on their core competencies and subcontract non-core tasks. In practice, the use of flexible arrangements available to contractors, such as labour employment agencies and temporary labour agencies, has eroded the stability of traditional employment. This increased job instability and the low entry barriers to the labour market are contributing factors in turn to low qualification levels – which by themselves are affected by the unstable employment relationships, thus creating a vicious cycle.

In many situations, workers do not have a labour contract, but merely a verbal agreement. This is especially true in small and medium enterprises and in the case of workers with short-term assignments. Poor employment conditions and verbal contracts exist in both developing and developing countries. For example, a survey in the United Kingdom found that 50 percent of construction workers in London had no written contracts and were uncertain about their terms, conditions, and rights (FLEX, 2018). Other consequences of precarious and unstable employment conditions are now discussed.

(Other) poor working conditions

Delayed payments and other wage violations – Unreasonably long payment terms, delayed payments, under-payment, and non-payment of wages constitute the biggest source of concern for construction workers. The phenomenon is the result of the movement towards the more flexible forms of employment mentioned before and the establishment of long subcontracting chains. The distance between clients and workers has become longer as there are several layers of commercial contract relationships, which negatively impacts payment processes as well as the bargaining power of labour. Delayed payments along the supply chain disproportionally affect the cash flows of small contractors employing workers and are a major cause of business mortality (FIEC, 2020). Lack of effective enforcement mechanisms, informality, and insufficient knowledge of basic rights (particularly among migrant workers) are also associated with this practice. Delayed payments to workers can be found both in projects involving several layers of subcontractors which depend on financing of the contracting chain as well as in small works by contractors operating with limited access to bank financing. Late payments can be found on both private- and public-sector projects and are considered an operational risk by contractors, who constantly need to monitor cash flows and operational expenses on both construction materials and wages.

Wage violations take place in both developing and developed countries. As an example, a study in the United Kingdom found that around one in ten employers pay their staff late due to late payment from clients (Federation of Small Businesses, 2018). Other examples of wage violations in Qatar, China, India, and the United States can be found in Chapter 7 of this book by Wells.

Low margins in the construction industry, limited access to finance, and non-existent or weak consequences for delays and non-payment mean that, in practice, employers tend to reoffend in a wide-array of malpractices including (EAP, 2018):

- Employers disappearing without paying wages they owe.
- Employers dissolving a company that owns wages and starting up a new company to avoid payments.
- Non-payment of accrued holiday.
- Workers regularly working an hour or two per week but not being counted.
- Disputes over the interpretation of contracts on issues such as travel time to construction sites.
- Charging workers for construction uniforms and necessary Personal Protective Equipment (PPE).

Many workers in the construction industry have long working weeks. For instance, in Slovakia, in the construction week, workers work for an average of 43.9 hours, making it the sector with the longest working hours (ILFR, n.d.). Whereas work beyond stipulated hours should be remunerated as extra hours, many times this is not accounted for. Working at night or on weekends usually entails additional charges which are either not accounted for or are paid for at rates below the stipulated figures. Workers under precarious employment contracts (such as temporary agency work and other multi-party employment arrangements, disguised employment relationships, and dependent self-employment) might have reduced bargaining power to claim non-remunerated hours.

Unsafe and unhealthy working conditions– Construction is considered a hazardous industry. Unhealthy working conditions with little or no culture of prevention nor mechanisms make the work in many construction sites precarious. Construction work requires demanding tasks to be undertaken by each worker alongside many other workers. For example, masons and carpenters work long hours in a highly physical job exposed to the weather. There are some technical occupations that require a specific know-how such as carpentry, tiling, or plumbing. Project managers and site managers work together to coordinate teams of workers in the most efficient way. Self-employed, casual, and fixed-term workers are less likely to have adequate qualifications and health and safety training opportunities than full-time workers of companies, yet they are the ones at highest risk of suffering injuries and accidents in construction sites (Fiorenza, 2021).

Health hazards in work on construction sites include respiratory diseases, hearing loss, and musculoskeletal disorders. Covid-19, in itself, represented an additional hazard to the lives of many construction workers (see Box 8.1). Extreme weather conditions are also an important element affecting workers' health, producing heat strokes. In projects requiring worker deployment away from their homes such as projects on remote sites, large civil engineering projects, and work in high-density urban areas with limited housing options, poor accommodation and living conditions affect workers. Architects and construction managers operating on site also have to work in conditions that are different from office work and they might have to endure particularly high temperatures.

Box 8.1 Health and safety aspects during the Covid-19 pandemic

The Covid-19 outbreak posed a unique challenge for the safety and health of workers in the built environment sector. Most countries tried to avoid the imposition of restrictions on the sector, but the pandemic still caused job losses in the construction industry. This meant that safety and health challenges needed to be addressed early on in the crisis. Worldwide, workers in the built environment were disproportionally affected because many cannot work from home; they cannot respect social distancing requirements in many operations of their day-to-day work; they do not possess minimal protective equipment (such as suitable masks and hand sanitisers); and the worksites do not have basic hygienic measures in place. Many are posted or migrant workers and quite often live in communal accommodation, with many workers sharing the same space. Furthermore, construction workers faced the challenge of working without having medical insurance or social protection rights.

There were different policy solutions implemented to minimise the effects of Covid-19 on construction workers. OSH interventions were prioritised in the short run.[1] In some countries, through social dialogue, several stimulus packages were rolled-out targeting workers and enterprises, but most construction workers remained unprotected. From a fiscal perspective, health and social packages were introduced in some countries. Short-term employment programmes, reduced social security contributions, and additional unemployment benefits were sometimes made available to workers in some developed and developing countries.

Workers in the built environment sector face a range of occupational risks associated with biological, chemical, physical, ergonomic, and psychosocial hazards. It is the sector with the most fatal accidents, and construction work is referred to as being *difficult, dirty, and dangerous*. Around 1 in 6 accidents takes place in construction, accounting for 60,000 lethal accidents per year (ILO, 2015). It is estimated that fatality rates in the least developed countries may be more than double the rates in developed economies, although reliable evidence is difficult to obtain (ICED, 2018). For instance, Patel and Jha (2016) estimated that in the Indian construction industry, which employs some 51 million workers, the number of fatal accidents could be anywhere from 11,614 to 22,080 each year. This represents around 24 percent of all occupational accidents occurring in India annually. The industry ranks first also in terms of non-mortal accidents (both in absolute terms and in relation to the number of workers) and the number of missed workdays resulting from workplace accidents (ILOSTAT, 2022).

Among many others, the most common working hazards at a construction site include falling from heights, handling materials, or being hit by equipment. The incidence and effects of workplace accidents can be broadly categorised depending on the type of accidents they cause (as shown in Table 8.1). Fatal accidents are mainly caused by falls from heights or being hit by moving objects, and non-fatal accidents by material handling and installation of drywalls.

Fraudulent, unlawful, and illicit work – Fraudulent and unlawful works are well known to many workers of the built environment proletariat. Types of unlawful work are presented in

Table 8.1 Categories of workplace accidents according to selected occupational prevalence and causes

Type of accident	*Occupations*	*Cause*	*Report and incidence*
Fatal accident	Roofers, construction operatives, elementary construction workers, carpenters, joiners	Falls from heights, being hit by moving objects such as vehicles and lifting equipment, contact with machinery, or electricity	Mostly reported. Risks are ten to a hundred times higher than other tasks
Non-fatal accident	Plumbers, pipefitters, steamfitters, carpenters, construction operatives	Material handling and installation of drywalls, piping, and ventilation-duct installation, or accidents while moving around construction sites	Often non-reported but believed to have much greater incidence than fatal accidents

Source: Authors adaptation of ILO (2017a).

Box 8.2. Compliance with labour law is generally lower in developing countries than in developed ones, and public institutions face greater challenges in enforcing the laws and ensuring compliance. However, there are issues such as modern slavery and human trafficking which are becoming major sources of concern in all countries around the world.

Box 8.2 Unlawful work in the construction industry

Lawful contracts are those where formal contracts exist and there is respect for the requirements. The formal requirements of an employment agreement refer to the conditions which, according to legislation, characterise the specific formal employment or contractual relationship to be used in the determined circumstances. Undeclared forms of work have long been present in the construction industry. Table 8.2 summarises these concepts. Around 30 percent of business in the maintenance and repair of buildings is not declared in Europe. In developed countries, new fraudulent and illicit contracting are also types of precarious employment.

In practice, it is difficult to distinguish between the four main forms of contracting work. In many instances, such arrangements are intended to give the impression that they are legitimate forms of contracting since they use legal employment relationships. However, under closer scrutiny, the apparent contract disguises a different employment relationship or a different employer from the contractual one. For instance, employment agencies may hire bogus self-employed workers where the worker appears to be self-employed, but the agency keeps time control and decides on the time schedules of the worker.

Table 8.2 Types of contract work

		Respect for requirements	
		Yes	*No*
Formal contracts exist	Yes	Lawful	Fraudulent
	No (of void)	Undeclared	Illicit

Source: Eurofound (2016).

The extensive use of short-term forms of employment, declining profit margins, insufficient labour supervision, and fierce competition (especially among small contractors) are behind the fraudulent contracting modalities or undeclared work (Eurofound, 2016). For instance, in Europe, the most common forms of fraudulent contracting work in the industry are as follows:

- *Bogus self-employment and freelance work*. Its use involves both low- and high-skilled workers. These "economically dependent workers" usually have a commercial or service contract rather than an employment contract and are therefore registered as self-employed when, in reality, their working conditions have a lot in common with those of employees. In bogus self-employment contracts, the relationship of subordination[2] and the open-ended contract duration are avoided. It is estimated that there are around 400,000 bogus self-employed construction workers in the United Kingdom (Jorens, 2008). Around 25 percent of workers in the Swedish construction industry could be in bogus self-employment. Similarly, 10 and 17 percent out of all self-employed workers could be in bogus self-employment in the Netherlands and Ireland, respectively.
- *Posting of workers and temporary agency work*. Posting of workers in other countries can disguise direct employment relationships, taking advantage of more flexible labour laws in the countries where temporary agencies are based. Fraudulent posting of workers takes place mainly in Europe, where workers and businesses are mobile. It is a complex task to distinguish between genuine posted workers and foreign workers.

It is commonly assumed that slavery is mostly concentrated in the developing countries. However, labour exploitation is increasingly being detected in the world's richest nations. In Europe, it is estimated that nearly three people in every thousand are victims of slavery (CIOB, 2018). Worldwide, approximately 4.5 million people are being forced to work in the construction industry, making it the second sector with the most modern slavery after domestic work (ILO, 2017d). It is present in small construction sites, which are rarely reached by labour inspectors. Forced labour is associated with migrant work, where there are additional risk factors such as the existence of credit-arrangement and debt schemes, restrictions on the ability of the workers to take full control of, and freely use, their wages (for example, a disproportionate portion of their wages is deducted for accommodation), workers do not have free access to their identity and residency documents, and workers may be forced to work more overtime hours than allowed by the national law (European Commission, 2017).

Examples of precarious workers of the built environment precariat

There are specific types of workers who are more likely to bear the brunt of precarious employment. These include uneducated men without qualifications, labour migrants, women, and the youth. The types of workers facing precarious employment tend to grow. Therefore, the list should be understood as examples rather than an exhaustive categorisation.

Uneducated men without qualifications in informal employment. Uneducated men without qualifications form the backbone of the construction workforce and the largest group of the built environment precariat. A male-dominated sector with unattractive stereotypes, construction provides employment to school dropouts because of its low entry barriers. In most countries, the worker does not need to have any qualification to enter many types of onsite jobs. Temporary and casual workers are commonly paid on a piecework basis (Estebsari and Werna, 2022). These workers often find themselves working long hours, either because they are forced to do so or because they want to earn as much money as possible while work is available. This is especially true for those who have migrated from rural areas or other countries. Temporary and casual workers typically work 10–12-hour days, 6 days per week. In the United Kingdom, research has shown that self-employed construction workers are paid per shift instead of per hour, and the typical shift is 10–12 hours per day, and they work 6 days a week. This highlights that the issue is not limited to developing countries. In sum, while piecework may be an efficient method for employers to manage temporary workers, it can also result in workers being overworked and underpaid. Many temporary and casual workers also form part of the informal sector. Workforce informality means unregulated and unprotected employment relationships. Informality is diverse, demand-driven, and locally specific, with geographical diversity among countries and regions. The latest available data shows that in 61 out of 101 countries, more than 70 percent of employment in the construction industry takes place outside the formal sector, mainly in self-employment (ILOSTAT, 2022). Informality puts workers at risk of vulnerability and precariousness. Other issues to be addressed include the informality practiced by and within the construction companies themselves through complex contracting chains and the organisation of construction processes. These aspects are intertwined and make it difficult to synthesise categories in the informal construction workforce. Informal employment can be considered under two categories (ICED, 2018):

- Employment in the informal sector, mainly self-employment and operating in the form of informal micro-enterprises.
- Informal employment outside of the informal sector, that is, formally registered enterprises that employ workers informally.

Self-employment and informal micro-enterprises are not a new phenomenon in the construction industry, but their prevalence has increased in recent years. Self-employment is partly explained by changing employment relations and the specialised nature of some trades. Many workers work directly for clients through these direct arrangements. However, self-employment is also associated with increased subcontracting by enterprises. In formally registered enterprises, workers can be employed informally so as to save on the taxes due and social security. Workers accept this as it may be a temporary opportunity for gain, a temporary stage awaiting formalisation, or a work experience whilst starting a business (Fiorenza, 2021). An EU-wide study from 2017 found that the main reason for construction workers to be self-employed was the lack of alternatives to work.

Informality is associated with SME development aspects. Employment surveys covering the informal sector show that there is an increasing number of construction workers employed in enterprises with less than five workers. In Europe, 50 percent of all construction workers work in micro companies and 39 percent do so in small- and medium-sized companies (Eurofound, 2017). Such individuals may work in the informal segment of the construction industry as employees, as self-employed workers, or as owners of small enterprises employing other workers (Wells, 2007). In India, much building activity is undertaken directly by small enterprises engaged directly by building owners. This is known as the Naka/Mandi section of the industry (ICED, 2018). Naka/Mandis are points in cities where workers gather in the morning to wait for customers, who come from the mass of individual house owners and petty contractors to hire them for a specific project and time. Although there is no reliable data, there is anecdotal evidence that this segment of the construction industry has increased in size in recent years. However, like all SMEs, the enterprises face some major barriers to their development and formalisation including low access to finance, productivity, skills development, and market access.

Labour migrants. Migrants have long been a structural component of the construction workforce in many countries. ILO (2016a) provides a full account of international migration issues in the construction industry. The intrinsically localised and employment-intensive nature of construction projects generates high demand for labour. Whenever the local workforce is insufficient, contractors hire labour agencies and temporary manpower companies operating in both the domestic and international labour markets. In countries with high urbanisation rates such as India and China, rural migrants constitute a large part of the construction workforce (RMMRU, n.d.). In Kyrgyzstan, a household income/expenditure survey in 2015 showed that migrants constitute around 29 percent of the total number of workers in the industry (compared with an average of 8 percent across all sectors) (ICED, 2018).

Migrant workers include both low-skilled workers (for example, in jobs in the construction industry and domestic services) and highly skilled professionals. A sizeable share of the low-skilled workers are in a vulnerable situation. International migrants in the construction industry are particularly vulnerable to fraudulent and abusive hiring practices and debt bondage, cases of abuse and exploitation, poor working and living conditions, as well as limited representation, voice, and access to justice.

The abuse of migrant construction workers exists in many countries. For instance, in China, a nation-wide study found that only 31 percent of domestic migrant workers receive their salaries monthly in accordance with Chinese law and that more than half work without contracts (Human Rights Watch, 2008). Unlike in other countries, Chinese domestic migrants are denied basic public services such as health and education, as these are linked to the household registration system (Hukou), which aims to prevent internal migration. Until recently, migrant construction workers who lack temporary household registration permits had been barred from seeking legal remedy against employers guilty of wage exploitation. The Chinese government has addressed the protection of wage payments through legal reforms and has piloted the use of an e-formalisation system that improves the efficiency of labour inspection (ILO, 2022).

Labour migration issues have come under renewed scrutiny in the last decade, particularly leading up to the execution of infrastructure projects for the Qatar 2022 World Cup. *Kafala* is a sponsorship system common in the countries in the Middle East whereby the employer acts as a sponsor of the migrant worker so that the worker can be issued with an employment and residence (ILO, 2017c). This is inherently problematic as it creates an imbalance between the rights of the employer and the worker, preventing the worker from changing jobs or even leaving the country without the permission of the employer. In the wake of the construction works of the 2022 FIFA World Cup, Qatar made significant progress towards reforming the *kafala* system and strengthening the rights of migrant workers. Most notably, in 2018, Qatar ended the "exit permit" requirement for most workers, making it possible for the worker to leave the country without the employer's permission. Qatar further ended the No-Objection Certificate requirement in 2020, enabling the worker to change jobs without the employer's permission. Other reforms included the introduction of a new mandatory minimum wage, the electronic wage protection system

(UAE, 2016), and prompt payment regulations. However, it is observed that, while legal reforms will allow construction workers to leave exploitative working relationships, they are unlikely to put an end to the abuse itself (Amnesty International, 2020). Employers continue to act as official sponsors; migrant workers cannot request their residence permits, nor cancel their residence permits.

Construction is witnessing the creation of an international labour market (ILO, 2016a). Although it started in the civil engineering segment of the industry, it is increasingly evident in the whole of the building and construction industry and includes new occupations. For example, during the great recession, 12 percent of Spanish architects migrated to other countries (including Germany and Switzerland, as well as others that had thriving industries such as China and Saudi Arabia), finding higher salaries and employment conditions than in Spain (El País, 2014). There is also anecdotal evidence of architects working in developing countries, where their skills are more appreciated and they have greater responsibilities than at home.

Women in construction. Despite concerted initiatives, in many countries, to recruit women, the construction industry remains one of the most gender-segregated industries in the world (BWI, 2021). There is a long-standing perception that the construction industry is male-dominated and is suited to only men. Women represent less than 10 percent of the total construction workforce globally and there are around 3 percent of women in trades (Eurofound, 2017; ICED, 2018). Women's work in construction can be broadly divided into three groups: women in professional and technical positions such as architects, engineers, and surveyors; women in administrative positions such as clerks, secretaries, and human resources; and women working on site. Table 8.3 shows the rate of women's participation in the formal construction industry and the percentage of all women paid workers in construction, by region. In most countries, the number of women in construction is low, but the figures are increasing moderately. The highest rates of female participation are found in Kazakhstan (35.7 percent), Singapore (29.1 percent), Kenya (22 percent), and Ethiopia (20.5 percent).

The rate of women's participation in construction is significant in South Asia and former Soviet countries, although in the latter, fewer young women are taking up STEP careers than during the Soviet period (BWI, 2021). Women face long-standing discrimination in the sector, which limits their ability to access blue-collar occupations and promote within businesses.

Table 8.3 Women's participation in the construction sector

Region	*Construction workers who are women (%)*	*Women paid workers in construction (%)*
North America	12.2	2.0
Latin America	5.4	0.9
Europe	9.2	1.4
Sub-Saharan Africa	6.5	0.8
Asia	10.1	1.5
MENA	2.0	1.8
Australia and Oceania	6.5	1.1

Source: ILOSTAT (2022), latest available figure (2010–2021) from 164 countries.

Note: "Women paid workers in construction" denotes the share of paid women working in the construction sector out of female workers in all sectors. "Construction workers who are women" denotes the share of women workers out of all workers in the construction industry.

Box 8.3 Women in informal construction work in India

Women form an important yet invisible part of India's construction industry. They constitute between 30 and 50 percent of the workforce, which is a notable exception in terms of women's participation in construction worldwide (BWI, 2021). Some 97 percent of the women work in the informal economy in low-skill occupations such as concrete mixing, digging, stone breaking, carrying of materials such as bricks, and cleaning of building sites. Day-labour and short-term verbal contracts are the most common forms of employment relationships. The women might also work as part of a family work unit, often without receiving direct payment. Working conditions are particularly poor; the working hours are rather long, and the women are paid lower wages than their male counterparts. A recent study by WIEGO (2020) found that among women informal construction workers, 30 percent nationally and 41 percent in urban India work 53 hours or more a week, while 40 percent and 27 percent, respectively, work 36 hours or less. The same report reported that among informal construction workers, the average hourly earnings of the men were 53 percent higher than that of the women. Other studies estimate that women are paid one-third to one-half of the wages of a male skilled worker.

As mechanisation increases in construction, many of the manual jobs on site may cease to exist, affecting women disproportionally (SEWA, 2005).

Although women face precarious conditions all around the globe, the situation of women participating in marginal blue-collar occupations in developing countries is particularly bad. Box 8.3 illustrates the situation of women participating in particularly difficult occupations in India. A study of informal construction workers in Dar es Salaam, Tanzania, revealed that only 4 percent were women (Jason, 2005). Their main roles included stone crushing, working in offices as storekeepers, and working as cleaners. Thus, very few women worked in skilled site construction occupations, such as masonry, carpentry, or electrical works. In South Africa, women make up 12.49 percent of the workforce; 53 percent work in unskilled trades, while 14 percent are in skilled work, 5 percent in technical and associated jobs, and 7 percent in management positions (BWI, 2021). In Uruguay, 5 percent of the workforce are women, working in administration or as professionals, and a small proportion undertake unskilled cleaning, gardening, and maintenance (BID, 2018).

Equal opportunities for women have not been achieved (WCB Initiative, 2020). Women often receive lower wages than men despite performing the same tasks and having the same responsibilities. Recruiters and companies that practice wage discriminations have cultural beliefs that associate labour productivity judgements with personal characteristics of employees, such as gender, race, and country of origin. The poor and inappropriate working conditions, discriminatory recruitment practices, gender stereotypes, and work culture not only constitute a barrier to employment, but also perpetuate the precarious conditions of employment for women in the construction industry. The excessive and irregular working hours and travel times often required in construction make it difficult for women to combine work with domestic and other responsibilities attributed to them in many countries.

Legal barriers to women's employment continue to limit the employment of women in the construction industry, as some trades are deemed too arduous, harmful, or unhealthy for them. For instance, Article 253 of Russia's Labor Code – Federal Law No. 197-FZ of 2001 and Resolution No. 162 of 25 February 2000 bars women from being employed in 456 types of work, many of which are construction trades. Similarly, Article 242 of the Colombian Labour Code establishes that women cannot work as industrial painters in building interiors (Diario Oficial de la Imprenta Nacional de Colombia, 1950). Thus, women who work in such trades are in the informal economy, with precarious working conditions and the threat of retrenchment. Indeed according to the World Bank Group's Women, Business and Law Reports (WBG, 2022), there are 33 countries that currently have laws that prevent women from working in at least one trade in the construction sector. Between 2019 and 2021, countries such as Benin, Montenegro, Saudi Arabia, Niger, the Democratic Republic of Congo, Mongolia, and Slovenia implemented reforms towards removing restrictions on the employment of women in specific trades in various sectors.

Child labour in construction sites and brick kilns. Most working children live in poverty conditions and are forced to work to help support their families and themselves (ICED, 2018). In most cases, they are forced by their parents, but they might also be forced by employers or middlemen. In most countries, participants in child labour on construction sites are mostly boys, as work on site is associated with lifting heavy and bulky materials and working at heights, often without safety equipment (ILO and UNICEF, 2020). In brick kilns, hazards include extreme temperatures and airborne ash created by the kilns. Children often younger than ten years old haul bricks all day, breathing air that is thick with dust. Table 8.4 presents a list of common construction tasks and the related hazards as well as their possible health consequences for working children.

Table 8.4 Child labour: Common tasks, hazards, and injuries

Location	*Tasks*	*Hazards*	*Injuries and health consequences*
Construction sites	Construction: hauling and stacking materials; carpentry; masonry	Heavy loads; dangerous heights; falling objects; sharp objects; power tools; live wires; moving vehicles; loud machines; exposure to extreme weather; dust	Joint and bone deformities; blistered hands and feet; lacerations; punctures from nails; back injury; muscle injury; head trauma; broken bones from falls; electrocution; noise-induced hearing loss; frostbite, sunstroke and other thermal stresses; dehydration; breathing difficulties
Brick kilns	Brick-making: toting, stacking	Heat from kilns and ovens; flying ashes; heavy loads; dropped bricks; dust; exposure to extreme weather; remote locations; poor sanitation; moving vehicles	Burns and heat stroke; dehydration; joint and bone deformities; musculoskeletal problems from repetitive motion; blistered hands; bruised feet from dropped bricks; lacerations; breathing difficulties; silicosis and other occupational lung diseases; heat and cold stress; insect bites; poor nutrition; bacterial and viral diseases; injury from moving vehicles

Source: ILO (2011b).

Box 8.4 Precarious working conditions of children in Ulaanbaatar, Mongolia

Half of the child labourers in the urban areas in Mongolia work in construction, making it the sector with the highest proportion (UCW, 2015). In 2015, the ILO conducted a rapid assessment of child labour in the construction industry covering construction sites, factories manufacturing building materials, and markets and shops selling building materials in seven districts of Ulaanbaatar (ILO, 2016b). The study found that:

- The majority of working children were boys, and 65 percent of them began working at the age of 15–18 years.
- Most of the workers did not have written contracts, as their parents made verbal agreements with the employers instead.
- Average working hours were 60 or more per week.
- Some of the labourers worked in jobs included in the list of occupations which were legally prohibited nationally.

Almost half of the labourers suffered injuries at work; many of them were not provided with PPE, and most had not received occupational health and safety advice.

Worldwide, there are about 160 million working children, with nearly 79 million of them trapped in hazardous occupations that endanger their health, safety, and moral development (ILO and UNICEF, 2020). It is pertinent to note that there have been important improvements since 2020. There were 245 million children working then, and 170 million of them did so in hazardous occupations. Many working children work in the construction supply chain, mostly in brick kilns, stone quarries, and cutting and carrying wood. As an example, the ILO (2011a) study reported that 56 percent of brickmakers in kilns in two districts in Afghanistan are children, with the majority of them being 14 years of age or younger. Others work directly in construction sites carrying building materials such as bricks and stones; cleaning; or in technical trades such as concrete placers and finishers. Child labour in construction is a phenomenon present in developing countries, mainly in Africa and Asia, particularly in India. Box 8.4 presents the particularly precarious working conditions of children in Ulaanbaatar, Mongolia.

The future of work and implications for the built environment precariat

It is useful to consider the major trends that are shaping the future of work in the construction industry to explore their potential consequences for the built environment precariat. These trends and their impact are now discussed.

New technologies and digitalisation of construction. The construction industry is going through a digitalisation transformation. With the implementation of smart building technologies and the development of new tools more adapted to the energy transition, policy efforts should address skills development and workforce adaptation to new technologies to avoid a digital divide. Construction 4.0, including Artificial intelligence, the Internet of Things, and the development of smart grids and homes, will likely have positive effects on the working conditions of highly skilled workers, who are likely to be better paid (Estebsari and Werna, 2022).

However, it is not clear how the unskilled and semi-skilled workers will be affected, as it depends on factors such as their ability to negotiate and whether there are social agreements between workers, employers, and governments. For example, there may be industry-wide effects of Building Information Modeling (BIM) as it is already being used in many developed countries to assess health and safety risks and generate a preventive work culture (Mordue and Finch, 2019). Blockchain technology can also increase traceability and uphold social safeguards in agriculture supply chains and has the potential to do the same in construction. Similarly, the use of digital technology can improve the efficiency of labour governance structures in tackling unlawful employment relationships and conditions. This will help to reduce occupational hazards for construction workers and generate employment opportunities for youth and women.

Automation of construction processes. Automation efforts in the construction industry include off-site construction as well as 3D printing (OECD, CEPAL and IDB, 2016). Off-site construction has gained renewed importance in the construction cycle in the last few years. In many countries, the volume of prefabricated housing has grown significantly following the Covid-19 pandemic. The global market for prefabricated buildings was estimated at USD 106.1 billion in the year 2020 and is projected to reach USD 153.7 billion by 2026, growing at a CAGR of 6.4 percent (Global Industry Analytics, 2021). The consequences of construction automation may differ between developed and developing countries as well as blue- and white-collar workers. For example, masons and bricklayers in developed countries could potentially become less useful if 3D printers attain commercial viability. The experience with off-site construction shows rather a shift towards assembly and framing tasks for builders and not exclusively substitution effects. Higher degrees of technology adoption will increase the demand for research and development and skilled workers.

Collaborative platforms. The emergence of collaborative platforms has opened up new marketplaces for goods and services from businesses and individuals. Such platforms allow workers in the construction industry to find business opportunities, often in maintenance and repair work (EFBWW and FIEC, 2020). Using (often bogus) self-employment staff, such platforms push down prices by paying workers below the legal minimum wage or what is stipulated in collective agreements applicable to workers in the construction industry. Ultimately, cost-reduction pressures contribute to the spread of short-term employment relationships.

Energy and circular economy transitions. The construction sector has a significant impact on climate change and environmental degradation. Construction work makes use of a large amount of resources. The sector is responsible for 25–40 percent of global energy use and 30–40 percent of global greenhouse gas emissions (ILO, 2010). In Europe, buildings are responsible for 40 percent of the continent's energy consumption and 36 percent of greenhouse gas emissions from energy, but only one in ten buildings undergoes energy-efficient renovation every year (European Commission, 2020). Another important aspect is the waste generated in construction activities. The emerging use of circular economy concepts in the construction industry aims at lowering CO2 emissions, making better use of natural resources, and producing less waste in the building life cycle. From new energy-efficient building to the refurbishment of existing buildings, the construction industry already plays a crucial role in the path towards sustainable development. The transition towards a decarbonised and circular building industry will have important job creation opportunities for construction workers (Mella and Werna, 2023). There are emerging skills which will require retraining and upskilling some of the existing tradespersons including roofers and installers of insulation materials and renewable energy elements. With regard to the circular economy, challenges emerge in the precarious conditions around the recycling and waste management of the built environment. The use of toxic materials in technologies such as solar photovoltaics may cause greater risk hazards for workers in waste management, a largely informal sector with atypical forms of employment. Building retrofitting will also come with challenges on how to remove asbestos from buildings without increased risks for workers.

Policy options improving work quality in the construction sector

Different approaches have been tried to address the precarious working conditions of construction workers. These can broadly be categorised under regulatory interventions, interventions on labour demand and supply, and interventions on labour market information. Some examples are as follows:

- *Labour market reforms.* There have been legal reforms of some key aspects of the labour market. For instance, the situations under which an employer can engage a worker in temporary employment are well defined and are increasingly being restricted (see Box 8.5),

Box 8.5 The Spanish experience in reducing workplace accidents

From 2007 to 2017, workplace accidents in the Spanish construction industry declined by 50 percent (European Platform Undeclared Work, 2019). Serious and very serious accidents have fallen by 77 percent. This is a result of the combined efforts between the government, employer organisations, and trade unions representing workers.[3]

In October 2006, the Spanish government approved law 32/2006 limiting subcontracting in the construction sector. The law prescribes that companies should have in their organisational structure a minimum of 30 percent of direct and permanent employment contract workers. The law limits to three levels of successive subcontractors for specialised workers and to one level of subcontracting for labour-intensive projects. This way, labour agencies cannot subcontract work further. The law stipulates the creation of a subcontracting registry. Firms have to provide quality and solvency guarantees, and the relationship between the main contractor and the subcontractors has to be written in a subcontracting book maintained at the worksite.

The work of the *Fundación Laboral de la Construcción* (FLC) has also helped to generate a preventive health and safety culture. Set up by employers and trade unions, the FLC delivers training in health and safety prevention and topic-specific sessions such as working at heights and working with mechanical anchors and chemicals. The FLC has a hotline which businesses and workers can call to seek advice on specific health and safety issues or to arrange a worksite visit. There is a wealth of manuals, videos, and gamified resources on the issue of health and safety and the use of ITCs to prevent accidents on site.

The financial model of the FLC is based on legally mandatory contributions by the industry and workers as well as funds from the government and charges for its training courses and other activities.

reducing the period of time an employee can work in successive contracts with the same employer or establishing a single type of employment contract that can address the dualisation of the labour market.

- *Promotion of equal access to decent employment for women and youth.* Governments' regulatory action can also address the inclusion of groups at a higher risk of precarious employment in the labour market, for example, by making it imperative that employers provide day nursery care at worksites where there are women workers with children. In India, the National Rural Employment Guarantee Act states that a day nursery should be provided at a worksite where there are more than five children under the age of six. The implementation of business-oriented TVET systems through apprenticeships and traineeships also helps the youth and women to gain access to employment in the construction industry.
- Social clauses in public procurement can address business competitiveness, skills upgrading, and health and safety risks (see, for instance, ILO (2018a) and Patterson Hurt-Suwana and Mahler (2021).
- Employment Intensive Investment Programmes (EIIPs) can improve job quality through temporary employment opportunities that combine local participation, resource use, and capacity building. EIIPs can be geared at civil work construction as well as building maintenance, and many of the programmes address gender and youth participation.
- Due diligence regulations of construction supply chains can help mitigate the presence of child and bonded labour. The European Parliamentary Research Service (2023) has recently (in September 2022) presented a proposal for a regulation to prohibit products made using forced labour, including child labour, on the internal market of the EU. The US Department of Labour has limitations on importing from countries with a high presence of child and bonded labour (see Box 8.6).

Box 8.6 Children working in brick kilns

Many children work in the brick kiln industry around the world. The US Department of Labor maintains a list of goods and their source countries, which it has reason to believe are produced by child labour or forced labour in violation of international standards. The updated list shows that 21 countries have a high presence of child labour in their brick and clay industries. These include as follows: Afghanistan, Argentina, Bangladesh, Bolivia, Brazil, Burma, Cambodia, China, Colombia, Ecuador, Egypt, India, Iran, Nepal, North Korea, Pakistan, Paraguay, Peru, Russia, Uganda, and Vietnam. A study on child labour estimated that brick kilns engage about 1.7 million children in India, at least 500,000 in Pakistan, and 110,000 in Bangladesh (FPRW-IPEC, 2014).

Normally, children are not directly employed by the brick kiln owners but start helping their parents from as young as five years old. Common tasks performed by children include preparing the soil, turning over cooked bricks, and creating piles of bricks. Wages are paid based on the piece-rate system, and hence, children continue to work as part of family labour units to enable the family to be paid more based on the quantity of bricks they produce.

National legislation sets out the rights and responsibilities of employers, workers, and labour inspectors for each country. Given the diversity of labour legislation and OSH initiatives, it is only possible to illustrate some efforts that have had a positive impact on reducing workplace accidents.[4] There is currently a change in the approaches of labour administrations. The traditional inspection model is based on fines and sanctions, which are reactive and deter, to a certain extent, non-compliance with labour laws. The new model is preventive, adaptive, and inclusive. The model requires changes in priorities, workplans, and resources, but also a shift in the mentality of labour inspectors to consider more strategic approaches and enhance the level and extent of the participation of employers and workers (ILO, 2017b).

The new approach has been fully embraced in many countries aiming to address labour compliance. The ILO released in 2017 a guide on Conducting Labour Inspections on Construction which highlights the different roles of employers, workers, and labour inspectors (ILO, 2017a). The model has been operationalised in different countries through workshops focusing on strategic compliance planning; some of the workshops were specifically for the construction industry (2017b).

There are few large Official Development Assistance (ODA) projects that address labour aspects in the construction industry. For instance, the Zambia Green Jobs Programme of the ILO promoted green job creation among MSMEs in the construction industry between 2013 and 2018. The programme addressed occupational safety and health, skills development, and social protection coverage of informal sector workers. Most ODA addresses multiple sectors in their interventions. The GIZ More Income and Employment in Rural Areas programme in Malawi for instance promotes demand-side interventions, including the development of scalable business models in ecological building material production (i.e., Tara Ecobricks) (GIZ MIERA, n.d.). There are many other national and regional initiatives and projects by governments, employers, trade unions, and civil society that address job quality in construction. An example of a regional initiative is the EU project Tackling Undeclared Work in the Construction Industry (ILO CINTERFOR, n.d.a). The project was rolled out in seven member states (Austria, Belgium, Bulgaria, France, Italy, Romania, and Spain). In each country, construction trade unions, employer federations, and enforcement authorities worked together to develop a range of policy initiatives spanning the spectrum of possible actions to tackle undeclared work in the construction sector. Similar projects can be found in ILO CINTERFOR (n.d.b), Hilti Foundation (n.d.), and Swiss Contact (n.d.).

The idea of a Universal Basic Income (UBI) (an unconditional cash transfer to every citizen of a country) has gained traction in the last few years. Although no country has rolled out a truly UBI for all citizens, Finland, Canada, and the Netherlands have piloted basic income schemes (ILO, 2018b). A UBI programme is a cross-sectoral approach that addresses the job and income insecurity associated with the growing precariousness and informality of employment. UBI advocates, such as Guy Standing, argue that a regular and predictable income reduces poverty and vulnerability more effectively than targeted schemes and protects the possible displacement of jobs by new technologies (ILO, 2018b). Critics of the UBI approach question the feasibility of a UBI considering its high economic costs and its capacity to reduce poverty and vulnerability.

Main findings and conclusion

In 2011, Guy Standing developed a provocative and influential new term: the *precariat* (Standing, 2011) which combines the concept of precarious work and the social class of the proletariat. A key element of the precariat is that precarious work is no longer an employment situation but rather a characteristic of the way a sizeable number of workers interact with the labour market throughout their lives. The underlying causes are the conformation to an international market system for construction labour, the structural technological transformation of construction, and the need for flexible hiring modalities in the industry. In a labour-intensive industry such as construction, the companies should be able to manage labour costs. Construction workers may go through successive temporary contracts, casual labour, and part-time employment without ever entering stable and open-ended work.

Some 50 years ago, construction contractors had many workers hired in stable full-time jobs under contracts and made efforts to maintain a continuous flow of projects to make use of their labour force. Globalisation and the rise of an international construction labour market made it necessary to adopt more flexible forms of employment. In many countries, labour markets became dualised in two silos, with one set of workers in protected employment conditions and another group in successive time-bound employment arrangements. The extensive use of flexible employment arrangements has given rise to a category of workers with unstable work prospects.

The chapter presented evidence on the nature and composition of the precariat in the built environment sector and their employment and working conditions. Concurrently, the construction industry is undergoing a deep technological and digital transformation, which will have deep impacts on the workers.

Various policy avenues exist for governments, employers, and workers to tackle this issue. These diverse options warrant in-depth exploration and thorough assessment of their effectiveness. A multitude of initiatives are in place to improve the working conditions within companies and for workers. These encompass labor market reforms, the advancement of equitable access to quality employment for women and youth, the integration of social clauses in public procurement, employment schemes, and the enforcement of due diligence regulations.

Notes

1 The ILO prepared an action checklist for the construction industry with practical measures designed to help stakeholders in the construction industry mitigate and prevent the spread of Covid-19, ILO (2020).
2 It is difficult to differentiate in practice between people who are really self-employed and running their own business and people who for example depend on a single employer for their income and thus have no real autonomy in running their “business”.
3 The decline in economic activity also contributed partially to the decline in accident rates. Increased activity in the last years however does not show an increase in workplace accidents.
4 The ILO Global Database on Occupational Safety and Health Legislation (ILO-LEGOSH, n.d.) provides a picture of the regulatory framework of the main elements of OSH legislation, including OSH management and administration, employers’ duties and obligations, workers’ rights and duties, and OSH inspection and enforcement.

References

Amnesty International (2020): Reality check: migrant workers rights with two years to qatar 2022 world cup. https://www.amnesty.org/en/latest/campaigns/2019/02/reality-check-migrant-workers-rights-with-two-years-to-qatar-2022-world-cup/

BID (2018): Estudio de genero de la industria de la construcción en Uruguay. https://publications.iadb.org/en/estudio-de-genero-en-la-industria-de-la-construccion-en-uruguay-sector-vial
BWI (2021): Women in Tardes, global research report. Avaialble at https://www.bwint.org/cms/bwi-launches-women-in-trades-global-research-report-2425
CIOB (2018): Construction and the modern slavery act tackling exploitation in the UK.
Diario Oficial (1950): Decree Number 2663 on the Colombian Labour Code published by the Imprenta Nacional de Colombia from 5 August 1950. Available at: https://leyes.co/codigo_sustantivo_del_trabajo/242.htm
EFBWW and FIEC (2020): Tackling undeclared work in the construction industry Toolkit. https://www.fiec.eu/application/files/8016/0127/8404/2020_-_TUWIC_Toolkit__EN__low_res.pdf
EIP (2018): Protecting the wages of migrant construction workers part two: Addressing the problem in Gulf Cooperation Council countries.
El País (2014): Arquitectos: 71% en paro. https://elpais.com/elpais/2013/10/09/del_tirador_a_la_ciudad/1381310848_138131.html
Estebsari and Werna (2022): Smart cities: grids, homes and the workforce: challenges and prospects. In Li, R., Chau, K.W. and Daniel Ho, C.W. Current state of art in artificial intelligence and ubiquitous cities. Singapore: Springer.
Eurofound (2016): Exploring the fraudulent contracting of work in the European Union, Publications Office of the European Union, Luxembourg.
Eurofound (2017): Sixth European working conditions survey – Overview report (2017 update), Publications Office of the European Union, Luxembourg.
European Commission (2017): Guidance on due diligence for EU businesses to address the risk of forced labour in their operations and supply chains https://trade.ec.europa.eu/doclib/docs/2021/july/tradoc_159709.pdf
European Commission (2020): Renovation Wave: doubling the renovation rate to cut emissions, boost recovery and reduce energy poverty, October 2020 Press Release. https://ec.europa.eu/commission/presscorner/detail/en/IP_20_1835
European Parliamentary Research Service (2023): Proposal for a ban on goods made during forced labour, February 17. https://epthinktank.eu/2023/02/17/proposal-for-a-ban-on-goods-made-using-forced-labour-eu-legislation-in-progress/#:~:text=On%2014%20September%202022%2C%20the,to%20promote%20decent%20work%20worldwide
European Platform Undeclared Work (2019): Regulating subcontracting in the construction sector (LSCS) Case study. Available at: https://www.ela.europa.eu/sites/default/files/2021-09/ES-LawSubcontractingChains.pdf
Federation of Small Businesses (2018): Pay it Forward - Lessons and recommendations for Europe from the UK payment landscape. https://amaiz.com/documents/pay_it_forward.pdf
FIEC (2020): Late payment in the Construction industry, Analytical report. https://www.fiec.eu/application/files/6616/0259/5873/2020_Analytical_Report_Late_Payment_in_Construction.pdf
Fiorenza (2021): Immigrant workers in the construction sector in Italy: impact on safety and health.
FLEX (2018): Shaky foundations: labour exploitation in London´s construction sector. https://labourexploitation.org/publications/shaky-foundations-labour-exploitation-londons-construction-sector.
FPRW-IPEC (2014): A health approach to child labour - Synthesis report of four country studies on child labour in the brick industry. International Labour Office.
GIZ MIERA (n.d.): GIZ KULIMA more income in rural areas of Malawi Programme. https://www.giz.de/en/worldwide/73930.html
Global Industry Analytics (2021): Prefabricated Buildings - Global market trajectory & analytics. https://www.strategyr.com/market-report-prefabricated-buildings-forecasts-global-industry-analysts-inc.asp
Human Rights Watch (2008): "One Year of My Blood" exploitation of migrant construction workers in Beijing. https://www.justice.gov/eoir/page/file/1276666/download
Hilti Foundation (n.d.): Construya project on affordable housing and technology in Latin America. https://www.hiltifoundation.org/projects/factsheet-construya
ILFR (n.d): Atypical Employment in Slovakia: Past and recent trends. https://www.ceit.sk/IVPR/images/IVPR/prezentacie/Presentation_Slovakia.pdf

ILO (2010): Green jobs in construction: Small changes – big effect, World of Work Magazine n°70, December 2010 - Promoting a recovery focused on Jobs. https://www.ilo.org/global/publications/world-of-work-magazine/articles/WCMS_151709/lang–en/index.htm
ILO (2011a): Buried under Bricks – A rapid assessment of bonded labour in brick kilns of Afghanistan. http://www.ilo.org/asia/publications/WCMS_172671/lang–en/index.htm
ILO (2011b): Children in hazardous work. What we know, what we need to know.
ILO (2015): Good practices and challenges in promoting decent work in construction and infrastructure projects. https://www.ilo.org/sector/Resources/publications/WCMS_416378/lang–en/index.htm
ILO (2016a): Migrant work & employment in the construction sector.
ILO (2016b): Mongolia policy brief: Child labour. https://www.ilo.org/wcmsp5/groups/public/—asia/—ro-bangkok/—ilo-beijing/documents/publication/wcms_491324.pdf
ILO (2017a): Conducting Labour Inspections on Construction A guide for labour inspectors. Available at https://www.ilo.org/wcmsp5/groups/public/—ed_dialogue/—lab_admin/documents/publication/wcms_570678.pdf
ILO (2017b): ILO approach to strategic compliance planning for labour inspectorates. https://www.ilo.org/labadmin/info/public/fs/WCMS_606471/lang–en/index.htm
ILO (2017c): Employer-migrant worker relationships in the Middle East: Exploring scope for internal labour market mobility and fair migration. https://www.ilo.org/wcmsp5/groups/public/—arabstates/—ro-beirut/documents/publication/wcms_552697.pdf
ILO (2017d): Global estimates of modern slavery: Forced labour and forced marriage. https://www.ilo.org/wcmsp5/groups/public/—ed_norm/—ipec/documents/publication/wcms_586127.pdf
ILO (2018a): Can we create better jobs in Africa's booming construction sector? Looking to market systems analyses to point us in the right direction, ILO The LAB. https://www.ilo.org/wcmsp5/groups/public/—ed_emp/documents/publication/wcms_652332.pdf
ILO (2018b): Universal Basic Income proposals in light of ILO standards: Key issues and global costing.
ILO (2020): COVID-19 action checklist for the construction industry. Available at: https://www.ilo.org/wcmsp5/groups/public/—ed_protect/—protrav/—safework/documents/instructionalmaterial/wcms_764847.pdf
ILO (2022): Technological solutions to guaranteed wage payments of construction workers in China.
ILOSTAT (2022): Data on formal, informal employment and self-employment; fatal and non-fatal accidents. Latest available data as of April 2022.
ILO-LEGOSH (n.d.): ILO global database on occupational safety and health legislation. Database available at: http://www.ilo.org/dyn/legosh/en/f?p=14100:1:0::NO
ILO and UNICEF (2020): Child labour: Global estimates 2020, trends and the road forward. https://www.ilo.org/wcmsp5/groups/public/—ed_norm/—ipec/documents/publication/wcms_797515.pdf
ILO CINTERFOR (n.d.a): Tackling undeclared work in the construction industry (TUWIC) project. https://www.oitcinterfor.org/en/experiencia/training-and-school-maintenance-uocra-foundation
ILO CINTERFOR (n.d.b): Fundación Escuela Taller de Bogotá (School-workshop foundation of Bogota). FETB. Colombia. https://www.oitcinterfor.org/en/experiencia/fundaci%C3%B3n-escuela-taller-bogot%C3%A1-school-workshop-foundation-bogota-fetb-colombia
ICED (2018): Construction sector employment in low income countries, infrastructure & cities for economic development. http://icedfacility.org/wp-content/uploads/2018/08/Report_Construction-Sector-Employment-in-LICs_Final.pdf
Jason, A. (2005): Informal construction workers in Dar es Salaam, Tanzania, Sectoral Activities Department, working paper no. 226. Geneva: ILO.
Jorens, Y. (2008): Self-employment and bogus self-employment in the European construction industry: A comparative study of 11 Member States. https://biblio.ugent.be/publication/1112697/file/1112792.pdf
Mella, A. and Werna, E. (2023): Skills and quality jobs in construction in the framework of the European green Deal and the post-covid recovery, JTC and EFBWW.
Mordue, S. and Finch, R. (2019): BIM for construction health and safety. Routledge.
OECD, CEPAL and IDB (2016): Economic prospects of Latin America 2017: youth, skills and entrepreneurship. Paris: OECD Publishing.

Patel, D. A. and Jha, K. N. (2016): An Estimate of Fatal Accidents in Indian Construction. In: P. W. Chan and C. J. Neilson (Eds.) Proceedings of the 32nd Annual ARCOM Conference, 5–7 September 2016, Manchester, UK, Association of Researchers in Construction Management, Vol 1, 539–548.
Patterson Hurt-Suwana, J. and Mahler, M. (2021): Social procurement to reduce precarious employment for Māori and Pasifika workers in the construction industry. Kōtuitui: New Zealand Journal of Social Sciences Onlin, Volume 16(Issue 1), 2021. https://www.tandfonline.com/doi/full/10.1080/1177083X.2020.1767164
RMMRU (Refugee and Migratory Movements Research Unit) (n.d.): Impact of labour migration to the construction sector on poverty: Evidence from India. Policy Brief. http://www.migratingoutofpoverty.org/files/file.php?name=rmmru-rp002-india-policy-brief-13-sep14.pdf&site=354
Standing, G. (2011): The precariat. The new dangerous class. London: Bloomsbury.
SEWA (Self-Employed Women's Association) (2005): At the Kadiyanaka: Challenges faced by construction workers in Ahmedabad. Ahmedabad: SEWA Academy.
Swiss Contact (n.d.): Skills to build project. https://www.swisscontact.org/en/projects/skills-to-build-s2b
UAE (2016): United Arab Emirates Ministry of Human Resources and Emiratisation, Ministerial Decree No. (739) On wage protection, Available at: www.ilo.org/dyn/natlex/natlex4.detail?p_lang=en&p_isn=104217&p_count=2&p_classification=12.02
UCW (2015): The twin challenges of child labour and educational marginalisation in the South-East and East Asia region: Preparing for a post 2015 world. Understanding Children's Work.
WCB Initiative (2020): Towards an equal construction industry. Available at: https://www.womencanbuild.eu/wp-content/uploads/2021/01/WCB_IO2_Joint-Report_immersive_experience-for-women.pdf
Wells (2007): Informality in the construction sector in developing countries.
WBG (2022): Women, business and the law. Several years. https://wbl.worldbank.org/en/reports
WIEGO (2020): Informal workers in India: A statistical profile. Available at: https://www.wiego.org/sites/default/files/publications/file/WIEGO_Statistical_Brief_N24_India.pdf

9

HUMAN RESOURCE MANAGEMENT AND DEVELOPMENT IN CONSTRUCTION

Strategic considerations

Yaw A. Debrah[1], *Aziz Christian Jabaru*[2], *Richard B. Nyuur*[3], *Florence Ellis*[4], *and Juliet Banoeng-Yakubo*[5]

[1]Swansea University, UK; [2]University of Ghana Business School, University of Ghana, Legon, Accra, Ghana; [3]University of Bradford, UK; [4]Kwame Nkrumah University of Science and Technology, Ghana; [5]De Montfort University, Leicester, UK

Challenges of human resource management methods in construction

A variety of human resource management (HRM)-based literature, approaches, and techniques have been written and put into practice to ensure that human resources in organisations are effectively managed and transformed in ways that are in synchronisation with the mission, goals, and strategies of the organisation or sector in which they are found. The majority of these HRM strategies have been successful across numerous sectors. However, research has shown that the construction industry presents particular difficulties that may make it challenging to apply critical components of the fundamental HRM strategies that appear to have been successful in other sectors which are more stable (Boselie, 2014; Warner, 2020). It is particularly challenging to adopt the fundamental principles of conventional HRM approaches due to the complexity and dynamism of the project-based culture prevalent in the construction industry (Assaad & El-Adaway, 2021; Li et al., 2015).

The construction industry, over the years, has contributed significantly to global economic activities. According to the Committee for European Construction Equipment (CECE), the construction industry is one of the largest in the world, and over 18 million people have been directly employed in the industry. This accounts for around 9 percent of the European Union's gross domestic product (GDP) (CECE, 2022). Additionally, construction generates new employment opportunities, promotes economic expansion, and offers some of the answers to social, environmental, and energy concerns (Akadiri et al., 2012; Ball, 2014; Haupt & Harinarain, 2016; Hildebrandt et al., 2017). In the United States, recent statistics from the Associated General Contractors of America (AGC) report that the industry employs more than 7.6 million people, has over 745,000 employers, and builds

 DOI: 10.1201/9781003262671-9

structures worth close to US$1.4 trillion annually (Associated General Contractors of America, 2022). Similarly, in Asia and Africa, the construction industry has been growing at a fast rate, with Asia, in particular, being tipped to become the key construction market of the global construction industry by 2030 (Robinson et al., 2021).

Despite technological advancements and investments in sophisticated equipment and methods, the construction industry remains one of the most labour-intensive industries (Hossain et al., 2020; Mosly, 2015). In most organisational initiatives in the industry, human resources account for the more significant part of expenditure (Rostoka et al., 2019). The industry employs a wide range of people from many occupational cultures and backgrounds, including unskilled, artisan, managerial, professional, and administrative occupations. Most of these diverse employees are itinerant and typically work on short-term projects. Unfortunately, the disparate backgrounds of the employees in the construction industry frequently disrupt the cohesion necessary for accomplishing project objectives (Jarkas et al., 2015). This can also result in employees having competing needs, making the construction industry a challenging area in which to manage people efficiently.

Over the years, there have been numerous demands for improvement in HRM practices in the construction industry (Aghimien et al., 2021; Ling et al., 2018). The unique challenges posed by the industry make addressing those demands essential. Across the world, there have been many reports on how employees on construction sites keep boycotting activities due to the poor treatment they face in their jobs (Adinyira et al., 2020; Creighton, 2014; Sishi, 2016). The risk factors employees deal with, especially in complex and challenging construction projects, are high (Debnath et al., 2016; Janani et al., 2017; Jayasudha & Vidivelli, 2016). This elevates the importance of employee protection in the industry. Evidence from all over the world has repeatedly supported the notion that the HR department is essential to the success of any construction project (Ling et al., 2018; Shamsuddin et al., 2015). The evidence also confirms the notion that the HR unit is not just a peripheral component of the construction industry but rather the cornerstone (Alzyoud et al., 2020). In order to avoid endangering the industry in the future, stakeholders must develop practical measures and mechanisms that will help minimise the numerous unique present and future problems the industry faces. It is also necessary to highlight the particular difficulties associated with some of the most prevalent HR practices used in the construction industry. This will act as a road map for implementing the appropriate actions to address these HR concerns in the construction industry.

Human resource development

Human resource development is one of the essential HR methods necessary for experiencing success in the construction industry. It is essential to develop a motivated workforce and an efficient, productive, innovative, and creative industry with a positive public image (Stone et al., 2020; Werner, 2021). It is also the most efficient means of maintaining, updating, and enhancing the intellectual capital of the industry's workforce and ensuring that its activities contribute positively to the general welfare of society (Werner, 2021). Nonetheless, investment in human resource development in the construction industry remains relatively low compared to other industries (Wilton, 2022). In the developing nations, notably in Africa and some parts of Asia and South America, the traditional culture of the construction industry, which places more importance on the brawn than the brains of employees, contributes to the low level of investment in human resource development (Boadu et al., 2020). With the aim of completing the majority of projects in these underdeveloped and emerging regions, personnel is hired based on the extent to which they can exert their physical energies (Boadu et al., 2020). This has led to employers making few efforts to increase their employees' intellectual potential.

Additionally, the advent and influx of high-end technological methods and equipment have heightened the need to extend the industry's training and human resource development capacity to ensure that the skills of the construction industry workforce in these regions remain current and relevant in the coming years (Koeleman, 2020; Zhou et al., 2013). However, there remains the challenge of high expenditure in operating some of this equipment. Employees also need to undergo the required training in some aspects of construction, which is highly technical and requires a high level of expertise (Orando & Isabirye, 2018; Windapo, 2016). All these serve as obstacles to firms' human resource development of employees in these underdeveloped and developing regions.

On the other hand, construction workers in the developed world and also China, are more educated and more skilled than ever due to the opportunities for growth they are presented with (Azeez et al., 2019). Therefore, they have higher expectations from their employers regarding supporting their professional development (Azeez et al., 2019; Johari & Jha, 2019). Hence, construction firms that fail to match the psychological expectations of their employees risk losing their most talented and ambitious employees to their rivals.

Globally, the construction industry continues to remain a hypercompetitive and highly dynamic arena. Firms that operate in this context, need to be proactive in managing the career development of their employees and ensuring that the individual's needs are aligned with those of the organisation.

Rewards / compensation

Every industry recognises that remuneration or rewards are an essential issue and a motivating factor for employees to perform well, contribute significantly to the firm's development and expansion, and help the firm achieve its objectives. Construction is one of the industries that require a robust compensation structure (Al-Kasasbeh et al., 2021; Salisu et al., 2015). This is because the construction employs a variety of workers, including engineers, architects, electricians, technicians, supervisors, and labourers, who are often required to work on holidays and weekends as construction projects have a specific deadline and must be completed according to client requirements despite any difficulties the operating environment and the weather might pose. In addition, construction employees are subject to accidents, injuries, and various other dangers; therefore, the reward system in this industry should be carefully considered.

In the construction industry, the components of the reward system are monetary and non-monetary (Druker, 2013). However, the reward system in the construction industry is fraught with several challenges. Some workers are dissatisfied because they believe they are not adequately compensated, while others do not find satisfaction in the work itself (Druker, 2013). A typical example is in Asia and some of the countries in the Middle East, where despite a persistent lack of construction workers, the increasing number of migrants competing for jobs in those countries is driving down compensation and reward schemes, leading to a considerable degradation in working conditions (Ling et al., 2013).

In Africa, delayed wage payment is a big issue for construction workers (Oke et al., 2017). The problem stems from the failure of the construction industry's payment system to adapt to significant changes in the employment of workers. In the past few decades, job relationships have been reconfigured due to the changing nature of employment, with labour externalised through subcontractors and temporary employment agencies. Principal contractors are growing more apart from construction workers, through their increasing reliance on extensive subcontracting chains. The key employers of the workforce are currently small and marginalised businesses with inadequate resources and financial capacity to pay workers' wages before receiving payment for their services from the main contractor, despite finishing their job and having already incurred expenses (Roy & Koehn, 2022).

Governments in Africa have been some of the most common causes of challenges like these. It is common to witness uprisings and strike actions from construction workers in many African countries over the issue of unpaid salaries and bonuses for construction work (Gashahun, 2020; Kamanga & Steyn, 2013). Employers have also had moments where they have been plunged into various financial difficulties and debts over the failure of the governments that engaged them for their services to pay them on time for the work they have done at the agreed points periodically, and even after the execution of the projects. In the first quarter of 2022 in Nigeria, construction employees from Qumecs Nigeria Limited, a foreign construction firm, who were the original contractors of a ring road project in Uyo, Akwa Ibom state, took to the streets, protesting the government's supposed debt to the firm (Punch News, 2022). The state government terminated the contract with Qumecs and awarded it to an indigenous construction firm, leaving a balance of approximately N572m to be paid to the original construction company. The purported incapacity of the state government to pay the amount resulted in unimaginable suffering for the workers, as the company's inability to pay workers was the result of the government's failure to pay the company what it owed. Over the years, construction workers, especially in other sub-Saharan African countries such as Ghana, Cameroon, and South Africa, have faced compensation challenges (Bowen et al., 2012; Gashahun, 2020).

In the European Union, Brexit will have significant repercussions for future construction industry employees, as the United Kingdom's withdrawal from the European Union will limit free movement and restrict migration through a point-based system based on salary and skill level (Brooks et al., 2020; Mohamed et al., 2017). The United Kingdom has one of the most robust construction industries in the world, employing over 150,000 from EU countries (Akhtar, 2021). The adoption of the Tier 2 system, which restricts the employment of migrants to 'highly skilled' or 'medium-skilled' individuals, will limit the mobility of workers from Eastern Europe who previously benefited from free movement (Akhtar, 2021).

Work-life balance

The ramifications of work-life balance in the construction industry have significantly propelled the rising interest in the concept in the industry (Tijani et al., 2020). In recent decades, poor work-life balance has been a pervasive and dangerous threat to the global construction industry. Studies in Australia, the United Kingdom, and other developed countries in the European Union have confirmed that the construction industry is bedevilled by acute poor work-life balance (Sang & Powell, 2013; Tijani et al., 2020). From empirical evidence, some of the causes of the phenomenon are common even in developing countries such as Nigeria, Kenya, and some parts of India.

The industry's competitive nature compels construction firms to operate with a slim profit margin, and adhere to stringent time limitations, and prompts them to reduce labour costs (Sunindijo & Zou, 2015). As a result, construction workers work longer hours than required by contract (Holden & Sunindijo, 2018). Other work-life balance challenges can be traced to rigid working hours and complicated tasks, which result from the features of construction projects (Holden & Sunindijo, 2018; Tijani et al., 2020). Zheng and Wu (2018) confirmed these challenges in their study on work-family conflict, perceived organisational support, and professional commitment among Chinese project professionals. Results from the study showed several dimensions of work-life balance and work-family conflict among construction workers. It was found that perceived organisational support from specific construction projects could mediate between work-family conflict and professional commitment.

The challenges of work-life balance being prevalent in the construction industry have raised much concern. Researchers have found some economic, social, and psychological problems linked to poor work-life balance in the construction industry (Holden & Sunindijo, 2018; Kotera et al., 2020; Tijani et al., 2020). In the United Kingdom and Australia, mental disorders, substance abuse, alcohol misuse, and marital dissatisfaction have all been stated as long-term effects that employees in the construction industry deal with (Tijani et al., 2020). In America, economic challenges such as a high turnover intention and poor organisational productivity have arisen due to an imbalance between work and life in the construction industry (Tijani et al., 2020).

Globally, increasing concerns about work-life balance in the construction industry have led to a proliferation of research aimed at identifying the precise origins and effects of this problem to find solutions. However, the challenge posed by workforce diversity and disparities in job characteristics among industry employees makes it difficult to identify the specific work-life balance issues that must be addressed (Lingard et al., 2015). A greater understanding of work-life balance among construction workgroups could inform the creation of policies suited to the needs of specific construction workgroups.

Occupational health, safety, and well being

Few methods of HRM can be as crucial as managing employees' health, safety, and well-being. It remains one of the most important tasks and obligations an HR manager must critically pay attention to due to it being a constant worry in the construction industry. Working in dangerous sites is occasionally necessary, and a robust training process that emphasises safety is essential to keeping everyone on the job safe. However, while protecting employees is not just the responsibility of an organisation, it can also help avoid financial challenges that firms that fail to adopt essential safety procedures face. Construction firms which neglect the occupational health, safety, and well-being of their workers may be exposed to high direct and indirect costs that arise from medical expenses, rising workers' compensation insurance premiums, lost time and associated cost while the site may be closed for the accident to be investigated, lost productivity while an injured worker recovers, the cost of finding and training substitute workers, the possibility of lawsuits from injured parties, and reputational damage to these construction companies.

Despite advancements in occupational health, safety, and employee well-being regulation, research, and management practices, construction remains one of the most hazardous industries to work in (Antwi-Afari et al., 2019; Lingard, 2013; Reese, 2018). Furthermore, in many countries, it is one of the few industries where occupational health and safety performance does not improve annually (Antwi-Afari et al., 2019). Although each country has its occupational health and safety legislation and enforcement procedures, on-site safety management must be tailored to each project's unique hazards. Also, the strategies that the management of construction companies adopt must be tailored to the unique combination of individual employees working on the project (Aminbakhsh et al., 2013).

Migrant workers are among the most vulnerable employees in the industrialised world, such as the United States, when it comes to poor occupational health, safety, and well-being (Hargreaves et al., 2019). They frequently perform tasks that are considered unclean, dangerous, difficult, degrading, and demeaning. They work for lower wages, longer hours, and in harsher conditions than the citizen workers, and they are frequently subject to human rights violations, abuse, and human trafficking (Buckley et al., 2016; Kumar, 2013). Moreover, migrant workers are more likely than their non-migrant colleagues to take considerable risks on the job and perform work without proper training and safety equipment and do so without complaining about harmful working conditions. They mostly cannot complain if they are undocumented migrants, due to not having work permits and other necessary residency requirements (Buckley et al., 2016).

The contribution of HRM to safe working conditions cannot be overemphasised. However, achieving safe working conditions is not a simple task. Health, safety, and well-being must be a top priority; it should be taken seriously and included in all management-system procedures to provide a unified approach. The challenge is that fragmentation and high levels of self-employment in the construction industry across the world necessitate workers to assume responsibility for their health, safety, and well-being.

Moreover, the construction industry's financially driven, male-dominated, and physically exertive milieu is unlikely to foster a safety-conscious mentality among the personnel (Clarke, 2014). On many construction sites, most employers and employees have readily embraced the idea of risk-taking and accidents occurring as part of their job. Attitudes such as these can only be minimised by the HR department when employers demonstrate that unnecessary risk-taking is undesirable and that safe working conditions are a non-negotiable requirement in the employment of workers in the construction industry. Also, the prioritisation of high safety standards and safety issues must be driven by senior managers, HR departments, effective systems, and continuous training and education programmes to change the attitudes and behaviours of the workers.

Recruitment

Recruitment is one of the HRM practices which can help the construction industry to compete with other industries for the limited number of willing and able workers. Historically, this has not been a challenge for the construction industry. However, it has become a key issue over the past few years. A protracted slowdown in population growth in many industrialised nations and changes in gender demographics affecting conventional recruitment into industries have made this market more competitive and increased the likelihood of skill shortages in the construction industry in the near future (Ho, 2016). In previous years, the proportion of men in the labour force of the industry was extremely high compared to that of women. This issue was prevalent in the developing world, leading to calls for gender balance in the construction industry (Adogbo et al., 2015; Hasan et al., 2021; Lewis & Shan, 2020). Although the gender balance is a little better than in previous years, the association of the industry with a male-dominated culture keeps serving as a stereotype-based barrier to the recruitment of women, particularly in the developing countries where women are not linked to highly intense environments like the construction industry (Adogbo et al., 2015; Hasan et al., 2021).

The construction industry employs over two million individuals in the United Kingdom, of which a little over 10 percent are females (Norberg & Johansson, 2021). This makes it a male-dominated sector among all the major sectors of the economy; this is despite the gender balance advocacy in the industry being increasingly more intense. Providing equal opportunities for all employees in the construction industry should be the foundation of excellent employment practices. Discrimination based on a person's gender, race/ethnicity, age, or disability results in the underutilisation of people's skills and talents and stifles worker diversity, innovation, and effectiveness.

Furthermore, the construction industry has gradually become less attractive for potential recruits. Knowledgeable persons are more likely to gravitate toward sectors that are perceived as having the best pay, working conditions, and opportunities for career advancement, and being the most glamorous and desirable to work in (Hom et al., 2017). Thus, the unattractiveness of construction as a career choice has been a source of concern for many groups and training organisations in the industry. Individuals are now leaning toward the technological world of robotics and data analytics compared to the more technical and physically intensive areas like construction (Lan et al., 2022).

Another reason the construction industry has a poor recruitment record is that it is often seen as one that overburdens its employees with workload (Brandt et al., 2015; Chen et al., 2016). Additionally, project- and site-based, and itinerant work patterns have often resulted in job insecurity (Ness & Green, 2013). Many construction workers often have to relocate in pursuit of new project opportunities, affecting their stability and stifling their career growth (Ness & Green, 2013). Challenges like poor on-site working conditions, health and safety record, and employee welfare provision within the industry all negatively affect employee recruitment (Lingard, 2013). Additionally, economic growth and increased construction demand are expected to result in significant shortages in traditional and emerging skills areas, given the shrinking labour market and the industry's image issues. For example, the Construction Skills Network (CSN) estimates that between 2023 and 2027, the UK construction sector will need an extra 224,900 workers, or over 45,000 per year (CSN, 2022).

Employee relations and conflict management

Employee relations in the construction industry involve developing and negotiating the conditions and goals of the employment relationship between employees and their employers. This process is crucial for workers in an industry known for its unethical and risky business practices. Historically, trade unions negotiated on behalf of their members through collective bargaining agreements (Conley, 2014). However, recent legislative reforms have offered greater flexibility to permit non-union collective bargaining, contract-based employment, and project-specific labour agreements (Conley, 2014). In many countries, collective bargaining agreements exist and are typically established voluntarily. Employer groups, which represent construction companies, and labour unions, which represent the interests of their members, are involved. These are high-level agreements on general working norms and pay rates, which are then suggested as national pay and conditions frameworks. In Ghana, one of the labour unions that work hard to ensure that ethical principles are applied in the construction industry for the benefit of both the employee and employer is the Construction and Building Materials Workers' Union (CBMWU) (Ghana

Trade and Union Congress, 2020). The Building, Construction and Allied Workers Union (BCAWU) plays a similar role for construction workers in South Africa (Hlatshwayo, 2018). In Australia, collective agreements in the construction industry are established by the Australian Industrial Relations Commission (AIRC) (Bray & Stewart, 2013).

Nevertheless, despite the presence of local labour unions in nearly every country, many vulnerable employees are exploited by employers in the pursuit of a quick income (Kern & Müller-Böker, 2015). In most cases, these individuals are unaware of their rights and others might be too intimidated to file a formal complaint with the unions that can pursue the case on their behalf. Also, the rapid growth of illegal immigrants in most developed nations and their focus on the construction industry as a place where they may find work without too many inquiries has contributed to poor relations between employees and employers (Abdul-Rahman et al., 2012; Bailey, 2021). This is especially evident in Asia, where labourers are treated inhumanely while undertaking their responsibilities (Bélanger, 2014; Kelly, 2013). Countries such as Kuwait, Qatar, Saudi Arabia, and the United Arab Emirates are often highlighted as being places where these shortcomings are evident. Most of the victims have been Africans and other Asians from developing nations seeking better employment. Since 2010, when Qatar was granted the opportunity to host the 2022 FIFA World Cup, human rights groups have drawn attention to the country's treatment of migrant workers (Beatrix, 2021). The preparations for the competition necessitated the expansion of the country's infrastructure. This resulted in a massive demand for human resources in the construction industry. However, in 2016, Amnesty International and major international labour organisations and unions such as the International Labour Organization alleged that Qatari businesses employed forced labour (BBC News, 2022). It was reported that numerous workers lived in deplorable conditions, were forced to pay exorbitant recruitment costs, had their income withheld, and had their passports confiscated. Other reports also alleged that since 2010, over 6000 migrants from other Asian countries like India, Nepal, Pakistan, Bangladesh, and Sri Lanka had died in various work sites in Qatar (Dorsey, 2014; Millward, 2017).

Moreover, it is reasonable to assume that in a sector, such as the UK construction industry, with a high proportion of self-employment and subcontracting, the formal importance of actively managing employee relations and, hence, HR departments, would be relatively low (Aronson, 2019). Nevertheless, the construction industry has a long history of collective bargaining institutions and industry-wide agreements, which continue to influence the character of employment relationships (Conley, 2014). Despite the diverse workforce, reports of employer and employee friction, and the complex nature of employee relations in the industry, labour unions continue to be powerful voices that need to be engaged with when advocating for proper HRM practices regarding employee relations and conflict management.

Performance management

A construction project is successful when it is completed on time, within budget, and meets performance objectives. There is a direct relationship between performance objectives and productivity, time management, and decision-making (Aeon & Aguinis, 2017). To achieve these objectives and execute construction projects in the most effective and structured manner possible, organisations must establish and retain strong teams, track and measure success, and guarantee the project's seamless operation. This is where the performance management of employees in the industry becomes crucial.

Performance management creates an atmosphere where employees and subcontractors can perform to the best of their abilities. It is how managers explain their team's requirements, measure achievement, provide feedback, and identify obstacles preventing their project team from performing its tasks (Aguinis, 2019). Performance management is essential in the construction industry because it enables managers to identify whether or not employees contribute to project success and the overall business plan. Through performance management, managers can immediately identify and fix problems and keep staff motivated, engaged, and on schedule, thereby enhancing productivity and profitability. For years, construction organisations have utilised the same performance management strategies (Aziz & Hafez, 2013; McGeorge & Zou, 2012). Some of these strategies are effective and should be retained in rotation, while others might need to be modified or eliminated. There is also room for improvement, no matter what the current record is, given the possible technological improvements which are available, changes in the construction industry, and societal shifts.

One of the challenges facing performance management in the construction industry is establishing a solid and cohesive team to perform the necessary duties on the project. Studies have proven over the years that construction firms are most successful when they have a solid and cohesive team of people working towards the same objective (Abylova & Salykova, 2019; Alias et al., 2014; Tripathi & Jha, 2019). The ideal team is comprised of individuals with complementary sets of skills. They should be motivated, determined, and devoted, capable of handling and solving complicated challenges with a level head. A successful team should also consist of people who have industry experience or are eager to learn. Creating a cohesive team can feel like a daunting task on a construction project. This is due to the highly diverse nature of job specifications and workforce in a construction project. Developing a team where employees' diversities in work and personality complement rather than contradict can be very challenging.

Another obstacle is the lack of clearly specified and attainable targets and key performance indicators (KPIs) for construction industry employees (Mesároš et al., 2021; Sibiya et al., 2015). This has the potential to create confusion among personnel and their employers as no one will be aware of the project's defined objectives and the primary measures that will be used to gauge performance. In addition, employers in the construction industry are adopting obsolete and ineffective performance management techniques such as profitability ratios, a narrow range of KPIs, balanced scorecard, and the European Foundation for Quality Management (EFQM) model, in today's rapidly evolving business environment (Holt et al., 2015; Muhsin & Nory, 2022). If a construction business does not apply the proper appraisal and onboarding tools, it risks having an unstructured and unmotivated workforce. This is a critical issue that can impede a project's success.

Feedback is one of the essential components of performance management in the construction industry. A worker who does not receive feedback will be unaware of whether or not their performance meets expectations. Employees' ability to achieve their utmost potential can be hindered by concerns that can be discussed openly with feedback. It also permits the problems of workers to be heard, acknowledged, and addressed, increasing overall job satisfaction and performance and promoting zeal, productivity, and expansion. Without employee feedback, organisations face a huge chance of employee disengagement. The most frequent problems with job feedback include: the input is inconsistent, infrequent, or not timely, not clear nor constructive, and unidirectional (Matthews, 2018). Due to the project-orientated nature of construction, continuous feedback can be challenging to implement. Some employees will work on various different projects during a specific period, while others will collaborate on a long-term project. This indicates that each employee will have unique strategic goals, objectives, and deadlines. Construction firms should avoid enforcing a strict performance evaluation procedure. The flow of performance management must correspond with how work is performed and should fit the speed and tempo of the construction project.

How has HRM evolved in response to these issues?

In the construction industry, everyday operations involve thousands of moving parts. However, none are more important than the humans who work on its projects. Whether dealing with a handful of employees or managing hundreds of people transitioning between jobs, human resource managers must always guarantee that everyone is educated, trained, appropriately compensated, and has their individual needs met. Without a strong and engaged staff, businesses tend to falter. In the aftermath of the COVID-19 pandemic, this is particularly true in the construction industry, where human resources professionals must manage not just the complicated projects of today but also a shifting workforce and employee management environment. Throughout the narrative in this chapter, many HR principles in the construction industry and their underlying issues are examined. Clearly, there are several HR concerns, but the crucial questions are as follows: What responsibilities do HR departments play in improving the performance of construction industry project teams? What steps have HR departments taken over the years to intervene and address the most pressing issues affecting the construction workforce?

Human resource development in the construction industry has always been regarded as one of the most significant HR challenges (Wilton, 2022). Human resource development in the construction business has often included staff training and growth. Employees in the construction industry demand continuous training and development (Fugar et al., 2013). In the traditional model of on-the-job training at construction sites, workers would receive pre-prepared courses on defined regulations, procedures, or processes, often at a location other than their place of employment, and be expected to apply this abstract knowledge later at their place of employment (Tabassi et al., 2011). Throughout the years, HR managers in construction companies have used a variety of strategies for on-the-job training. Most often, HR departments and training companies dispatch trainers to the construction site to provide on-the-job training to construction personnel (Mathis et al., 2016). When construction companies employ a significant number of skilled bricklayers, carpenters, plumbers, steel workers, welders, and so on, they may use apprenticeship training, a unique kind of on-the-job training (Tabassi et al., 2011). This training is usually conducted in accordance with standards that have been set and are mostly regulated.

With the emergence of new technology, HR managers in the construction industries of the majority of developed nations are assisting employees to learn how to utilise these technologies to stay competitive (Harris et al., 2021; Vrontis et al., 2022). Construction companies are using HR technology and current HR software designed specifically for the construction industry to prepare, train and develop their staff for the implementation of modern ways of going about construction work (Vrontis et al., 2022). Some of this software not only aid in storing documentation, but also track upcoming training and certification needs, eliminate guesswork regarding deadlines, and transform what could be a jumbled mess of paperwork into an easily navigable digital library, complete with notifications to alert employees and employers of upcoming certification renewal as well as other essential information related to their work (Dachner et al., 2021). Modern software being pushed out by HRM departments enable contemporary construction workers to be provided with new training courses, educational materials, and other paperwork required to expedite their learning of the new skills, practices and ideas (Kurek, 2021).

The problem of recruitment in the construction industry is another HR principle that modern HRM is helping to solve in recent years. Successful construction companies' HR managers continue to offer avenues to recruit new employees, and workers to learn new skills, raise new leaders and managers, and get extensive industry experience (Banfield et al., 2018; Stewart & Brown, 2019). Construction firms that appeal to women and younger people can attract a more diverse and highly-talented workforce (Hasan et al., 2021). To attract these individuals, cutting-edge technology is required. Younger, tech-savvy individuals seek reassurance that they can remain at the leading edge of technology as their careers advance. Top construction firms attract such tech-savvy workers by promoting the usage of drones, wearables, artificial intelligence, and tech-connected working locations (Abioye et al., 2021; Courtway, 2020). These technologically driven enticements also make it simpler to recruit individuals from different industries. HR managers in the construction industry are rapidly shifting from conventional recruiting methods and embracing technology to streamline everything from onboarding to training, and from educational opportunities to employees' daily needs (Pattanayak, 2020). They are also applying new elements of recruitment, infusing AI and data analytics into recruitment processes, establishing new efficiencies for the organisation and fostering new talents and employee buy-in (Vrontis et al., 2022). These modern ways also help in adjusting recruitment methods, schedules and responsibilities, reduce the stress that manual procedures and sometimes even email messaging, may generate.

In recent years, HR managers have been constantly admonished to assist in transforming the construction industry into a career path for workers (Torraco & Lundgren, 2020). The construction industry's high turnover rate continues to make recruiting and training new employees an expensive endeavour for employers (Nkomo & Thwala, 2016). Even though the human resources department will never be able to completely eradicate turnover, it is of the highest significance to make efforts to mitigate the problem. By placing a focus on construction as a career, employees take a new focus with them each day.

Occupational health, safety, and well-being is another essential HR value that has emerged in the construction industry over the last several decades. To guarantee a safer working environment for employees, construction companies and HR managers are using contemporary procedures to ensure compliance with legislation and safety protocols (Harris et al., 2021). As the construction industry's usage of highly specialised and technologically advanced equipment increases, such precautions have become crucial. HR specialists are implementing occupational health, safety and well-being systems for workers (Kowalski & Loretto, 2017). They include health and safety management wearables such as tech-controlled helmets and contemporary PPEs (Gaur et al., 2019; Holt et al., 2015). In developed countries, the construction industry is progressively adopting these sorts of technologies to help promote workplace health, safety, and well-being (Gaur et al., 2019). Smartwatches are being used to monitor body temperature and heart rate and alert employees if anything is found to be out of the ordinary (Choi et al., 2017; Mardonova & Choi, 2018). Some of the wearables potentially take the shape of exoskeletons that could be worn while operating heavy equipment, standing for extended periods, or safeguarding the body's posture (Choi et al., 2017).

In the construction industry, robots are also becoming increasingly widespread (Saidi et al., 2016). HR professionals are enabling the deployment and integration of robots and AI (Saidi et al., 2016; Vrontis et al., 2022). Human employees are being aided, and in some cases completely replaced, while doing dangerous or injury-causing tasks such as carrying large or bulky goods, travelling to inaccessible areas, performing repetitive tasks, conducting inspections, and many more (Saidi et al., 2016). Sites and areas are also being inspected using drones (Elghaish et al., 2021; Entrop & Vasenev, 2017). These drones are fitted with laser scanners, sensors, and cameras to undertake tasks which ensure that the work environment is safe for people and that there are no possible hazards (Entrop & Vasenev, 2017). This device reduces the likelihood of employees encountering dangerous conditions.

Health and safety management applications and software are also being used to maintain contact with data for analyzing trends, assessing the hazards of activities, and ensuring compliance with rules (Reese, 2018; Zou et al., 2017). This has made it easier for HR managers to keep up with the myriad of industry rules, health and safety practices, and policies governing the maintenance of a safe work environment. These updated systems and software may facilitate accurate tracking, efficient management of staff training, risk assessment for a specific activity, workflow optimisation, accident reporting, and inspection management, among other things (Zou et al., 2017).

Another one of the most important HR areas in which modern HRM is significantly changing in order to address it is performance management and the provision of incentives and pay to construction employees. Due to the diverse character of the construction industry's workforce, these two particular concepts continue to pose challenges. Nonetheless, modern HR technology has played a crucial role in tackling some of the most significant challenges faced by construction companies. Many HR managers have used performance management technology that facilitates the collection of input from different sources for a 360-degree system of performance measurement and management (Baig et al., 2022).

In the area of rewarding and compensating workers in the construction industry, HR managers in construction companies are moving toward deploying more complex HR technology systems that monitor each employee's hours worked, ascertain their competitive base pay, bonuses, sick leave, and vacation pay, profit sharing entitlement, company-sponsored health plans, and retirement plans (Harris et al., 2021; Kurek, 2021). By using HR technology systems, payrolls are more efficient and workers are compensated more fairly in construction businesses. Excessive expenditures that come from rewarding and compensating staff who are not working or are taking longer than required to accomplish assignments are being avoided through time monitoring using these complex technologies. Modern human performance management and reward systems are helping construction companies to operate more effectively and allowing them to concentrate on meeting their goals (Harris et al., 2021). It makes it simpler for them to manage their staff and maximise the productivity of their employees.

Can HRM assist in resolving the HR problems in the construction business, or do we need other solutions?

Over the years, HRM has played a variety of crucial roles in tackling the challenges presented in the narrative in this chapter about HR principles in the construction business. HRM in the construction business is experiencing major changes at the moment. Globally, HRM is shifting from old and traditional practices to more creative, technologically integrated, and efficient methods of operation, as has been discussed. Modern technology is simplifying human resources operations in the construction industry, where managing people is a difficult job. For their HR practices and company model to function well in the current market, HR managers are embracing new strategies. Modern HRM continues to assert that strong and successful HR managers and systems are the most efficient means and options for handling HR concerns in the construction industry. Top HR construction and management software like Procore Technologies, OpenSpace, Dusty Robotics, Built Robotics, and BuildStream, among many others, are currently leading the way and proving how modernised forms of HRM can truly propel construction firms into making the most impact and being highly effective in the construction industry. While these new HRM procedures are more widespread in developed nations, emerging nations throughout the globe are progressively recognising the necessity for this HR revolution (Singh et al., 2022).

Conclusion

The research aimed to clarify our knowledge of HRM and explain why it is challenging to adopt basic ideas and strategies that have worked in other sectors of the construction industry. Several crucial HRM strategies were examined to pinpoint obstacles that make using these strategies in the construction industry challenging. Some of the strategies covered were in the areas of recruitment; employee relations and conflict management; performance management; rewards and compensation; work-life balance; and occupational health, safety and well-being. It was observed that some of the main barriers in human resource development were the high running costs for specific construction equipment and people undertaking highly specialised training in various elements of construction. Also, it was found that the rewards system is plagued with difficulties, such as workers who are dissatisfied with their jobs because they feel they are not being paid enough, while others in developing nations also struggle with late payments. Poor work-life balance in the construction industry is also associated with an array of economic, social, and psychological issues such as mental illnesses, substance abuse, alcoholism, and excessive labour turnover.

Additional significant issues were found in the area of occupational health, safety, and well-being, where construction is still one of the most dangerous job sectors. Migrant workers and workers in developing countries faced the highest forms of risks due to the poor working conditions they are often subjected to. On the issue of recruitment in the industry, poor on-site working conditions, safety records, the provision of employee welfare, and the perception of the industry as being dominated by males all have a detrimental impact on staff recruitment. The construction industry has had a long history of collective bargaining institutions and industry-wide agreements that continue to shape the nature of employment relationships in terms of employee relations and conflict management. Although most nations have collective bargaining and other employment laws and mechanisms for the construction industry, most workers, especially migrant workers, end up being abused by unscrupulous employers because the workers are ignorant of their rights. Finally, difficulties in creating a strong and cohesive team to carry out tasks and the absence of clearly defined and reachable objectives and KPIs for employees in the construction industry continues to be some of the most urgent performance management difficulties that the industry faces today.

The research was concluded by examining how contemporary HRM has changed in response to some of the problems mentioned in the HRM methods studied. In implementing the HRM techniques described in the study, it was found that HRM is moving away from antiquated and traditional practices and toward more innovative, technologically advanced, and effective methods of operation.

Contributions to public policy and future research directions

The analysis has one important practical implication. There is still space for improvement, especially in emerging nations, despite the recent progress in the inclusion of technology in administering these HRM approaches. In emerging nations, it is important to promote a culture and environment that encourages technical advancement. To assist in building practical and useful HRM solutions for construction businesses, it is necessary to increase the capacity and competencies of construction firms. Construction companies in developing nations may be able to grow through several technological phases and catch up with the developed nations as a result of technological breakthroughs.

Future research could proceed along the following lines. First, studies of HRM practices in construction businesses in developing nations are required to shed more light on the sector-based and firm-level issues that appear to hinder the development of contemporary HRM technologies in the construction industry in developing nations. In addition, it is important to look at how contemporary HRM technologies may impact. Ideally, this study will encourage more research into HRM and growth in the construction industry in both developed and developing countries.

References

Abdul-Rahman, H., Wang, C., Wood, L. C., & Low, S. F. (2012). Negative impact induced by foreign workers: Evidence in Malaysian construction sector. *Habitat International*, *36*(4), 433–443.

Abioye, S. O., Oyedele, L. O., Akanbi, L., Ajayi, A., Delgado, J. M. D., Bilal, M., Akinade, O. O., & Ahmed, A. (2021). Artificial intelligence in the construction industry: A review of present status, opportunities and future challenges. *Journal of Building Engineering*, *44*, 103299.

Abylova, V., & Salykova, L. (2019). Critical success factors in project management: A comprehensive review1, 2. *PM World Journal*: https://Pmworldlibrary.Net/Wp-Content/Uploads/2019/06/Pmwj82-Jun2019-Salykova-Abylova-Critical-Success-Factors-in-Project-Management2. Pdf.

Adinyira, E., Manu, P., Agyekum, K., Mahamadu, A.-M., & Olomolaiye, P. O. (2020). Violent behaviour on construction sites: Structural equation modelling of its impact on unsafe behaviour using partial least squares. *Engineering, Construction and Architectural Management*, *27*(10), 3363–3393.

Adogbo, K. J., Ibrahim, A. D., & Ibrahim, Y. M. (2015). Development of a framework for attracting and retaining women in construction practice. *Journal of Construction in Developing Countries*, *20*(1), 99.

Aeon, B., & Aguinis, H. (2017). It's about time: New perspectives and insights on time management. *Academy of Management Perspectives*, *31*(4), 309–330.

Aghimien, L. M., Aigbavboa, C. O., Anumba, C. J., & Thwala, W. D. (2021). A confirmatory factor analysis of the challenges of effective management of construction workforce in South Africa. *Journal of Engineering, Design and Technology*, *ahead-of-print*.

Aguinis, H. (2019). *Performance management for dummies*. John Wiley & Sons.

Akadiri, P. O., Chinyio, E. A., & Olomolaiye, P. O. (2012). Design of a sustainable building: A conceptual framework for implementing sustainability in the building sector. *Buildings*, *2*(2), 126–152.

Akhtar, Z. (2021). Eastern European workers: Exploitation in the construction industry and enforcement by regulatory agencies. *Construction Law International*, *16*(2), 42–50.

Al-Kasasbeh, M., Abudayyeh, O., Olimat, H., Liu, H., Al Mamlook, R., & Alfoul, B. A. (2021). A robust construction safety performance evaluation framework for workers' compensation insurance: A proposed alternative to EMR. *Buildings*, *11*(10), 434.

Alias, Z., Zawawi, E. M. A., Yusof, K., & Aris, N. M. (2014). Determining critical success factors of project management practice: A conceptual framework. *Procedia-Social and Behavioral Sciences*, *153*, 61–69.

Alzyoud, A. A. Y., Ogalo, H. S., & ACDMHR. (2020). Strategic Management of Health and Safety at Work: Critical Insights for HR Professionals in the Construction Sector. *Annals of Contemporary Developments in Management & HR (ACDMHR), Print ISSN*, 2632–7686.

Aminbakhsh, S., Gunduz, M., & Sonmez, R. (2013). Safety risk assessment using analytic hierarchy process (AHP) during planning and budgeting of construction projects. *Journal of Safety Research*, *46*, 99–105.

Antwi-Afari, M. F., Li, H., Wong, J. K.-W., Oladinrin, O. T., Ge, J. X., Seo, J., & Wong, A. Y. L. (2019). Sensing and warning-based technology applications to improve occupational health and safety in the construction industry: A literature review. *Engineering, Construction and Architectural Management, 26*(8), 1534–1552.

Aronson, R. L. (2019). *Self employment: A labor market perspective*. Cornell University Press.

Assaad, R., & El-adaway, I. H. (2021). Impact of dynamic workforce and workplace variables on the productivity of the construction industry: New gross construction productivity indicator. *Journal of Management in Engineering*, *37*(1), 4020092.

Associated General Contractors of America (AGC). (2022). *Construction Data*. Retrieved on August 26 from https://www.agc.org/learn/construction-data

Azeez, M., Gambatese, J., & Hernandez, S. (2019). What do construction workers really want? A study about representation, importance, and perception of US construction occupational rewards. *Journal of Construction Engineering and Management*, *145*(7), 4019040.

Aziz, R. F., & Hafez, S. M. (2013). Applying lean thinking in construction and performance improvement. *Alexandria Engineering Journal*, *52*(4), 679–695.

Baig, U., Khan, A. A., Abbas, M. G., Shaikh, Z. A., Mikhaylov, A., Laghari, A. A., & Hussain, B. M. (2022). Crucial causes of delay in completion and performance management of the construction work: Study on the base of relative importance index. *Journal of Tianjin University Science and Technology*, *55*(6), 75–102.

Bailey, T. R. (2021). *Immigrant and native workers: Contrasts and competition*. Routledge.

Ball, M. (2014). *Rebuilding construction (Routledge revivals): Economic change in the British construction industry*. Routledge.

Banfield, P., Kay, R., & Royles, D. (2018). *Introduction to human resource management*. Oxford University Press.

Beatrix, I. (2021). The 2022 FIFA World Cup in Qatar: Turning the spotlight on workers' rights. *European Parliamentary Research Service,* PE 698.856.

Bélanger, D. (2014). Labor migration and trafficking among Vietnamese migrants in Asia. *The Annals of the American Academy of Political and Social Science*, *653*(1), 87–106.

Boadu, E. F., Wang, C. C., & Sunindijo, R. Y. (2020). Characteristics of the construction industry in developing countries and its implications for health and safety: An exploratory study in Ghana. *International Journal of Environmental Research and Public Health*, *17*(11), 4110.

Boselie, P. (2014). *EBOOK: Strategic human resource management: A balanced approach*. McGraw Hill.

Bowen, P. A., Edwards, P. J., & Cattell, K. (2012). Corruption in the South African construction industry: A thematic analysis of verbatim comments from survey participants. *Construction Management and Economics*, *30*(10), 885–901.

Brandt, M., Madeleine, P., Ajslev, J. Z. N., Jakobsen, M. D., Samani, A., Sundstrup, E., Kines, P., & Andersen, L. L. (2015). Participatory intervention with objectively measured physical risk factors for musculoskeletal disorders in the construction industry: Study protocol for a cluster randomized controlled trial. *BMC Musculoskeletal Disorders*, *16*(1), 1–9.

Bray, M., & Stewart, A. (2013). From the arbitration system to the 'fair work act': The changing approach in Australia to voice and representation at work. *The Adelaide Law Review, 34*(1), 21–41.

British Broadcasting Corporation (BBC). (2016). *Qatar 2022: 'Forced labour' at World Cup stadium*. Retrieved on August 26 from https://www.bbc.com/news/world-middle-east-35931031

Brooks, T., Scott, L., Spillane, J. P., & Hayward, K. (2020). Irish construction cross border trade and Brexit: Practitioner perceptions on the periphery of Europe. *Construction Management and Economics*, *38*(1), 71–90.

Buckley, M., Zendel, A., Biggar, J., Frederiksen, L., & Wells, J. (2016). Migrant work & employment in the construction sector. *International Labour Organization*, Geneva.

Chen, J., Song, X., & Lin, Z. (2016). Revealing the "Invisible Gorilla" in construction: Estimating construction safety through mental workload assessment. *Automation in Construction*, *63*, 173–183.

Choi, B., Hwang, S., & Lee, S. (2017). What drives construction workers' acceptance of wearable technologies in the workplace?: Indoor localization and wearable health devices for occupational safety and health. *Automation in Construction*, *84*, 31–41.

Clarke, J. (2014). Consultation doesn't happen by accident. *Management*, *22*(3), 316–331.

Conley, H. (2014). Trade unions, equal pay and the law in the UK. *Economic and Industrial Democracy*, *35*(2), 309–323.

Construction Industry Training Board (CITB). (2023). *Construction Skills Network Industry Outlook - 2023-2027*. Retrieved on August 26 from https://www.citb.co.uk/about-citb/construction-industry-research-reports/construction-skills-network-csn/

Courtway, J. E. (2020). Wearables, augmented and virtual reality, integrated project delivery, and artificial intelligence. *CONSTR. LAW*, *40*, 25.

Creighton, B. (2014). Why can't "they" leave our labour laws alone?'. *Phillipa Weeks Lecture, ANU*, 15 October 2014.

Dachner, A. M., Ellingson, J. E., Noe, R. A., & Saxton, B. M. (2021). The future of employee development. *Human Resource Management Review*, *31*(2), 100732.

Debnath, J., Biswas, A., Sivan, P., Sen, K. N., & Sahu, S. (2016). Fuzzy inference model for assessing occupational risks in construction sites. *International Journal of Industrial Ergonomics*, *55*, 114–128.

Dorsey, J. M. (2014). The 2022 World Cup: A potential monkey wrench for change. *The International Journal of the History of Sport*, *31*(14), 1739–1754.

Druker, J. (2013). Reward management in construction. In *Human Resource Management in Construction* (pp. 276–304). Routledge.

Elghaish, F., Matarneh, S., Talebi, S., Kagioglou, M., Hosseini, M. R., & Abrishami, S. (2021). Toward digitalization in the construction industry with immersive and drones technologies: A critical literature review. *Smart and Sustainable Built Environment*, *10*(3), 345–363.

Entrop, A. G., & Vasenev, A. (2017). Infrared drones in the construction industry: Designing a protocol for building thermography procedures. *Energy Procedia*, *132*, 63–68.

Fugar, F. D. K., Ashiboe-Mensah, N. A., & Adinyira, E. (2013). Human capital theory: Implications for the Ghanaian construction industry development. *Journal of Construction Project Management and Innovation*, *3*(1), 464–481.

Gashahun, A. D. (2020). Causes and effects of delay on African construction projects: A state of the art review. *Civil and Environmental Research*, *12*, 41–53.

Gaur, B., Shukla, V. K., & Verma, A. (2019). Strengthening people analytics through wearable IOT device for real-time data collection. *2019 International Conference on Automation, Computational and Technology Management (ICACTM)*, 555–560.

Ghana Trade Union Congress (2020). Trade Unions and Industrial Relations in Ghana. *Labour Relations Manuals.*

Hargreaves, S., Rustage, K., Nellums, L. B., McAlpine, A., Pocock, N., Devakumar, D., Aldridge, R. W., Abubakar, I., Kristensen, K. L., & Himmels, J. W. (2019). Occupational health outcomes among international migrant workers: A systematic review and meta-analysis. *The Lancet Global Health*, *7*(7), e872–e882.

Harris, F., McCaffer, R., Baldwin, A., & Edum-Fotwe, F. (2021). *Modern construction management*. John Wiley & Sons.

Hasan, A., Ghosh, A., Mahmood, M. N., & Thaheem, M. J. (2021). Scientometric review of the twenty-first century research on women in construction. *Journal of Management in Engineering*, *37*(3), 4021004.

Haupt, T., & Harinarain, N. (2016). The image of the construction industry and its employment attractiveness. *Acta Structilia*, *23*(2), 79–108.

Hildebrandt, J., Hagemann, N., & Thrän, D. (2017). The contribution of wood-based construction materials for leveraging a low carbon building sector in Europe. *Sustainable Cities and Society*, *34*, 405–418.

Hlatshwayo, M. (2018). Building workers' education in the context of the struggle against racial capitalism: The role of labour support organisations. *Education as Change*, *22*(2), 1–24.

Ho, P. H. K. (2016). Labour and skill shortages in Hong Kong's construction industry. *Engineering, Construction and Architectural Management*, *23*(4), 533–550.

Holden, S., & Sunindijo, R. Y. (2018). Technology, long work hours, and stress worsen work-life balance in the construction industry. *International Journal of Integrated Engineering*, *10*(2), 13–18.

Holt, E. A., Benham, J. M., & Bigelow, B. F. (2015). Emerging technology in the construction industry: Perceptions from construction industry professionals. *2015 ASEE Annual Conference & Exposition*, 26–595.

Hom, P. W., Lee, T. W., Shaw, J. D., & Hausknecht, J. P. (2017). One hundred years of employee turnover theory and research. *Journal of Applied Psychology*, *102*(3), 530.

Hossain, M. A., Zhumabekova, A., Paul, S. C., & Kim, J. R. (2020). A review of 3D printing in construction and its impact on the labor market. *Sustainability*, *12*(20), 8492.

Janani, R., Kalyana, P., & Yazhini, S. (2017). Investigation and control of major risks on construction sites. *Chemical Science*, *14*, 3087–3096.

Jarkas, A. M., Al Balushi, R. A., & Raveendranath, P. K. (2015). Determinants of construction labour productivity in Oman. *International Journal of Construction Management*, *15*(4), 332–344.

Jayasudha, K., & Vidivelli, B. (2016). Analysis of major risks in construction projects. *ARPN Journal of Engineering and Applied Sciences*, *11*(11), 6943–6950.

Johari, S., & Jha, K. N. (2019). Challenges of attracting construction workers to skill development and training programmes. *Engineering, Construction and Architectural Management*, *27*(2), 321–340.

Kamanga, M. J., & Steyn, W. J. v. d. M. (2013). Causes of delay in road construction projects in Malawi. *Journal of the South African Institution of Civil Engineering= Joernaal van Die Suid-Afrikaanse Instituut van Siviele Ingenieurswese*, *55*(3), 79–85.

Kelly, L. (2013). A conducive context: Trafficking of persons in Central Asia. In *Human trafficking* (pp. 73–91). Willan.

Kern, A., & Müller-Böker, U. (2015). The middle space of migration: A case study on brokerage and recruitment agencies in Nepal. *Geoforum*, *65*, 158–169.

Koeleman, T. (2020). *The construction industry of tomorrow implemented today: An introduction of robotic machines on the construction site for parametric city densification*. Student theses, Delft University of Technology.

Kotera, Y., Green, P., & Sheffield, D. (2020). Work-life balance of UK construction workers: Relationship with mental health. *Construction Management and Economics*, *38*(3), 291–303.

Kowalski, T. H. P., & Loretto, W. (2017). Well-being and HRM in the changing workplace. In *The International Journal of Human Resource Management* (Vol. 28, Issue 16, pp. 2229–2255). Taylor & Francis.

Kumar, M. (2013). Inimitable issues of construction workers: Case study. *British Journal of Economics, Finance and Management Sciences*, 7(2).

Kurek, D. (2021). Use of modern IT solutions in the HRM activities: Process automation and digital employer branding. *European Research Studies Journal, 24*(1), 152–170.

Lan, J., Yuan, B., & Gong, Y. (2022). Predicting the change trajectory of employee robot-phobia in the workplace: The role of perceived robot advantageousness and anthropomorphism. *Computers in Human Behavior*, *135*, 107366.

Lewis, A. K., & Shan, Y. (2020). Influencing factors on recruitment and retention of women in construction education: A literature review. *Construction Research Congress 2020: Safety, Workforce, and Education*, 717–725.

Li, H., Lu, M., Hsu, S.-C., Gray, M., & Huang, T. (2015). Proactive behavior-based safety management for construction safety improvement. *Safety Science*, *75*, 107–117.

Ling, F. Y. Y., Dulaimi, M. F., & Chua, M. (2013). Strategies for managing migrant construction workers from China, India, and the Philippines. *Journal of Professional Issues in Engineering Education and Practice*, *139*(1), 19–26.

Ling, F. Y. Y., Ning, Y., Chang, Y. H., & Zhang, Z. (2018). Human resource management practices to improve project managers' job satisfaction. *Engineering, Construction and Architectural Management*.

Lingard, H. (2013). Occupational health and safety in the construction industry. *Construction Management and Economics*, *31*(6), 505–514.

Lingard, H., Turner, M., & Charlesworth, S. (2015). Growing pains: Work-life impacts in small-to-medium sized construction firms. *Engineering, Construction and Architectural Management*, *22*(3), 312–326.

Mardonova, M., & Choi, Y. (2018). Review of wearable device technology and its applications to the mining industry. *Energies*, *11*(3), 547.
Mathis, R. L., Jackson, J. H., Valentine, S. R., & Meglich, P. (2016). *Human resource management*. Cengage Learning.
Matthews, G. (2018). Employee engagement: What's your strategy? *Strategic HR Review*, *17*(3), 150–154.
McGeorge, D., & Zou, P. X. W. (2012). *Construction management: new directions*. John Wiley & Sons.
Mesároš, P., Behúnová, A., Mandičák, T., Behún, M., & Krajníková, K. (2021). Impact of enterprise information systems on selected key performance indicators in construction project management: An empirical study. *Wireless Networks*, *27*(3), 1641–1648.
Millward, P. (2017). World Cup 2022 and Qatar's construction projects: Relational power in networks and relational responsibilities to migrant workers. *Current Sociology*, *65*(5), 756–776.
Mohamed, M., Pärn, E. A., & Edwards, D. J. (2017). Brexit: Measuring the impact upon skilled labour in the UK construction industry. *International Journal of Building Pathology and Adaptation*, *35*(3), 264–279.
Mosly, I. (2015). Safety performance in the construction industry of Saudi Arabia. *International Journal of Construction Engineering and Management*, *4*(6), 238–247.
Muhsin, I. F., & Nory, Z. K. (2022). Extended balanced scorecard approach for evaluating the performance of Iraqi construction companies. *Journal of Management Information & Decision Sciences*, *25*(S7), 1–10.
Ness, K., & Green, S. (2013). Human resource management in the construction context: Disappearing workers in the UK. In *Human Resource Management in Construction* (pp. 42–74). Routledge.
Nkomo, M. W., & Thwala, W. D. (2016). Mentoring on Retention of employees in the construction sector: A literature review. *Creative Construction Conference*, 305–310.
Norberg, C., & Johansson, M. (2021). "Women and 'ideal' women": The representation of women in the construction industry. *Gender Issues*, *38*(1), 1–24.
Oke, A. E., Ibironke, O. T., & Bayegun, O. A. (2017). Appraisal of reward packages in construction firms: A case of quantity surveying firms in Nigeria. *Journal of Engineering, Design and Technology*, *15*(6), 722–737.
Orando, M., & Isabirye, A. K. (2018). Construction workers'skill development: A strategy for improving capacity and productivity in South Africa. *International Journal of Economics and Finance Studies*, *10*(1), 66–80.
Pattanayak, B. (2020). *Human resource management*. PHI Learning Pvt. Ltd.
Punch News. (2022). *Construction workers protest company debts by Akwa Ibom govt*. Retrieved on August 26 from https://punchng.com/construction-workers-protest-company-debts-by-akwa-ibom-govt/
Reese, C. D. (2018). *Occupational health and safety management: A practical approach*. CRC press.
Rostoka, Z., Locovs, J., & Gaile-Sarkane, E. (2019). Open innovation of new emerging small economies based on university-construction industry cooperation. *Journal of Open Innovation: Technology, Market, and Complexity*, *5*(1), 10.
Roy, S. K., & Koehn, E. (2022). Construction labor requirements in developing countries. *2006 GSW*.
Robinson, G., Leonard, J., & Whittington, T. (2021). Future of construction. A global forecast for construction to 2030. London, UK: Oxford Economics.
Saidi, K. S., Bock, T., & Georgoulas, C. (2016). Robotics in construction. In *Springer handbook of robotics* (pp. 1493–1520). Springer.
Salisu, J. B., Chinyio, E., & Suresh, S. (2015). The impact of compensation on the job satisfaction of public sector construction workers of Jigawa state of Nigeria. *The Business & Management Review*, *6*(4), 282.
Sang, K., & Powell, A. (2013). Equality, diversity, inclusion and work-life balance in construction. In *Human resource management in construction* (pp. 187–220). Routledge.
Shamsuddin, K. A., Ani, M. N. C., Ismail, A. K., & Ibrahim, M. R. (2015). Investigation the safety, health and environment (SHE) protection in construction area. *International Research Journal of Engineering and Technology*, *2*(6), 624–636.
Sibiya, M., Aigbavboa, C., & Thwala, W. (2015). Construction projects' key performance indicators: a case of the South African construction industry. *Proceedings of the 2015 International Conference on Construction and Real Estate Management*, Lulea, Sweden, 11–12.

Singh, R. K., Agrawal, S., & Modgil, S. (2022). Developing human capital 4.0 in emerging economies: an industry 4.0 perspective. *International Journal of Manpower*, *43*(2), 286–309.

Sishi, K. K. (2016). *Factors influencing labour unrest: A case of construction employees residing in Quarry Road West informal settlement,* (Doctoral dissertation), University of KwaZulu-Natal.

Stewart, G. L., & Brown, K. G. (2019). *Human resource management*. John Wiley & Sons.

Stone, R. J., Cox, A., & Gavin, M. (2020). *Human resource management*. John Wiley & Sons.

Sunindijo, R. Y., & Zou, P. X. W. (2015). *Strategic safety management in construction and engineering*. John Wiley & Sons.

Tabassi, A. A., Ramli, M., & Bakar, A. H. A. (2011). Training and development of workforces in construction industry. *Ángel F. Tenorio, Prof. Dr.*, 150.

The Committee for European Construction Equipment (CECE) (2022). Internal Market, Industry, Entrepreneurship and SMEs. *European Commission*. Retrieved on August 26 from https://single-market-economy.ec.europa.eu/sectors/construction_en

Tijani, B., Osei-Kyei, R., & Feng, Y. (2020). A review of work-life balance in the construction industry. *International Journal of Construction Management*, 1–16.

Torraco, R. J., & Lundgren, H. (2020). What HRD is doing—What HRD should be doing: The case for transforming HRD. *Human Resource Development Review*, *19*(1), 39–65.

Tripathi, K. K., & Jha, K. N. (2019). An empirical study on factors leading to the success of construction organizations in India. *International Journal of Construction Management*, *19*(3), 222–239.

Vrontis, D., Christofi, M., Pereira, V., Tarba, S., Makrides, A., & Trichina, E. (2022). Artificial intelligence, robotics, advanced technologies and human resource management: a systematic review. *The International Journal of Human Resource Management*, *33*(6), 1237–1266.

Warner, M. (2020). Human resource management in China revisited: introduction. In *Human Resource Management in China Revisited* (pp. 1–18). Routledge.

Werner, J. M. (2021). *Human resource development: talent development*. Cengage Learning.

Wilton, N. (2022). An introduction to human resource management. *An Introduction to Human Resource Management*, 1–100.

Windapo, A. O. (2016). Skilled labour supply in the South African construction industry: The nexus between certification, quality of work output and shortages. *SA Journal of Human Resource Management*, *14*(1), 1–8.

Zheng, J., & Wu, G. (2018). Work-family conflict, perceived organizational support and professional commitment: A mediation mechanism for Chinese project professionals. *International Journal of Environmental Research and Public Health*, *15*(2), 344.

Zhou, Z., Irizarry, J., & Li, Q. (2013). Applying advanced technology to improve safety management in the construction industry: A literature review. *Construction Management and Economics*, *31*(6), 606–622.

Zou, Y., Kiviniemi, A., & Jones, S. W. (2017). A review of risk management through BIM and BIM-related technologies. *Safety Science*, *97*, 88–98.

10

CULTURE IN CONSTRUCTION

A driver as well as a barrier for the improvement of labour situations within construction industry

Wilco Tijhuis

University of Twente, Faculty of Engineering Technology, Department of Construction Management & Engineering, Enschede, The Netherlands

Introduction

The built environment is in continuous development because of several influences, which do lead to changes and moving of its users. Think for example about urbanisation, supported by infrastructural developments, etc. But also other (actual) influences like for example pandemics (e.g. Covid-19), conflicts (war), and disasters (climate change, food-shortages, etc.) make people move to other regions, thus (re)starting human settlements.

These developments are not new; throughout history, there were these kinds of movements of groups of people, leading to abandoning existing environments and/or (re)use of regions elsewhere. What all these developments do have in common is the need for (new/refurbished) shelters and adjacent infrastructure. These areas of work are delivered by the construction industry, making it a sector which is continuously under pressure for delivering 'value for money'. And that is not always easy, because several problems within society do influence the ability of successful delivery of their products/assets (i.e. housing, roads, offices, bridges, etc.).

This chapter focuses on the impact of differences in business-cultures, making construction processes not that easy, or at least 'challenging'. It highlights the organisation-related aspects as results of earlier partly participative research within construction processes, especially from a labour-oriented viewpoint and within an international scope. Important aspect of this research into human behaviour within construction processes was its focus on so-called 'critical incidents'. This made it an interesting theme, leading to a better understanding and insight into business-cultural (often abbreviated as 'cultural' or 'culture's') differences, influencing also nowadays construction processes and projects. It describes difficulties and solutions, trying to give an outlook for improvements to avoid critical situations within the construction process.

For this outlook, a model is used here, symbolising culture's influence on labour situations, which should lead to better working circumstances for individuals and organisations by stimulating a better understanding of each other, resulting in an improved delivery of the built environment to its owners and users.

DOI: 10.1201/9781003262671-10

Some basics of culture: Dimensions of human behaviour

Definition of culture

Speaking about 'culture' in general, Schneider has defined it as '*how we do things around here*' (Schneider, 2000). However, it is not only about *how* we do things, but it also relates to *what, why, when, whom,* etc. Nevertheless, it is related to the *behaviour* of individuals and groups. In an earlier definition by Kroeber and Klukhohn (1952) which is used in this chapter, 'culture' is set out more explicitly as follows:

> *Patterns, explicit and implicit, of and for human behaviour acquired and transmitted by symbols, constituting the distinctive achievements of human groups, including their embodiment in artefacts; the essential core of culture consists. traditional (i.e. historically derived and selected) ideas and, especially, their attached values; culture systems may, on the one hand, be considered as products of action, on the other as conditioning elements of future action.*

Especially their underpinning of the *products of action* and the *conditioning elements of future action* indicates an interesting value for industry as well as research from the viewpoint of the following aspects:

1 *The culture-theme can be analysed by focusing on the (results of) projects/organisations and the behaviour of people within it.*
2 *Such analysis results can be used to 'predict' behaviour of people within future projects/organisations.*

Within an industry setting, for example, within a construction company or project, these aspects might be used to prevent further negative outcomes of actions and/or improve results based on earlier experiences. From this viewpoint, the knowledge of 'culture' within organisations/projects is undoubtedly an important theme in labour circumstances, for example, because it can be a barrier as well as a stimulus for better performance of the individual, functioning stand-alone or within groups. Nevertheless, it should be kept in mind that, according to Hatch (1993), cultures are *dynamic* and *change* over time, usually gradually and predictably but subject to occasional perturbations. This dynamic aspect of culture makes it not only a difficult subject to analyse, but also a very interesting subject, especially because:

> *'Culture' is there where people are, being part of society, influencing and being influenced by the society.*

And considering the fact that the construction industry is still a people's business, 'business-culture' still plays a prominent role within it, especially because of the *differences* between business-cultures, leading to often unforeseen and/or unpredicted situations.

Research into differences between business-cultures

Investigations of differences between business cultures within the construction industry have existed for a long time, not in the least because construction industry as such is, on the one hand, a specific branch of industry, but on the other hand, it operates within the larger context of the industry as a whole. Earlier research projects from for example Hofstede from the early 1980s gave interesting insights on how differences between human behaviour influence organisations; this can be for example project organisations as well as companies. And although many other researchers after him focus on this area of research, for example, Cameron and Quinn, Trompenaars and Hampden-Turner, etc. (Cameron & Quinn, 1999; Trompenaars & Hampden-Turner, 1998), these approaches are mainly focusing on operationalising and detailing it further. Therefore, the 'basic Hofstede-approach' has still stayed one of the most widely used and cited results within the international research and business society. As an important result, Hofstede distinguished four so-called 'dimensions' of human behaviour, i.e. (Hofstede, 1980):

- *Short---------------Power-distance---------------Long Individualism*
- *Collectivism ------------------------------ Masculinity*
- *Femininity ------------------------------*
- *Weak---------------Uncertainty-avoidance--------------------Strong*

These four dimensions do characterise in a certain way the 'drivers' for human behaviour from a labour-perspective during their daily work within organisations.

Interesting was that Hofstede did not have access to China because of the political situation in those days. Later on, during the mid-1990s, when China had become a more open society, he expanded his research into that part of the world, which lead to the following two extra dimensions (Hofstede, 2001):

- *Short-term-------------Future-orientation---------------Long-term*
- *Indulgence ---Restraint*

And although many researchers have further evolved on this basic 'four-plus-two-dimensions' of human behaviour, the basics of this analysis still stay intact and are useful for analysing 'patterns' within human behaviour.

However, although it opened up the way of thinking more systematically about business-cultures by focusing on human behaviour, it also became clear that it was not possible to predict behavioural-patterns strictly, simply because human beings still do have their own will, making them also able to behave differently and thus staying quite 'fluid'.

Construction industry and culture: Analysing human behaviour

Considering the dimensions of human behaviour within organisations, it is also clear that construction industry is also being influenced by these distinguished dimensions of human behaviour. However, in the international construction research as well as business community during the mid-1990s, there hardly existed such systematically analysed human behaviour, despite the fact that many construction projects still led to human conflicts and bad results. For example, by cost-overruns, lack of quality, non-fulfilment of clients' expectations. Nevertheless, an interesting and challenging moment in history made way for starting such kind of systematic research, leading to a long-term partly participative project in comparing human behaviour in construction processes, being carried out by the author (Tijhuis, 2022): The break-down of the Berlin Wall in 1989, leading to a suddenly 'melting together' of the earlier split-up East and West parts of Germany. This led in the early 1990s to a huge demand for renovation and expansion, especially in the East German region, thus attracting many German and international individuals and companies within construction industry. In fact, a big 'melting-pot' of human behaviour, which suddenly were influencing each other within construction processes for realising projects with an urgent need for re-building the German society as one country again. A real challenging period for German and international construction industry!

In the period between 1993 and 1996, a partly participative research project was carried out by the author as part of his Ph.D. research, merely focusing on international construction processes from the viewpoint of a Dutch-German contractor and project-development company acting as a German subsidiary and using German construction processes, mainly run by Dutch employees. The research project took place during and parallel to the researcher's daily activities within construction industry. Especially the comparison between Dutch and German construction organisations and the behaviour of individuals in their daily labour situation were analysed. Due to the partly participative approach, there was open access to information and experiences which, according to Spradley, is a necessary factor when doing this type of research (Spradley, 1980).

Important aspect of this research into human behaviour within construction processes was its focus on so-called 'critical incidents'. This made it an interesting theme, leading to a better understanding and insight into business-cultural (often abbreviated as 'cultural' or 'culture's') differences, influencing also nowadays construction processes and projects. The main results have been published in several national and international publications (Tijhuis, 1996; 1999) and are still being used to develop further insight and experiences into international construction practices and its research, supporting the improved functioning of multi-cultural processes and projects in the national and international construction industry.

Research into human behaviour within international construction processes: An example

Background and structure of the considered research project

Due to the fall of the Berlin Wall in 1989, a large building market had arisen in the countries behind the former Iron Curtain. Various parties were particularly active in areas in former East Germany. Dutch enterprises were represented there too, as well as in several areas in former West Germany. Various investigations in the building sector had confirmed that it was necessary for Dutch building companies to enter into these markets to maintain a competitive position. However, entering into these markets is not always that simple, which was made very obvious by the fact that several foreign companies including Dutch ones were withdrawing. It would seem that operating on an international building market would require more than just working with and under the 'guidance' of different technical regulations or standards. The partly participative research project was therefore primarily aimed at the situation and experiences of building companies and project developers that started activities in the new German provinces with their Dutch backgrounds.

In the preliminary research, it turned out that these parties had quite some problems with the rather fragmented German building process, which they were not used to. In the Netherlands, the common building process has a relatively integrated approach. The research was therefore especially oriented on differences between the Netherlands and

Germany in the organisation of the building process, in which special attention was paid to the possible backgrounds of the differences and focusing on the (resulting) human behaviour related to them. To analyse a number of matters within several cases, partially participative research was carried out within building companies and project developers. This was combined with the analysis of various theories, such as those of Hofstede, Mintzberg, and Sanders and Neuijen (Hofstede, 1980; Mintzberg, 1979; Sanders & Neuijen, 1987). Besides that, there was participation in other international research projects, where comparisons were made between project organisations and processes within building companies in France, Germany, the Netherlands, and Finland, among others.

Basic results of the considered research project

The described research activities collectively contributed to the ascertaining of the fields of interest, in which differences and their possible backgrounds were brought to light. Mainly from in-depth interviews with for example several practitioners in the field, it appeared that the three areas *contact*, *contract*, and *conflict* were of major importance within the business processes and the communication-structure therein. These three areas, being prominent fields of attention, have been depicted in Figure 10.1, also known as the 3C-Model™.

In this model, causal relations are used against the background of the categories *culture*, *project organisation*, and *technology*, respectively, which together served as a basis for a hypothesis (Tijhuis, 1996; 2001).

Against the background of the categories involved, the prominent fields of attention were incorporated in a hypothesis:

- *If Dutch builders in Germany have a conflict with the other parties involved during the building process, it is the consequence of incorrectly formulated contracts that are a result of misconceptions when contacts were made.*

This hypothesis was then tested by means of an analysis of the various case studies. The main goal was to determine where things went wrong during the building process.

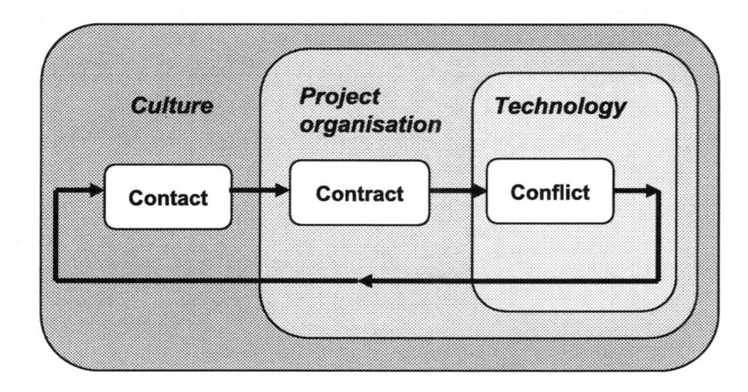

Figure 10.1 Representation of the causal relations between the three areas 'contact, contract, and conflict', being prominent fields of attention, and their major importance within the business processes and the communication-structure therein, also known as the 3C-Model™ (Tijhuis, 1996; 2001).

The results, which are portrayed as lessons for each case study, did not lean to the rejection of the hypothesis. After that, a validation of the lessons and the hypothesis was carried out.

For this purpose, use was made of a panel of experts, with experience in the Dutch and German building trade.

This validation provided the author with the result that most of the lessons can generally be declared valid. The hypothesis was therefore declared to be generally valid within the context of the research. Finally, conclusions were drawn and certain recommendations were included too. These conclusions and recommendations have been placed in a so-called NEDU-Matrix™ in the shape of keywords. The matrix gives the conclusions and recommendations for the Dutch parties that were actively involved or wished to become active on the German building market.

The primary focus was on contractors and project developers, and more especially on the project organisations they – or others – used or in which they participated.

Following the vertical axis of this matrix, there can be distinguished three prominent fields of attention, while the horizontal axis contains the three related background categories.

Figure 10.2 represents the NEDU-Matrix™ (Tijhuis, 1996), in which the role of each of the respective background categories is given per prominent field of attention, summarised by the keywords placed in the matrix fields.

NEDU Matrix	*Context* Culture	*Organisation* Project organisation	*Concequences* Technology
Contact	(a) Avoiding uncercertainty Negotiating skills	(b) Orientationon the process Orientationon getting results	(c) Regulations Customs
Contract	(d) Reliability Balance of interests	(e) Integration Accountability	(f) Reach of the contract Experts
Conflict	(g) Orientationon claims Orientationon solutions	(h) Durability of the relationship Customer orientation	(i) Professional knowledge Complexity of regulations

Figure 10.2 The NEDU-Matrix™ with conclusions and recommendations given as keywords, placed in the matrix fields (Tijhuis, 1996).

The main point that can be drawn from the conclusions and recommendations is as follows, keeping in mind that it was drawn in 1996:

- *The present national and international building trade will have to undergo a cultural revolution that is more customer friendly.*
  Having some ideas of what the customer wants and especially how a customer will assess a particular achievement (e.g. in the shape of a building or building product) plays a vital role. This especially holds when conflicts arise as a result of flaws in the product or building.
- *The prevention of such conflicts can mainly be achieved by a greater degree of trust between and reliability of all the parties involved.*

In addition, clearer communication on the one hand and proper information on the other, particularly in the shape of written professional knowledge and insight into unwritten customs, are helpful.

The reduction of the surplus regulations by the proper authorities may be one of the most important conditions for the creation of a good foundation for the realisation of these changes.

In the end, these were useful results, having several Dutch companies prevented from 'going into the wrong direction' by not choosing for internationalisation in the German market because of the fact that differences in human behaviour (i.e. business-culture) significantly play an important role in the emergence and rise of conflicts; knowing them and/or being aware of those has helped and will still help to prevent for and/or solve conflicts within construction processes.

Results of the considered research project and follow-up projects

Since 1996, when the results of the research project were presented in a Ph.D. thesis and defended, several initiatives have been taken to taken to further analyse and operationalise the results. For example, the developed 3C-Model™ has been used during the analysis of several case studies and led to interesting lessons learned for the prevention of conflicts within construction processes by considering the (differences between) business-cultures and the human behaviour therein.

And because research into 'culture' (i.e. business-culture) within organisations and/or companies can be of the most added value if carried out from within that organisation and/or company itself, this can be done by the researcher by 'being part of the action' in a so-called 'partly participative action approach'.

However, if doing so, one should always stay aware not to become too close to the action, i.e. continuously switching between analysing 'inside-out' and 'outside-in'. This aspect is described by Sanders as 'being a "third culture man"' (Sanders, 1995).

It was also one of the important characteristics of the way the above-described research project was executed, therefore reaching for reliable as possible results.

Several years later, in 2011, and with a first reprint in 2017, the author, Dr. Tijhuis, together with Dr. Fellows, jointly published a book consisting of detailed theoretical analyses plus several international case studies, which were described and analysed within the period after completion of the above-described considered research project. These case studies considered business cases within the construction industry, located for example in The

Netherlands, Germany, Poland, Turkey, The United Arab Emirates, and China. A brief selection of the several interesting outcomes of these case studies is now presented (Tijhuis & Fellows, 2011 & 2017).

The Netherlands: Developing a complex inner-city project

A serious attention to the culture (behaviour, reputation, etc.) of possible parties involved when establishing the project team should therefore be recognized within the selection process. The 'match' with its project environment should be taken seriously. Trying to understand team members 'hidden agendas' as well as their 'official agendas' is of great importance.

Germany: Construction of rationalised terraced housing

Although one is more comfortable with a personal way of handling conflicts, one should be prepared to accept that the juridical way will often become the final means; when working with professional as well as private parties. This may also be an extra reason to be very selective during acceptance or rejection of projects and their accompanying contracts and business partners.

Poland: Subcontracting infrastructural and foundations works

During negotiations for getting problems solved, one should especially keep the focus on those issues which connect parties, instead of focusing only on those issues which are just separating parties. This will increase the chance of finding solutions by routes and/or issues of mutual interest between the parties involved.

Turkey: Tendering for developing a production factory

Although parties do have their national background in common, their behaviour within a multi-national group or consortium is not automatically an equal one; this is because every individual party will work according to its own agenda, which may be different from the other parties involved, regardless their own or other's nationality. Therefore, the matching of team members should be done especially with the focus on matching the goals and agendas.

United Arab Emirates: Designing a production factory

Working together with different specialist parties in a tram needs a strong sense or awareness of the weak signals related to the behaviour of the different parties involved. In particular, awareness of the other party's professional and/or private backgrounds and drive can reduce the risk of unpredicted behaviour during the construction process.

China: Organising an international distribution structure for special building materials

Existing contracts between parties do not mean that the arrangements made in such contracts are always followed clearly. Especially in business cultures with a focus on short-term goals, an opportunistic approach is often used to follow alternative approaches outside the contracts. However, if parties can balance this method by a continually focusing on long-term goals and keeping the value of reputations, this still can lead to a satisfactory situation for all parties involved. Nevertheless, this will often need a business-culture in which having and keeping of a good reputation plays an important role.

Parallel to these challenging outcomes of case studies and follow-up projects, the growing awareness of the influence of human behaviour on construction processes, often called a 'soft factor', was also positioned more specifically as a special field of attention within the internationally renowned CIB-platform, which for already many years connects academic and industry experts for improving construction industry by exchange of knowledge and experience. Not in the least since the establishing in circa 2004 of a so-called 'CIB W112 working commission Culture in Construction', the research theme 'culture in construction' has gained increasing attention from the international construction industry and academic institutions, studying and teaching construction-related research themes (Tijhuis, 2011).

After all and in general, throughout the recent years, it has become clear that the theme of 'business-culture' (often abbreviated as 'culture') in construction industry really has become 'part of the deal'. And this is not only within the boundaries of the described research project, but also elsewhere. Especially because nowadays the construction industry is increasingly globalising, leading to a growing 'mix' of multi-cultural processes and projects, creating a further challenging business-environment.

Lessons learned: Some symbolic considerations regarding culture in construction

Introduction

After being many years within international construction industry, from both the viewpoint of academic research as well as from a business-perspective, it all has to do with the way if and how interactions between parties are taking place, i.e. if it really leads to 'doing business' (creating value) or if it leads to 'conflicts' (destroying value).

To explain this more in detail, a symbolic representation of the meaning of culture in construction is being used. This is represented by graphics based on the shape of the so-called 'One Za'abeel' building in Dubai, the United Arab Emirates (under construction in 2022).

Remark

Without diving deeper into the construction-processes themselves, in this case the amazing building-design represents in a symbolic way some of the characteristics of 'culture', compared and described within the following paragraphs.

Comparison 1: Individuals might be connected but are not the same

When observing the One Za'abeel building from a larger distance, one can see that the building, consisting of two high-rise towers, is interconnected at approximately halfway their height. See Figure 10.3 for an impression of the building and its surroundings from a larger distance, representing these two high-rise buildings and landscaping with crossing infrastructure connections.

Altogether this building with two towers plus an in-between connection is still quite impressive but – considering Dubai's quite spectacular urban environment – it still does not look that very special compared to other buildings in the same region, at least when one

Figure 10.3 Impression of the One Za'abeel building and its surroundings in Dubai, United Arab Emirates (Tijhuis, 2022a).

looks at it from a longer distance. Nevertheless, the building has a very interesting shape if comparing it with some 'culture' characteristics … .

Comparing the towers of the building from a distance with some 'culture' characteristics

One can consider the two interconnected towers from a longer distance as two individual people standing beside each other, i.e.:

- They might look the same, thus expect to behave the same or even be 'connected', but they still will not have the same kind of mindsets.
- However, from a distance, one cannot see this difference immediately.

In Figure 10.4, this is represented schematically.

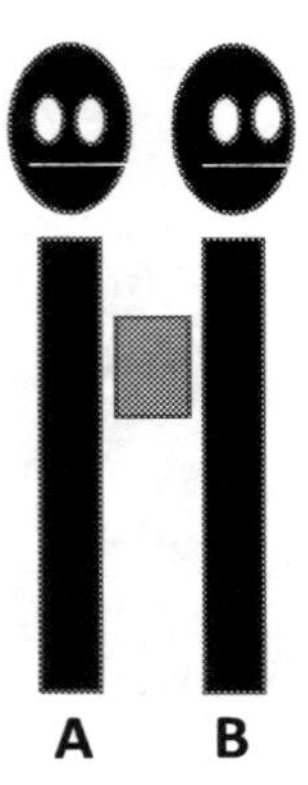

Figure 10.4 Schematic representation of individuals A and B standing beside each other; they might behave differently or the same or could even be connected, but will still have different mindsets.

Figure 10.5 Impression of the typical shape of the One Za'abeel building and its world-record cantilever-beam design structure in Dubai, United Arab Emirates (Tijhuis, 2022b).

Comparison 2: Individuals are connected as well as separated by their culture

If looking from a different angle towards the building and from a shorter distance, it suddenly becomes clear that it has a very impressive design by having a world-record cantilever-beam design structure. See Figure 10.5 for a detailed impression of this typical shape of the building (Tijhuis, 2022b).

Comparing the cantilever-beam of the building from a different angle and a distance with some 'culture' characteristics

One can consider the interconnected towers from a distance as two individual people standing beside each other, but explicitly interconnected by their 'culture' (i.e. represented by the cantilever-beam). However, this interconnection also creates a kind of distance, which symbolises the multi-faceted aspects of 'culture', i.e. 'culture' not only connects individual people, but also distinguishes them as individuals. See Figure 10.6, which represents this schematically.

Comparison 3: Individuals are not that what they might look like

When after all focusing more into detail at one of the individual towers, one can see that it has facades with specific surfaces and edges. This can for example be seen by the way it reflects (i.e. reacts) in a particular way the sunlight, because of their different shapes and materials at the surface. See Figure 10.7 for an example of this (Tijhuis, 2022c).

Comparing the surface of a tower of the building with some 'culture' characteristics

Although both of the two towers do have more or less the same facades, their structures and specific surfaces are somewhat different, depending on the use of the floors. That is also the case with different people: They might even be a kind of look-a-like, but there will always be differences, characterising them as individuals with their specific qualities, i.e. talents, knowledge, and skills. See Figure 10.8, which represents this schematically.

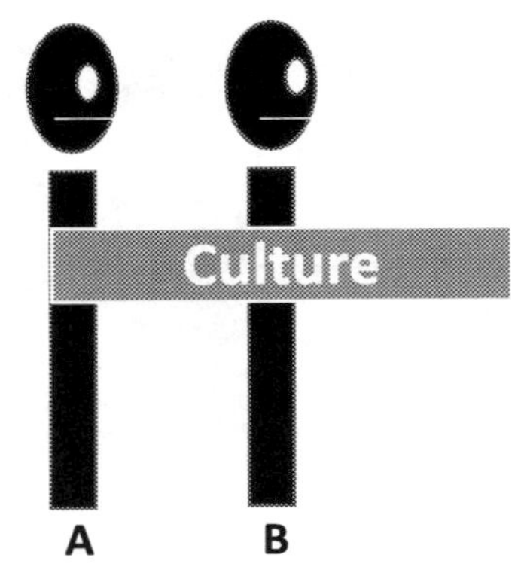

Figure 10.6 Schematic representation of individuals A and B standing beside each other but explicitly interconnected by their 'culture'. However, this interconnection also creates a kind of distance, which symbolises the multi-faceted aspects of 'culture' by also distinguishing individuals, too.

Figure 10.7 Impression of the typical facade and its reflection of sunlight of one of the towers of the One Za'abeel building in Dubai, United Arab Emirates (Tijhuis, 2022c).

Figure 10.8 Schematic representation of an individual A or B, being individuals with their specific qualities, i.e. talents, knowledge, and skills.

Future outlook: Culture's influence on labour

Having analysed culture in construction in detail in this chapter and making symbolic comparisons with the described 'One Za'abeel' building, one can say that:

- *'culture' acts as a kind of 'force' which connects as well as distinguishes individual people, thus of immediate influence on the labour situation within construction industry.*

These two aspects can even exist parallel to each other at the same time. Based on the analyses in this chapter, a model is used, representing the 'forces' which emerge when there are cultural differences between people, for example, within labour situations, seeing the following influences of culture:

1 *Having the same 'culture' makes it easier for people to communicate, understanding not only the spoken/written communication but also the informal/weak signals, i.e. especially by sharing the same values, attracting people towards each other. In this case, culture is 'helpful' because it eases their communication, which might lead to stimulate their collaboration and better working circumstances.*
2 *Whereas when having a different 'culture', makes it more difficult for people to communicate, although still understanding the spoken/written communication, but with a risk of misunderstanding the informal/weak signals. In this case, culture 'hurts' because it distracts people from each other and worsens their communication, which might lead to conflicts in their collaboration and bad working circumstances.*

These both influences of culture are symbolised within the model, represented schematically in Figure 10.9, and explained as follows.

In this model, the (1) 'helpful' role of culture for individual A is represented by 'force' F2 (the dotted arrow upwards, i.e. culture's influence), lifting the individual A upwards, thus leading to a lighter weight (i.e. 'force' F1, the full arrow downwards) for the individual A to carry, resulting into a smiling face:

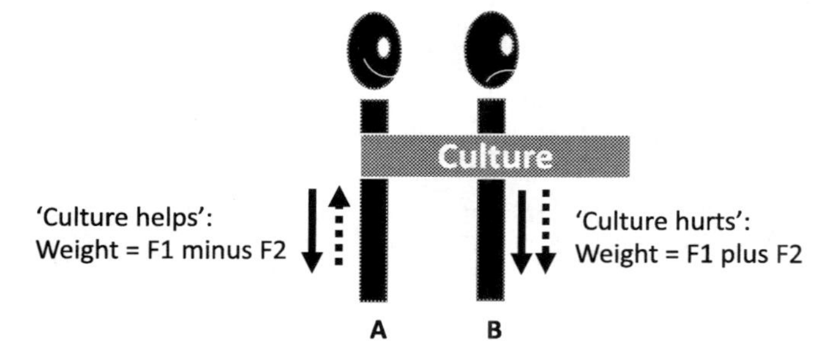

Figure 10.9 Model, representing individuals A and B, standing beside each other within a labour environment, being influenced by different aspects of 'culture', i.e.: Culture is 'helpful' or culture 'hurts', leading to happy or angry people.

This is the situation one wants to create within a labour environment, i.e.: Smiling, happy people, supporting each other, leading to a better collaboration and stimulating working environment.

Parallel to this, the (2) 'hurting' role of culture for individual B is represented by 'force' F2 (the dotted arrow downwards, i.e. culture's influence), pushing the individual B downwards, thus leading to a heavier weight (i.e. 'force' F1, the full arrow downwards) for the individual B to carry, resulting into an angry face:

This is the situation one wants to avoid within a labour environment, i.e.: Angry people, not supporting each other, leading to a bad collaboration and depressing working environment.

See Figure 10.9, representing this model schematically.

These both forces of 'helping' and 'hurting' individuals are important aspects of 'culture', making this an increasingly important theme, influencing the execution of construction processes and leading towards successful or failing projects. From this viewpoint, it can be stated as follows:

'*Construction is a people's business, thus making culture part of the deal*'.

Based on lessons from the past, this is an increasingly important and challenging outlook for nowadays and future construction industry and the processes therein, with their increasingly globalising characteristics, leading to a growing 'mix' of multi-cultural processes and projects, creating an increasingly challenging business-environment and labour situation with its working circumstances therein. This leads to a need for people with increasing awareness and learning the competencies and skills needed to handle these challenges properly. Indeed, challenging times are to come![1]

Note

1 Because Tijhuis was the organiser of the workshop and editor of the proceedings, the 'et al' are all the authors of the individual papers within this proceedings. Because there are more than three authors, it is written 'et al', as a common rule.

References

Cameron, K.S. & Quinn, R.E. (1999) *Diagnosing and Changing Organizational Culture: Based on the Competing Values Framework*; Addison-Wesley; New York.

Hatch, M.J. (1993) *The dynamics of organizational culture*; Academy of Management Review; Vol.18, Nr.4; pp.657–693.

Hofstede, G. (1980) *Culture's Consequences – International Differences in Work-Related Values*; Vol.5, Cross-Cultural Research and Methodology Series; Sage Publications; Beverly Hills, London, New Delhi.

Hofstede, G. (2001) *Culture's Consequences: Comparing Values, Behaviors, Institutions and Organizations Across Nations*, 2nd Edition. Sage Publications: Thousand Oaks, London, New Delhi.

Kroeber, A.L. & Kluckhohn, C. (1952) *Culture: A critical review of concepts and definitions*; Papers of the Peabody Museum of American Archaeology and Ethnology; Vol.47, Nr.1; p.247.

Mintzberg, H. (1979) *The Structuring of Organizations*; Prentice Hall; Englewood Cliffs, New Jersey.

Sanders, G.J.E.M. (1995) *Being "a third culture man"*; Cross Cultural Management: An International Journal; Vol.2, Nr.1; p.507.

Sanders, G.J.E.M. & Neuijen, B. (1987) *Bedrijfscultuur: Diagnose en Beïnvloeding*; Van Gorcum; Assen, Maastricht.

Schneider, W.E. (2000) *Why good management ideas fail: The neglected power of organizational culture*; Strategy and Leadership; Vol.28, Nr.1; pp.24–29.

Spradley, J.P. (1980) *Participant Observation*; Holt, Rinehart and Winston; New York, Chicago.

Tijhuis, W. (1996) *Bouwers aan de Slag of in de Slag? – Lessen uit Internationale Samenwerking: Onderzoek naar Nederlandse Ervaringen in het Duitse Bouwproces*; Ph.D.-Thesis; with English and German Summary: 'Contractors at Work or in Conflict? – Lessons from International Collaboration'; Eindhoven University of Technology, Eindhoven; Kondor Wessels Berlin GmbH, Berlin.

Tijhuis, W. (1999) *Focussing at the Client's wishes and behaviour in construction management – (Re) starting at the front-end of construction-process*; University of Twente, Enschede, WT/Beheer BV, Rijssen; paper on conference: 'Customer Satisfaction: A Focus for Research & Practice in Construction'; proceedings; 5th–10th September; pp.148–156; Ed.: P.A. Bowen and R.D. Hindle; University of Cape Town, Cape Town.

Tijhuis, W. (2011) *Report – Developments in construction culture research: Overview of activities of CIB W112 'culture in construction'*; Journal of Quantity Surveying & Construction Business; Vol.1, Nr.2; pp.66–76; Edited by: Khairuddin Abdul Rashid & Christopher Nigel Preece; Published by Procurement and Project Delivery System Research Unit, International Islamic University Malaysia; ISSN 2229-8339; Malaysia.

Tijhuis, W. (2022) *3C-model & NEDU-matrix - Research outcomes*; Course Material; Part of the Lecture Series: 'Culture in International Construction 2021–2022'; Version 24th of March 2022; University of Twente, Faculty of Engineering Technology, Department of Construction Management & Engineering (CM&E), Enschede (NL).

Tijhuis, W. (2022a) *Impression of the One Za'abeel building and its surroundings in Dubai*; picture; October 2022; Dubai, United Arab Emirates.

Tijhuis, W. (2022b) *Impression of the typical shape of the One Za'abeel building and its world-record cantilever design structure in Dubai*; picture; October 2022; Dubai, United Arab Emirates.

Tijhuis, W. (2022c) *Impression of the typical facade of one of the towers of the one Za'abeel building in Dubai*; picture; October 2022; Dubai, United Arab Emirates.

Tijhuis, W. & Fellows, R. (2011 & 2017) *Culture in international construction*; book; published by: Spon Press – an imprint of Taylor & Francis; ISBN13: 978-0- 415-47275-3 (hbk) & ISBN13: 978-0-203-89238-1 (ebk); 1st Edition published in 2011 (hardcover & e-book); 2nd Edition published in 2017 (paperback); Published by Spon Press, 2 Park Square, Milton Park, Abingdon, Oxon (UK); Simultaneously published in the USA and Canada by Spon Press, 711 Third Avenue, New York (USA).

Tijhuis, W. et al. (2001) *Culture in Construction – Part of the Deal?*; Proceedings of 2- day workshop at University of Twente, Enschede; 22nd and 23rd of May 2001; Task-group TG23 of CIB - Conseil International du Batiment; Rotterdam.

Trompenaars, F. & Hampden-Turner, C. (1998) *Riding the Waves of Culture: Understanding Cultural Diversity in Global Business*; Second edition; McGraw Hill; New York.

11
GENDER, CONSTRUCTION WORK, AND ORGANISATION

Maria Johansson and Kristina Johansson
LULEA UNIVERSITY OF TECHNOLOGY, SWEDEN

Introduction

A gender perspective on construction work makes it visible that it is a *gender-segregated* arena dominated by *men* and cultures and notions associated with *masculinity*. That men and women tend not to work in the same industry, in the same occupations and/or with the same work tasks not only applies to the construction sector. The division of labour in most Western and non-Western societies tends to be organised along gender lines. For example, there is an overrepresentation of women in agriculture in developing countries and health and social work, education, and other service work in developed countries.[1] These industries tend to have lower incomes than many other sectors (International Labour Organization, 2016). Nevertheless, it seems that construction work is particularly gender-segregated, as women workers in most construction contexts are close to none existing. Why is that so?

The aim of this chapter is to add empirical, theoretical, and practical insights to issues concerning gender and equality in construction work. We start with a closer interrogation of the gender profile of construction and the representation of men and women as trade workers and managers in various geographical contexts based on statistical sources. We then continue the exploration with the help of the available insights from the literature on gender in construction. Based on the varying focus and theoretical starting points, we critically examine and analyse the literature divided into three themes, that we have titled *the barriers that hinder women in construction*, *the gendered character of construction work and organisations*, and *men and masculinity in construction*. The chapter ends with implications for change, including some examples of related practical measurements, and conclusions.

The gender profile of construction work

International statistics indicate that female participation in the construction industry is low and that the distribution of women is generally similar regardless of country (International Labour Organization, 2016). It is important to be aware that variations in statistical measurements and other differences between countries, such as differences in the work

DOI: 10.1201/9781003262671-11

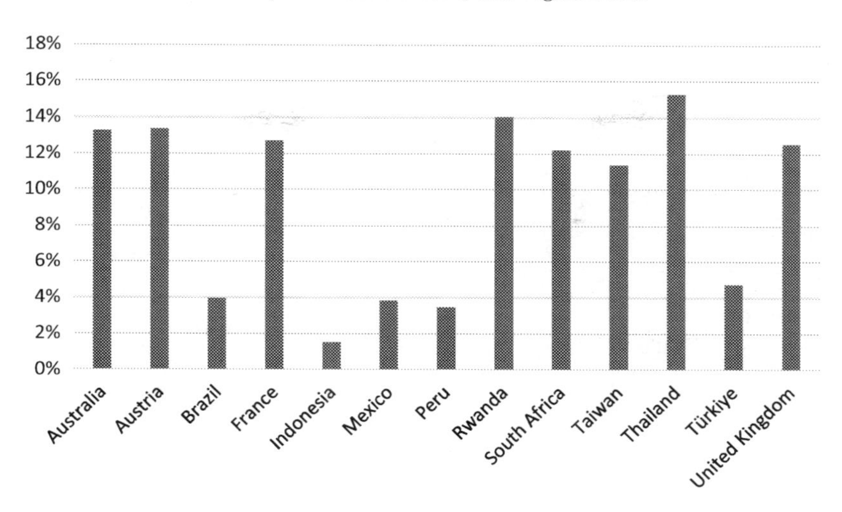

Figure 11.1 Female participation in the construction industry in a selection of countries.
Source: International Labour Organization (2022). ILOSTAT database.[2]

content, make it difficult to compare statistics between countries. In Figure 11.1, statistics from a selection of countries are presented mainly to illustrate the low female participation. Here, particularly low percentages of women are reported for Brazil, Indonesia, Mexico, Peru, and Türkiye, while Australia, Austria, France, Rwanda, South Africa, Sweden, Taiwan, Thailand, and the United Kingdom report slightly higher female participation between 11% and 15%. Overall, approximately 10% of women worked in the construction industry in Europe in 2021 (Eurostat, 2022), and just over 11% were reported in the US construction industry in 2021 (U.S. Bureau of Labor Statistics, 2022).

Women working in the construction industry, are mainly assigned to administrative positions, such as sales, office work, and service occupations. Additionally, women are present in professional management positions (see International Labour Organization, 2016; The National Association of Women in Construction, n.d.). However, statistics on trade workers such as carpentry and masonry indicate that only a small percentage are women (International Labour Organization, 2015; The National Association of Women in Construction, n.d.; Statistics Sweden, 2022). India stands out as it is estimated that female participation is up to 50%. However, in the Indian construction industry, women are reported to predominantly be hired in positions at the lowest levels of the organisational hierarchy to perform semiskilled work tasks such as blending concrete (International Labour Organization, 2015). Hence, to make sense of the gender profile across national contexts, variation in work organisation and degree of rationalisation must also be taken into consideration. It also makes evident the need to consider not only the representation of male and female construction workers, but also additional patterns of gender segregation within the industry.

Taking a closer look at statistics on women in occupations related to construction in Sweden, a country generally viewed as fairly gender equal, reveals a pattern that is consistent with other Western countries where there is low overall female participation, especially in regard to manual trades. As shown in Table 11.1, overall, female participation in the construction industry in Sweden has slightly increased in recent years. Female

Table 11.1 Craft and related trades workers: Building and manufacturing workers in Sweden (by sex and average age 2018)

	Total	*Women*		*Men*	
	n	*%*	*Average age*	*%*	*Average age*
Woodworkers, carpenters	49 800	1	36	99	40
Plumbing and central heating fitters	21 500	1	35	99	39
Rail and road construction workers	18 400	4	38	96	41
Building frame and related trades workers not elsewhere classified	18 100	3	33	97	36
Painters and related workers	15 000	9	31	91	41
Concrete placers, concrete finishers and related workers	9 900	2	34	98	41
Bricklayers and related workers	8 700	2	32	98	40
Painters and industrial painters	5 400	9	38	91	40
Building structure cleaners and related workers	4 700	12	35	87	37
Floor layers	4 199	1	33	99	41
Heating and air conditioning mechanics	3 900	1	35	99	41
Roofers	3 700	1	36	99	38
Scaffold builders	3 400	2	38	98	27
Glaziers	2 700	4	39	96	40
Insulation workers	2 100	2	41	98	40
Chimney sweepers	1 200	8	32	92	42
Construction diver	200	3	32	97	38

Source: Statistics Sweden.

participation was 8% in 2010, 9% in 2015, and 10% in 2020 (Statistics Sweden, 2022). A slight increase in female participation is also visible among students enrolled in construction education in upper-secondary schools, which also shows a slight increase from 8% in 2015 to 9% in 2020. Although change is occurring slowly, this might be cautiously interpreted as an improving trend.

In manual trades, statistics show that approximately 1%–2% of the workforce is women in occupations such as carpenters, plumbers, concrete workers, bricklayers, roofers, floor layers, and scaffold builders. Women who labour as construction workers related to roads and railways and as glaziers are slightly higher, at 4%. Painting appears to be an area where female participation is significantly higher, approximately 9%, and female participation in construction cleaning is 12%. The table also makes evident that women are generally younger than male workers. This example highlights the extremely low number of women engaged in manual construction occupations.

The barriers that hinder women in construction

In general, studies that address work in the construction industry from a gender perspective often focus on women and take women's experiences and situations as a starting point. Typically, these studies result in the identification of barriers or obstacles that are found to hinder women, including both barriers that hinder women from entering construction work and barriers that hinder the few women who do work in construction from being promoted. A similar focus also dominates the literature on gender in other male-dominated industries, such as mining (cf. Baruah & Biskupski-Mujanovic, 2021; Salinas & Romaní, 2014; Benya, 2017;

Musonda 2020). When women are placed in the focal point of a male-dominated organisation, analyses specifically deal with persistent "minority problems" (cf. Kanter, 1993) for women, including both open and passive resistance (from male coworkers, managers, unions, and society) to women and gender equality in the workplace, which becomes extra work environment problems for women.

Two literature reviews relating to gender and construction work with a particular focus on the situation of women have been published, reaching similar conclusions. Navarro-Astor et al. (2017) investigated barriers to women's career development in the construction industry from an international perspective. The review presents findings of 12 categories of career barriers found in 60 scientific articles produced between 2000 and 2015. The barriers found are related to work-family balance, gender stereotypes, allocation of posts and activities, promotion, working conditions, sexist culture, harassment and lack of respect, recruitment and selection, lack of recognition, pay, social networks and others. That various, interconnected circumstances within the construction industry and organisations tend to hinder women is further described by Hasan et al. (2021) in their review of 128 scientific articles on women's situation globally in the construction industry. The results were divided into four different areas, spanning gender roles and work culture, gender segregation in the form of "glass walls" and "glass ceilings" keeping women out, job satisfaction, and gender diversity initiatives. It identified how cultural norms stipulating that women take the main responsibility for children and home life, create a dilemma for women construction workers attempting to combine paid labour and home life. The second area revolves around barriers women tend to face either to enter the construction industry or to be promoted. Several barriers are reported, such as sexism and harassment, an unsafe work environment, poor work-life balance, and biased recruitment practices based on informal networks. Furthermore, extra costs, such as locker rooms and other facilities requiring separation between women and men, are also viewed as preventing construction companies from hiring women. Related to job satisfaction, studies showing that female employees leave their jobs in the construction industry more frequently than male employees are presented. This is reported to be caused by discrimination, harassment, lack of career opportunities, unequal pay, long working hours, and unhealthy work situations. Finally, gender diversity initiatives, which include policy and legislation reforms, are considered necessary to increase gender equality in the construction industry. It is reported, however, that the impacts of policies are limited (Hasan et al., 2021).

The literature further shows that the barriers identified for women in the construction industry are both general and vary with geographical context. Vijayaragunathan and Rasanthi (2019) investigate why female participation in the Sri Lankan construction industry is low and the three most important barriers found are commitments to family, sexist attitudes, and lack of female role models. An investigation of the situation of women with a university degree working in the construction industry in Peru found similar barriers as other studies, such as a lack of career opportunities (Barreto et al., 2017). Although several reported barriers seem to be similar, there are also studies making visible situations where women working in the construction industry are extra precarious and exposed to hardship. Choudhury (2013) investigates the experiences of poor female construction workers in Bangladesh. Choudhury (2013) argues that although women working in the construction industry in countries labelled "developed" are also reported to meet inequality at work, the situation of poor women working at construction sites in Bangladesh (and other countries in the global south) is not comparable. It is explained that Bangladesh is highly patriarchal, where the dominant ideal of men as breadwinners means that women's participation in paid labour tends to reflect badly on her husband, as he is not seen to be able to provide for his family. The results of the study show that women worked without a work schedule where working hours differed and could be rather long. They had no working clothes and often worked barefoot. Some also had their youngest children with them. Similarly, Bowers (2019) investigates how female migrant construction workers experience their work situation in India. It is explained that women are often a part of a family unit that all come to work at the same construction site. They also live there while they work. Women tend to be seen as helping their husbands at work rather than being productive workers. They are paid less than their husbands, and often women's pay is given to their husbands. Furthermore, it is reported that they do not have any possibilities for career progression as well as a lack of legal rights and a lack of protection against sexual harassment and poor working conditions.

The gendered character of construction work and organisation

Studies focusing on women and their profile as minorities provide insights into the barriers that hinder women from participating in construction work. However, starting and ending gender analyses with women without a thorough consideration of the structural and cultural contexts that women occupy, risks reproducing a notion of construction work and organisation as neutral and logical entities. A more comprehensive understanding of the gendered character of the construction industry is found in the literature that analyses construction work and organisation as the result of gendered processes and practices. The concept of "gendered organisations" refers to organisations and their activities as intertwined with gender and distinguish between men, women, masculinities, and femininities. Rather than gender being an external phenomenon, that in a later phase affects (or does not affect) the organisation, the concept thus helps to emphasise how gender is an integral part of organisations (Acker, 1990; 2006). Such an approach presupposes an understanding of organisations as social and ongoing processes shaped by the everyday practices that make up their activities. It also presupposes an understanding of gender not as a stable category but as the result of socially situated "doings" (cf. West & Zimmerman, 1987).

Shifting the focus from women to the gendered character of construction work and the workplace provides further insights needed to understand the complexity of gender in construction. In their investigation of the informal aspects of recruitment, retention, and progression in two multi-international construction companies in Australia, Galea et al. (2020) analyse construction as a "masculine space". Departing from feminist institutional theory, they study interactions between formal and informal rules by conducting observations and interviews with both men and women at six construction sites. Within such spaces, and due to sexism in language, images, and the allocation of tasks, women tended to be positioned as different and deviant. Part of this masculine space was a workplace culture that sexualised women, with women subjected to being filmed in the shower, receiving comments on their breasts, being called "babe" and other derogatory remarks by coworkers and managers, and being exposed to pornography at construction sites. Such a gendered workplace culture also shaped men's and women's career prospects. The findings indicate that informal networks were an advantage for men through retention and progression, while women had less access to such networks. Men typically get their jobs based on their access to informal networks, resulting in work opportunities, while women typically apply for their jobs through formal channels by answering job applications and going to one or more job interviews. The authors also found examples where managers tended to encourage women to choose career paths leading towards design instead of project management, which made it more difficult to be promoted. Instead, men were more likely to be given opportunities for project management roles. Additionally, a very high workload and long working hours were evident in the investigated companies, up to more than 80 hours per week. In addition, informal rules and a workplace culture that privileged long working hours meant that it was considered "bad" to not stay on the job. Working long hours, even when the work was done, was also identified as making it difficult to achieve a balance between work and family life. Although there were formal rules concerning parental leave and flexible work hours in the studied organisations, informal rules resisted this flexibility. It was also found that it was easier for men to relocate geographically because their families then moved with them; women, however, had a harder time when there was less interest from their families in also moving with them (Galea et al., 2020).

Norberg and Johansson's (2021) analysis of how women in the construction industry are discursively represented and how they talk about their workplaces provides further insights into the gendered character of construction work and organisation. On a general note, women were represented as sought-after labour and welcomed by the construction industry. However, if analysed more closely, women who enter the industry are represented in a problematising manner. The findings show that stereotypical perceptions of women as passive and more emotional than men lead to an image that women need to gain competencies traditionally seen as male to fit into construction workplaces. For example, women are advised to be calm, not emotional, listen more, and speak less. Such stereotypical notions of femininity (and masculinity) thus mean that women are perceived as not truly belonging in the industry. It also suggests that women need to change rather than the construction industry. Women who have already taken up positions in the industry are described as role models, as successful, and as the ones to follow. However, demands on these women are found to be very high, with expectations to work harder than their male colleagues. Above that, they are also left with the responsibility to act as change agents by transforming the industry culture. In conclusion, the analysis shows that women entering the industry tend to be met by gender-biased attitudes, discrimination, and unrealistic demands.

Norberg and Johansson (2021) make evident how the representation of women draws on taken-for-granted assumptions about men and women as inherently different, as well as about construction work and organisation as associated with men and masculinities (cf. Cockburn, 1991). Such assumptions of gender differences add to men's privilege by positioning women as deviants in relation to a preconceived male norm (cf. Brown, 1995). Similar conclusions have been made in related industries. For example, Eveline and Booth (2002) found that an increased proportion of women in the Australian mining industry was assumed to have a positive impact on men's behaviour in the workplace and thus, by extension, on productivity. Similarly, Mayes and Pini (2014) saw that gender equality was assumed to be "good for business" for the Australian mining industry. This was motivated by the fact that women were assumed to bring something different to the workplaces compared to the men who were already there, such as communication skills, which would help to "civilise" workers (cf. Johansson & Ringblom, 2017; Norberg & Fältholm, 2018). Guerrier et al. (2009) show that more women were considered particularly positive for the IT industry because the work has to a greater extent come to be about communication, relationship building and empathy, that is, abilities and qualities associated with women and femininity.

Men and masculinity in construction

A third and final approach to gender found in the literature on the construction industry has a particular focus on men and masculinity. As argued by Collinson and Hearn (1994), there is a particular need to "name men as men" in the critical analysis of gendered power relations in organisations. While the perspective of men constitutes the implicit starting point for most conceptualisation of organisations, masculinity tends to remain taken for granted and hidden. Shifting the focus to men and masculinity in construction work and organisation thus constitutes a means to further nuance the relevance of gender.

Similar to studies of other male-dominated high-risk occupations (cf. Stergio-Kita et al., 2015), a recurring theme in studies that do focus on masculinity in construction work is risk and safety. It is typically found that when (male) trade workers do their job, they engage in behaviour that glorifies risk-taking, toughness, physical endurance, and endurance of physical pain as virtues of "real" men (Fielden, 2000; Iacuone, 2005; Lu & Sexton, 2010; Ness, 2012). For example, using safety equipment and scaffolding is interpreted as incompatible with the (gendered) ideals of trade workers, fostering a conception of breaking safety regulations as being an expression of manliness (Gherardi & Nicolini, 2002; Ness, 2012). Drawing on interviews and observations, Iacuone (2005) and Baxter and Wallace (2009) relate the gendered culture of construction sites to the behaviour of men and how hierarchies of masculinities in construction are established and maintained during daily interactions between men. These studies all suggest that gender hierarchies are created and manifested not only in the relations of men and women but also among men and various forms of masculinities. The role of multiple masculinities in the creation of gender hierarchies and that these various forms of masculinities in specific times and places are ordered in relation to a specific "hegemonic" form of masculinity was introduced by Connell (2005). Hence, drawing on the concept of hegemonic, subordinated, and marginalised masculinities it becomes evident that the promotion of risk-taking in construction becomes a way to simultaneously subordinate other forms of masculinities, perceived as "unmanly".

Gherardi and Nicolini (2002) link trade workers' tendency to treat their own safety lightly with the creation of gendered identities and site managers' tendency to explicitly question the manliness of trade workers who request safety equipment or tools to reduce the physical strain of their work. They also found that despite the high-risk nature of the jobs and poor application of safety regulations, when accidents occurred, the injured individuals were blamed, implying that the accidents were due to the workers' incompetence rather than to the organisation and management of work. Similar insights are provided by Ajslev et al. (2016) in an analysis of Danish trade workers' tendency to "trade health for money", describing how the workers' ability to work hard and endure severe and/or chronic pain constituted a means for them to construct a positive (masculine) work identity. This was also manifested in the firing of workers who took time off due to physical strain more than once or twice. According to the interviewed workers, the rapid pace of work caused the most harm, and their perception that the pace of work was linked to both productivity and remuneration levels further contributed to the workers' neutralisation and acceptance of physical strain (Ibid.). By situating analyses of gender in relation to the organisational context of construction, Gherardi and Nicolini (2002) and Ajslev et al. (2016) highlight, in different ways, how gender identities and norms relate to overarching organisational structures and ideals. Hence, rather than in terms of the behaviour of individual men, such an approach enables scrutiny of the context that structures such behaviour.

The literature on men, masculinity and risk-taking in male-dominated industries tends to revolve around the traditional, "macho" masculinity that tradesmen perform and embody while doing their job. More recent studies have identified changes in the expression of masculinity in construction in ways that challenge previous insights into male trade workers' celebration of risk-taking. Hanna et al. (2020) investigate health and well-being at construction sites, drawing on a case study of stakeholders relating to the UK construction industry. All interview participants agreed that construction has a "macho" male-dominated nature that affects health and well-being in the industry. It was "macho" due to the domination of a workplace culture that celebrated physical strength, toughness, and stoicism. However, structural factors built on competition meant that productive pressure and profit were prioritised over occupational health and safety (cf. Stergiou-Kita et al., 2015). However, when examined more closely, there were also various examples of generational differences, suggesting a more complex relationship between gender, health and well-being and a more pluralistic expression of masculinity in construction. Whilst construction workers seem to perceive help-seeking and healthy diets as contradicting images of tough masculinity, younger trade workers embraced a more "hybrid" form of masculinity that involved body enhancement and aesthetic health. Additionally, the case study showed a sense of comradery between men who offered informal peer support to workers that enabled, rather than prevented, the promotion of positive health and well-being within the workplace.

The notion that contemporary men and notions of masculinity are undergoing change is evident in the literature beyond construction work. Bridges and Pascoes (2014) promote the concept of "hybrid masculinities" to make sense of the emergence and effects of recent transformations in masculinities. "Hegemonic masculinities" (Connell, 2005) attempt to understand the power struggles and hierarchies between men leading to the subordination of certain forms of masculinities perceived as "unmanly" and certain form of masculinities domination over others. In contrast, "Hybrid masculinity" refers to "men's selective incorporation of performances and identity elements associated with marginalised and subordinated masculinities and femininities" (Bridge and Pascoe, p. 246). Placed in the focal point in such studies are the ways in which men with power currently tend to incorporate "bits and pieces" of behaviour and identities traditionally subordinated and coded as "gay", "black", or "feminine", into their identity project as "modern men". While some scholars argue that this transformation is a sign of the emergence of more egalitarian masculinity and a significant change in gender equality, others, including Bridges and Pascoes (ibid.), suggest that hybrid masculinity is a sign of the flexibility of masculinity and the ability to maintain power and privilege in new faces.

Implications for policy and change

Thus far, this chapter has responded to the question of why female participation in the construction industry is so low that it was raised in the introduction. By insights from research analysing the situation of women specifically, the analysis of organisations as continuously formed by gendered processes and practices and the analysis of the role of masculinities in construction workplaces, the intertwined patterns shaping construction work and construction workplaces that affect female participation are evident. This section describes and discusses some recommendations made by researchers on how to increase gender equality in the construction industry. Some practical examples are also provided. Overall, rather than making changes on specific levels, studies make visible the need to

Box 11.1 Women becoming skilled trade workers, an example from India

The Self-Employed Women's Association (SEWA) is a national trade union in India. In 2003, the Karmika School for construction workers started to offer training for individuals working in the construction industry, focusing on female workers. Vocational training, training for trade competence in construction trades such as carpentry, plumbing, and masonry, and others as well as certification of skills, are offered. More than 5,000 people have been educated through the programme, which is reported to have increased the salary of the participants (Chen, 2008).

implement changes in several aspects simultaneously, targeting the systemic, structural, and cultural aspects of the industry (Hasan et al., 2021).

Legislation, welfare systems, and policy reforms

The state of welfare systems and legal rights affect women working in the construction industry. In countries with less developed welfare systems and a lack of legal rights for women, a precarious situation for women in the construction industry is reported (cf. Bowers, 2019; Choudhury, 2013). There is certainly a need for development of legal rights to protect women at work as well as for development of welfare systems to ensure good health and education for all. Related to the construction industry, inequalities in the access to basic education and construction-related training is reported to contribute to women's difficulties for career progression in the industry. Box 11.1 shows a practical example of a programme aimed to contribute to education and training in the construction industry in India.

Legislation and policy reforms are a way in which political decisions are made to improve gender equality. In an analysis of equal employment policies in the Australian construction industry, French and Strachan (2015) found that policies tend to be designed mainly to meet minimal legislative demands, such as to overcome harassment and discrimination. The only impact found in the implementation of policies in the article is a slight increase in the number of women working in clerical and sales roles. An area where women often work if they are hired in the industry. Furthermore, it is reported that policies tend to be overridden by masculine workplace culture and practices (Hasan et al., 2021). Galea et al. (2015) state that policies that aim to increase the number of women are less favourable since they tend not to lead to changes in norms and gender practices in the construction industry. Instead, it is recommended that policies should be aligned with company values because then they are more likely to be perceived as relevant for both women and men at the workplace.

The image of the industry

The construction industry is also described to have an image of being dangerous and high in physical strain. This image is described to make the industry seem less interesting to some groups thus hindering gender equality and diversity. Changes to break gender segregation and to make more people curious about working in the various occupations in the industry is recommended. For example, Naoum et al. (2020) mean that reaching out to young girls to promote working in the construction industry is a good way to make more women enter and stay in the construction industry.

Working conditions

A common barrier addressed concerns women's tendency to have the main responsibility for home and family life, which conflicts with construction work that entails long working hours (Bryce et al., 2019; Hasan et al., 2021; Naoum et al., 2020). It is found that women tend to leave the industry following maternity leave due to traditional family roles; if they do stay, it can negatively affect their career planning (Bryce et al., 2019; Navarro-Astor, 2017). Hence, recommendations to make it possible with flexible working conditions are put forwards. It is suggested that employers should discourage employees from working on weekends. It is also recommended that employers should advertise more part-time positions, offer other forms of flexible arrangements such as job sharing where two employees split one full-time job, and enable employees to work part of the week from home (Bryce et al., 2019). However, these suggestions seem to be directed more towards increasing the number of women in the industry rather than leading towards gender equality. Additionally, as noted by Galea et al. (2020), within construction companies that offer flexible worktime solutions, norms and values that glorify long working hours hinder implementation. This indicates that it is not sufficient to only change the working conditions. It is also important to change norms, values, and culture at construction sites.

Norms, values, and culture

Since a masculine culture is put forwards as hindering women from being included in the industry, challenging norms and values and changing the culture, striving for an inclusive culture, is also highlighted as important for gender equality in the industry. Hasan et al. (2021) report on studies showing that where women are treated fairly, have career opportunities, and have a safe work environment, women employed in the construction industry were positive about remaining. There is a lack of recommendations on how to proceed with changing masculine cultures, and thus a practical example is provided (see Box 11.2).

Formal and informal social networks

The findings emphasise the need for the individual to be chosen by colleagues and managers, both before entering the industry and during employment, which requires access to informal social networks that open opportunities for employment in a construction company. Furthermore, informal social networks open possibilities for individuals to prove their abilities, which is important for promotion. With less access to these networks, it is argued that women tend to be neglected. Men, however, seem to be included in these networks to a greater extent, even before entering the industry (Galea et al., 2020). One way suggested to increase women's social networks is to develop mentoring programmes targeting both genders to form social networks where the individual gets to know the organisation and increases possibilities for career progression (Bryce et al., 2019). Having access

Box 11.2 Stop Macho Culture, a Swedish example

To change the culture in the construction industry and to increase gender equality and diversity, the union of workers in the construction industry (Byggnads) together with the association of construction managers (Byggcheferna) started a three-year project called *Stop Macho culture* in 2015. It was motivated by an upcoming labour shortage presumably caused by macho culture. The aim was to create a discussion about jargon, social norms, and behaviour commonly characterised as macho culture. It was launched as a media campaign with ads containing short statements in the subway, newspapers, and social media, as well as visits to schools, companies, and authorities. Targeting the public as well as construction workers and construction managers specifically to discuss and to question social norms and values. After three years, it was decided that the project was too important to end, instead the project continued indefinitely without an end date.

On the project's website, information about macho culture is provided as well as several tools developed to change the culture. These tools are directed to construction workers, construction managers, and students.

One of the tools is called *MachoIndex* which consists of eleven questions where construction managers can identify the prevalence of macho culture in their company. Then, they can move on with other tools to change the culture. Another example of a tool is a conversation exercise called *Talk so that everyone can listen*. Here, cards with short statements gathered from Swedish construction sites are provided digitally. They are then discussed by construction crew members (Stop Macho Culture, n.d).

to female role models is also proposed as a way to increase women's job satisfaction and career advancement. However, Hasan et al. (2021) found conflicting results about the effects of female role models. Furthermore, in Norberg and Johansson's (2021) study, female role models were identified as risking being largely responsible for changing the masculine culture in the industry towards one of greater gender equality. Hence, if the industry or construction organisations choose to highlight female role models (see Box 11.3) it is important to make sure that they have a good work environment and that the organisation is responsible for its processes and culture, not female role models.

Box 11.3 Women's Awards

In several countries, *Women's awards* have been introduced to celebrate accomplishments made by women working in the industry. A goal of the awards is to encourage women and to strive for gender equality and diversity in the industry, to promote female role models and motivate construction employers to hire women. For example, the European Women in Construction & Engineering Awards (WICE Awards, n.d.) and the Empowerment and Recognition of Women in Construction Awards in South Africa (Erwic Awards, n.d.).

Concluding remarks

This chapter aimed to add empirical, theoretical, and practical insights into issues concerning gender and equality in construction work. First, it can be concluded that female participation in the construction industry is low, and that the pace of change concerning gender distribution in the industry is slow. This particularly applies to skilled trades.

Through three themes, the barriers that hinder women in construction; the gendered character of construction work and organisations, and men and masculinity in construction, the chapter provides insights into why female participation is low by making different aspects affecting gender equality in construction work organisations visible. The first theme provides examples of barriers faced by women through the lens of being in the minority in the workplace. Studies of women's experiences as minorities in construction organisations have identified several barriers that women tend to face. These barriers concern for example biased recruitment and promotion processes, sexual harassment, and inequalities in the allocation of work tasks based on perceptions of traditional gender roles (Hasan et al., 2021; Navarro-Astor et al., 2017). However, although similar barriers have been reported in studies in various countries, differences in geographical locations also need to be addressed. For example, Barreto et al. (2017) and Choudhury (2013) report that women in the construction industry in Bangladesh and Peru face overt resistance to their participation in the labour market, and in the construction industry specifically. Such open resistance to female participation tends not to be reported in studies taking place in the global north. Studies further reveal that women working in the construction industry in some developing countries are extra exposed to risks due to a lack of legal rights (Bowers, 2019; Choudhury, 2013). Additionally, work tasks are described to be different, as women are described as "helpers" and mainly revolve around unskilled work. These, often poor, women's work situations cannot easily be compared to the situation of women working in the construction industry in the global north.

In the second theme, the focus is placed on the work organisation where construction work and workplaces are analysed based on the view of organisations as gendered and construction workplaces as connected to notions of masculinity. Hence, the focus turned to the context of construction work and workplaces rather than on women per se. It enabled the inclusion of studies analysing construction work as masculine spaces and made visible how stereotypical notions of femininity and masculinity shape the workplace structure and processes and place women as outsiders in that context (cf. Galea et al., 2020; Norberg & Johansson, 2021). More specifically, studies show that women are seen to bring something different compared to men, such as communication skills, or the ability to change the culture of the industry, which at the same time contributes to the separation and coding of characteristics as female or male (Norberg & Johansson, 2021). Construction workplaces thus present a masculine space where women are described as not belonging. The theme emphasises organisations as continuously shaped by gendered processes and practices.

The third theme puts men and masculinity as the focal point by demonstrating how masculine hierarchies are developed and, to some extent, challenged in the workplace. It is argued that understandings of various forms of masculinities as they play out in male-dominated organisations such as the construction industry provides important insights into understanding the conditions for gender equality at work. For example, it is shown how risk-taking, and endurance of pain shape hegemonic hierarchies where masculine identities are ranked according to perceived "manliness" (cf. Gherardi & Nicolini, 2002; Ness, 2012).

Masculine identities are described as closely linked to organisational processes and macho culture. Hence, it stipulates how masculinities are shaped in relation to their context rather than in relation to specific individuals. Whether changes towards hybrid masculinities are development towards gender-equal workplaces or just showing the changing nature of masculinity within remaining power structures remains to be seen.

Finally, comparing gender analyses concerning the construction industry with analysis of other male-dominated industries makes visible that there are several similarities. Reports on discrimination, a lack of opportunities for career progression and sexual harassment are evident in many studies. However, the construction industry stands out because of its particularly low female participation. In general, other industries such as mining and forestry also seem to have moved from physically demanding work tasks towards new forms of work. These tendencies are not reported in relation to construction work. Instead, construction work is still described as physically demanding, and a common argument for the lack of women in the industry relates to women being perceived as not physically strong enough to do the job. However, working conditions are described to be harsh also for men (cf. Ajslev et al., 2016). Similar results have been found in studies of the construction industry in different countries and over the last 20 years which puts an emphasis on the dire need for change if women and men are to flourish as employees in gender-equal healthy workplaces in the construction industry.

Notes

1 See International Labour Organization, 2016, p. 98, for their division between developing and developed economies.

2 Female participation in the formal construction industry reported in 2019, 2020, or 2021.

References

Acker, J. (1990). Hierarchies, jobs, bodies: A theory of gendered organizations. *Gender & Society*, 4(2), 139–158.

Acker, J. (2006). Inequality regimes: Gender, class, and race in organizations. *Gender & Society*, 20(4), 441–464.

Ajslev, J.Z., Møller, J.L., Persson, R., & Andersen, L.L. (2016). Trading health for money: Agential struggles in the (re) configuration of subjectivity, the body and pain among trade workers. *Work, Employment and Society*, 31(6):887–903.

Barreto, U., Pellicer, E., Torres-Machí, C., & Carrión, A. (2017). Barriers to the professional development of qualified women in the Peruvian construction industry. *Journal of Professional Issues in Engineering Education and Practice*, 143(4). 10.1061/(ASCE)EI.1943-5541.0000331

Baruah, B., & Biskupski-Mujanovic, S. (2021). Gender analysis of policy-making in construction and transportation. Denial and disruption in the Canadian green economy. In Magnusdottir, G. L. & Annica Kronsell, A. (eds.)(2021) *Gender, Intersectionality and Climate Institutions in Industrialised States*. London: Routledge.

Baxter, J., & Wallace, K. (2009). Outside in-group and out-group identities? Constructing male solidarity and female exclusion in UK builders' talk. *Discourse and Society*, 20(4), 411–429.

Benya, A. (2017). Going underground in South African platinum mines to explore women miners' experiences. *Gender & Development*, 25(3): 509–522.

Bowers, R. (2019). Navigating the city and the workplace: Migrant female construction workers and urban (im)mobilities. *Global Labour Journal*, *10*(1). 10.15173/glj.v10i1.3406

Bridges, T., & Pascoe, C.J. (2014). Hybrid masculinities: New directions in the sociology of men and masculinities. *Sociology Compass*, 8(3), 246–258.

Brown, W. (1995). *States of injury: Power and freedom in late modernity*. Princeton, NJ: Princeton University Press
Bryce, T., Far, H., & Gardner, A. (2019). Barriers to career advancement for female engineers in Australia's civil construction industry and recommended solutions. *Australian Journal of Civil Engineering*, 17(1), 1–10, 10.1080/14488353.2019.1578055
Chen, M. (2008). *Skills, employability, and social inclusion: Women in the construction industry*. WIEGO network, Harvard University.
Choudhury, T. (2013). Experiences of women as workers: A study of construction workers in Bangladesh. *Construction Management & Economics*, 31(8), 883–898. 10.1080/01446193.2012.756143
Cockburn, C. (1991). *In the way of women: Men's resistance to sex equality in organizations*. Basingstoke, UK: Macmillan. 10.1007/978-1-349-21571-3
Collinson, D.L., & Hearn, J. (1994). Naming men as men: Implications for work, organization and management. *Gender, Work and Organization*, 1(1), 2–22.
Connell, R. (2005). *Masculinities*. (2nd ed.). Cambridge: University of California Press.
Erwic Awards. (n.d.). Awards overview. Downloaded from https://erwicawards-cidb.co.za/
Eurostat. (2022). Share of women by economic activity. Downloaded from: https://ec.europa.eu/eurostat/web/products-eurostat-news/-/edn-20220304-1
Eveline, J., & Booth, M. (2002). Gender and sexuality in discourses of managerial control: The case of women miners. *Gender, Work and Organization*, 9(5), 556–578. 10.1111/1468-0432.00175
Fielden, S.L., Davidson, M.J., Gale, A.W., & Davey, C.L. (2000). Women in Construction: the Untapped Resource. *Construction Management & Economics*, *18*(1), 113–121. 10.1080/014461900371004
French, E., & Strachan, G. (2015). Women at work! Evaluating equal employment policies and outcomes in construction. *Equality, diversity and inclusion: An International Journal*, 34(3), 227–243. 10.1108/EDI-11-2013-0098
Galea, N., Powell, A., Loosemore, M., & Chappell, L. (2015). Designing robust and revisable policies for gender equality: Lessons from the Australian construction industry. *Construction Management and Economics*, *22*(5-6), 375–389. 10.1080/01446193.2015.1042887
Galea, N., Powell, A., Loosemore, M., & Chappell, L. (2020). The gendered dimensions of informal institutions in the Australian construction industry. *Gender, Work & Organization*, *27*(6), 1214–1231. 10.1111/gwao.12458
Gherardi, S., & Nicolini, D. (2002). Learning the trade: A culture of safety in practice. *Organization*, 9(2):191–223. 10.1177/1350508402009002264
Guerrier, Y., Evans, C., Glover, J., & Wilson, C. (2009). 'Technical, but not very ... ': Constructing gendered identities in ITrelated employment. *Work, Employment and Society*, 23(3), 494–511. 10.1177/0950017009337072
Hanna, E., Gough, B., & Markham, S. (2020). Masculinities in the construction industry: A double-edged sword for health and wellbeing? *Gender, Work and Organization*, *27*(4), 632–646. 10.1111/gwao.12429
Hasan, A., Ghosh, A., Mahmood, M. N., & Thaheem, M. J. (2021). Scientometric review of the twenty-first century research on women in construction. *Journal of Management in Engineering*, *37*(3). 10.1061/(ASCE)ME.1943-5479.0000887
Iacuone, D. (2005). "Real men are tough guys": Hegemonic masculinity and safety in the construction industry. *Journal of Men's Studies*. 13(2):247–266.
International Labour Organization. (2015). *Good practices and challenges in promoting decent work in construction and infrastructure projects*. Geneva: International Labour Office.
International Labour Organization. (2022). *ILO modelled estimates database, ILOSTAT* [Labour Force Statistics]. Available from: ILO Data Explorer.
International Labour Organization. (2016). *Women at work: Trends 2016*. Geneva: International Labour Office.
Johansson, M., & Ringblom, L. (2017). The business case of gender equality in Swedish forestry and mining—Restricting or enabling organizational change. *Gender, Work & Organization*, 24(6), 628–642. 10.1111/gwao.12187
Kanter, R.M. (1993). *Men and women of the corporation*. (2. ed.) New York: BasicBooks.
Lu, S.L., & Sexton, M. (2010). Career journeys and turning points of senior female managers in small construction firms. *Construction Management & Economics*, 28(2), 125–139. 10.1080/01446190903280450

Mayes, R., & Pini, B. (2014). The Australian mining industry and the ideal mining woman: Mobilizing a public business case for gender equality. *Journal of Industrial Relations*, 56(4), 527–546. 10.1177/0022185613514206

Musonda, J. (2020). Undermining gender: Women mineworkers at the rock face in a Zambian underground mine, in *Anthropology Southern Africa*, 43(1), 32–42.

Naoum, S.G., Harris, J., Rizzuto, J., & Egbu, C. (2020). Gender in the construction industry: Literature review and comparative survey of men's and women's perceptions in UK construction consultancies. *Journal of Management in Engineering*, *36*(2), 12. 10.1061/(ASCE)ME.1943-5479.0000731

Navarro-Astor, E., Román-Onsalo, M., & Infante-Perea, M. (2017). Women's career development in the construction industry across 15 years: Main barriers. *Journal of Engineering Design and Technology*, *15*(2), 199–221. 10.1108/JEDT-07-2016-0046

Ness, K. (2012). Constructing masculinity in the building trades: 'Most jobs in the construction industry can be done by women'. *Gender, Work & Organization*, *19*(6), 654–676. 10.1111/j.1468-0432.2010.00551.x

Norberg, C., & Fältholm, Y. (2018). "Learn to blend in!": A corpus-based analysis of the representation of women in mining. *Equality, Diversity and Inclusion*, 37(7), 698–712. 10.1108/EDI-12-2017-0270

Norberg, C., & Johansson, M. (2021). "Women and 'Ideal' Women": The Representation of Women in the Construction Industry. *Gender Issues*, *38*(1), 1–24. 10.1007/s12147-020-09257-0

Salinas, P., & Romaní, G. (2014). Gender barriers in Chilean mining: A strategic management, *Academia Revista Latinoamericana de Administración*, 27(1): 92–107.

Statistics Sweden. (2022). *Craft and related trades workers: Building and manufacturing workers in Sweden* by sex and average age 2018. Downloaded from: https://www.scb.se/en/finding-statistics/

Stergiou-Kita, M., Mansfield, E., Colantonio, A., Garritano, E., Lafrance, M., Lewko, J., & Travers, K. (2015). Danger zone: men, masculinity and occupational health and safety in high risk occupations. *Safety Science*, 80, 213–220. 10.1016/j.ssci.2015.07.029

Stop Macho Culture. (n.d.). Stop Macho Culture. Downloaded from: https://stoppamachokulturen.nu/english/

The National Association of Women in Construction (NAWIC). (n.d.). Statistics of women in construction. Downloaded from: https://www.nawic.org/statistics

U.S. Bureau of Labor Statistics. (2022). Labor force statistics from the current population survey. Downloaded from: https://www.bls.gov/cps/cpsaat18.htm

Vijayaragunathan, S., & Rasanthi, T. (2019). An insight to women in construction for fostering female careers in Sri Lankan construction industry. *Journal of International Women's Studies*, 20(3), 168–173.

Watts, J.H. (2007). 'Porn, pride and pessimism: Experiences of women working in professional construction roles'. *Work, Employment and Society*, 21(2), 299–316.

West, C., & Zimmerman, D. (1987). Doing gender. *Gender & Society*, 1(2), 125–151. 10.1177/0891243287001002002

Wice Awards. (n.d.). Women in construction and engineering awards. Downloaded from: https://www.womeninconstructionawards.co.uk/

12

MISTREATMENT OF MIGRANT CONSTRUCTION WORKERS

Trajectory from the past to the present and into the future

Abdul-Rashid Abdul-Aziz[1], *AbdulLateef Olanrewaju*[2], *and Poline Bala*[3]

[1]Wawasan Open University, Penang, Malaysia; [2]Universiti Tunku Abdul Rahman, Malaysia; [3]Universiti Malaysia Sarawak, Malaysia

Introduction

As Qatar began constructing stadia, infrastructure and other development projects for World Cup 2022, the global media gave coverage to the abuse of construction migrant workers engaged for these projects. Institutions such as the International Labour Organisation, Amnesty International and Human Rights Watch were relentless in uncovering the plight of these migrants who came mainly from South Asia. Exploitation of construction labour migrants has been going on for decades, even centuries, but never before had so much global attention been aroused on this issue. The reality is that labour abuse of migrant construction workers is a global issue which has historical roots.

This chapter can be divided into two parts. The first begins by charting construction labour flows between countries from ancient times to more recent times. Cross-border construction labour flow will continue unabated, especially to developed economies, due to the twin effects of an ageing domestic construction workforce in host nations and high levels of poverty in labour-abundant countries. Even labour-saving technologies cannot arrest this movement of labour. Construction migrant trails criss-cross the globe. Benefitting from foreign remittances, among others, labour-abundant countries put in place various measures to facilitate migration flows. However, their desire to liberalise the cross-border movement of labour under the aegis of the multilateral free trade agreement failed.

The second part of this chapter focuses on the ill-treatment of in-migration construction workers. Four main aspects are elaborated upon: exploitative wages, insecure employment and forced labour, low safety standards and poor living conditions. The chapter ends by posing the question whether construction labour migrants will ever get to enjoy fair, safe and decent employment given that abuse of migrant labour is deeply entrenched in the construction industry of host countries and that their legislation and the judicial systems are intrinsically discriminatory. This chapter concludes by making certain recommendations for international organisations and non-government organisations to adopt.

DOI: 10.1201/9781003262671-12

Construction labour flows

Historical and more recent construction labour migration

Construction labour migration is as old as humankind. Impressive ancient edifices that stand until today, sometimes with nearby settlement remains, bear testimony to large-scale mobilisation of workers either by compulsion or voluntarily. An army of people was mobilised to work on the Great Wall of China and other monumental structures during the reign of the first Qin Emperor of China, Qin Shi Huangdi (259–210 BCE), famous for unifying ancient China (Yates, 2001). In ancient Egypt, peasant labour was conscripted during the floods to help build the pyramids (Allen, 1997). Evidence of purpose-built villages for these workers near the pyramids attests to their home away from home (Booth, 2015). The Mayan religious centre of Teotihuacan, built around the second and third centuries, was the result of compulsory movement of population (Manzanilla, 2014). Archeological settlement remains at Durrington Walls, 3 kilometres from Stonehenge, suggest that migrants worked on the megalith (Pearson et al., 2011). Even though migration and forced labour are separate issues, they are closely intertwined. A recent tri-partite publication from the International Labour Organisation (ILO), Walk Free and International Organisation for Migration (IOM) (2022) estimated that migrant workers are three times more likely to be in forced labour than non-migrant workers.

Fast forward to more recent times, urban growth and physical infrastructure development in many nations owe in large part to domestic as well as international construction migrants. In the seventeenth and eighteenth centuries, seasonal construction migrant workers constituted part of the Dutch labour market (van Lottum, 2011). During the Industrial Revolution, European countries such as France and Germany experienced internal migration whereby rural construction workers established themselves in cities (Moch, 2011). By the nineteenth century, Irish construction workers were present in large numbers in England, constructing canal and railway networks as well as other infrastructure projects (Mulvey, 2018), working alongside English workers as well as the Scots and Europeans. The 'navvy' made up of specialised labour force provided the muscle power and practical expertise for the great infrastructure projects of England's Industrial Revolution (Morris, 1994).

The westward expansion of nineteenth-century America, characterised by canals and railroad construction, owed in large measure to the barely visible and forgotten Irish, Chinese, Mexican, Mormon and black immigrants as well as native-born working-class Americans (Sadowski-Smith, 2008; Dearinger, 2015). Railways were a prerequisite to Canada's 'wheat boom' which began in the late 1800s (Lewis & Robinson, 1984). Their construction was aided by workers of various nationalities, including Ukrainians, Poles, Czechs, Slovaks, Serbs, Hungarians, Croats, Bulgarians, Macedonians, Turks and Chinese (Selby, 2012).

South Africa became the leading recipient of migrants in the southern African region following the discovery of diamonds in Kimberley in the 1860s, and gold in the Witwatersrand in the 1880s (Moyo, 2018). Construction craft workers such as plumbers and electricians from Europe, mainly from the United Kingdom, also arrived during that time (Mukora, 2008). Following the abolition of the trans-Atlantic slave trade in 1850 which eventually led to the abolition of slavery in 1888, Brazil was forced to find alternative sources of workers, especially during the 1870s when the country experienced a 'railway mania' (Lamounier, 2000). The workers comprised Brazilians (usually migrant farmers working off-season), Europeans (including Portuguese, Germans, Italians, English, French and Swiss), Chinese and former slaves. During the 1800s, pre-independence Singapore owed all its roads, canals and public works to convicts shipped from India, so claimed Edmund Augustus Blundell, the Governor of the Straits Settlement in 1855–1859 (Turnbull, 1970).

Construction labour migration is as old as history. As pointed out at the end of this chapter, the few historical records point to labour abuse having historical origins. It can be extrapolated that mistreatment of construction migrants took place even farther back in time.

Present international construction labour flows

Many countries, including the ones mentioned earlier, still depend on domestic and cross-border migrant workers. Again here, only a few examples are highlighted. Canada's $180 billion infrastructure plan will stall unless the shortage of skilled construction workers is addressed (Cook, 2021); one source points to a likely shortage of more than 300,000 construction workers by 2030 (Buildforce, 2021). South Africa and Kenya are among the countries in Africa that are currently facing a shortage of local skilled construction artisans (Tshele & Agumba, 2014; Kiganda, 2022). The United Kingdom is also facing a construction labour shortage exacerbated by the COVID-19 pandemic which disrupted the inflow of immigrant labour (CITB, 2021). Compounding the problem is the departure of the United Kingdom from the European Union (EU) which ended the automatic right of EU citizens to enter and reside in the United Kingdom with no restrictions on the types of jobs they could hold. The RICS Global Construction Monitor Report for the fourth quarter of 2021 states that construction labour shortage is an issue in many other countries including Australia, Canada, the United States and many countries in the Middle East such as Oman (RICS, 2022). The entire southern Europe, comprising Italy, Spain, Greece and Portugal are also facing construction humanpower shortfalls (Varvitsioti and Kazmin, 2022). These shortages in these countries and more provide the stimulus for labour in-migration.

Construction labour shortage in Canada (Buildforce, 2021), the United Kingdom (CIOB, 2012; ONS 2018), the United States (Sokas et al., 2019), Republic of Korea (Seol, 2018), Australia (Sivam et al., 2017) and many other developed countries stems from the ageing local construction workforce with youths joining the industry in insufficient numbers. Even labour-saving technologies are unlikely to free labour-starved nations from the need to import labour. While many construction projects already include limited amounts of basic pre-manufactured and offsite fabricated components, such as windows, balustrades and external fencing, larger offsite and pre-manufactured members have yet to gain full traction (Ofori-Kugaru & Osei-Kyei, 2021) in many countries.

On the supply side, despite the high migration costs (Abella, 2018), poverty and financial obligations provide the push factor for the migration of unskilled and semi-skilled construction workers (Human Rights Watch, 2012). Their migration could either be in the form of temporary residency (i.e., circular) or permanent (BWI, 2013). Some may even use certain countries as a stepping stone to acquire new skills for their eventual destinations (Buckley et al., 2016).

Construction labour flows are likely to continue in the years to come as the push and pull factors described in this section will continue to persist (BWI, 2013). Construction work, which covers the gamut of new construction, refurbishment, demolition and reconstruction, will always take place the world over. The next section explains that construction labour migration traverses the globe, thus giving labour abuse a global significance.

Direction of contemporary labour flows

The media coverage often highlights the contentious issue of contemporary labour migration, often clandestinely, from developing to developed countries. However, the trails also take place between developed countries and between developing countries.

Developing-developed labour flows

The hardline stance on immigration adopted by Donald Trump during his 2016 presidential campaign and subsequent presidency brought worldwide attention to the scale of illegal movement of people into the United States (Baker & Bader, 2022), many of whom ended up working in the construction sector. A study by a research organisation showed that the United States is heavily reliant on immigrants for construction work (i.e. roughly a quarter of the total workforce) with certain occupations being heavily dependent on them (see Table 12.1). More than 10 percent of construction workers are estimated to be undocumented immigrants (Svaljenka, 2021). Mexicans make up the largest immigrant group in the US construction industry (Spindler-Ruiz, 2021). Other immigrant groups come from Central America, Latin America and Eastern Europe (Erlich & Grabelsky, 2005).

There is a long tradition in Western Europe of construction jobs being filled by migrants (Wells, 2016). Around 1974, most Western European countries abandoned migrant labour recruitment, and introduced restrictive entry rules (Castles, 2006). The construction sector is one of the biggest employers for undocumented migrant workers in Europe (Honsberg, 2004). These people come from the South (as well as East). For example, in Portugal, construction labour shortage created the space for African immigrants since the late 1960s (Carreira, 1982; Mendoza, 2000). Since the late 1980s, numerous unauthorised migrants from developing countries as well as Eastern Europe have been entering Italy, Greece and Spain (Reyneri, 2012). Specifically, the workers originated from Africa (including from Nigeria, Senegal, Morocco and Tunisia); Asia (from the Philippines and Sri Lanka) and Eastern Europe (from Albania and Romania). Construction is one of the sectors in which these migrants secure irregular employment. In the United Kingdom, the latest figures from the Office for National Statistics indicate that, between 2014 and 2016, there were almost 50,000 regularised non-EU immigrants working in the construction industry (ONS, 2018).

Table 12.1 Immigrants in selected construction-related occupations in the United States

Workers	*Number of foreign-born workers*	*Share of all workers, foreign born (%)*	*Number of undocumented workers*	*Share of workforce that is undocumented (%)*
Construction workers	686,000	38	445,800	23
Carpenters	382,000	31	225,600	19
Painters and paperhangers	264,000	45	167,300	29
Roofers	98,000	46	75,600	32

Source: American New Economy (2020) and Svajlenka (2021).

Note: The data on foreign-born workers was based on the 2018 American Community Survey while the data on undocumented workers was based on the 2018 and 2019 American Community Survey.

Table 12.2 Number of foreign construction trainees under the EPS, March 2015

Country of origin	*Number*
Vietnam	2,797
Cambodia	1,383
Myanmar	1,353
Thailand	921
Indonesia	338
Others	275
Total	7,067

Source: Seol (2015), cited in (ADBI, ILO and OECD, 2016).

Note: 'Others' covers Sri Lanka, the Philippines, Uzbekistan, Bangladesh, Mongolia and Pakistan.

In Singapore, the roughly 300,000 migrant construction workers originated mainly from Bangladesh, India and China (Wee et al., 2022). A tripartite publication from the Asian Bank Development Bank Institute, ILO and Organisation for Economic Co-operation and Development (2016) reported that in 2004, the Republic of Korea initiated the Employment Permit System (EPS) to admit low-skilled foreign workers for the benefit of small companies that have difficulty recruiting local workers. The cumulative number of unskilled construction workers in the Republic of Korea from developing countries in 2015 is shown in Table 12.2.

Developed-developed labour flows

Construction labour movement also takes place between developed nations. The UK Office for National Statistics data (mentioned earlier) indicated that between 2014 and 2016, there were roughly 37,000 construction workers from EU 15 countries (i.e., Austria, Belgium, Denmark, Finland, France, Germany, Greece, Ireland, Italy, Luxembourg, the Netherlands, Portugal, Spain and Sweden) (ONS, 2018). Portugal became a country of emigration in the second half of the nineteenth century (Pereira & Azevedo, 2019). The fourth wave of Portuguese emigration started around 2001. Emigration flowed to the traditional destinations of Germany, France, Switzerland and one new country, Spain. Most of the Portuguese emigrants worked in civil construction. Many Portuguese emigrants who lost their jobs during the financial crisis of 2008 either returned to Portugal or re-emigrated to England or France.

In New Zealand, while the majority of people who received short-term visas to work in construction-related occupations from 2011 to 2020 were from developing countries, there was a sprinkling from the United Kingdom, Ireland and the United States (Schiff, 2022). During the reconstruction work in the wake of the Canterbury earthquakes in 2010 and 2011, the New Zealand government relaxed its immigration rules, including decoupling visas from specific jobs (INZ 2015; Woodhouse 2015). The United Kingdom (25%) and Ireland (25%) contributed half of migrant workers in Canterbury in 2014 (Statistics New Zealand as cited in Searle et al. 2015). Australian citizens and permanent residents do not need a visa to work in New Zealand, and vice versa (Schiff, 2022). There was a net outflow of around 9,800 technicians and trades workers from New Zealand to Australia between 2010 and 2013, and a net outflow of 4,100 construction workers between 2010 and 2014. Subsequently, there was a net inflow of around 1,600 technicians and trades worker from

2014 to 2017 and a net inflow of around 300 workers from 2015 to 2017. This coincided with the increase in residential construction activity in New Zealand.

Developing-developing labour flows

Russia is home to one of the world's largest migrant population, second only to the United States (Human Rights Watch, 2009). The majority come from the countries of the ex-Soviet Union such as Takijistan, Uzbekistan, Ukraine and Belarus. Forty percent of the migrants work in the highly unregulated construction sector. Asia has played a major role in global migration patterns, and will continue to do so, involving construction labour, with flows to developed as well as developing countries (ADBI, ILO and OECD, 2016). Migrants from neighbouring Laos, Myanmar and Cambodia occupy the labour vacuum in Thailand created by Thai construction workers seeking employment overseas (Kongchasing & Sua-iam, 2021). Malaysia is one of the destination countries that Thai construction workers migrate to (Sakolnakorn, 2019). Malaysia is also host to Bangladeshi site operatives (Abdul-Aziz, 2001a), as is the Maldives (Heslop & Jeffery, 2020). South Asians (i.e., Indians, Bangladeshis, Pakistanis, Nepalis, Afghans and Sri Lankans) make up the bulk of overseas migrants in the Gulf Cooperation Council (GCC) that is comprised of Bahrain, Kuwait, Oman, Qatar, Kingdom of Saudi and United Arab Emirates (Abdul-Aziz et al., 2018). There were around 41,000 Indian construction workers even in the small Kingdom of Bhutan in June 2019, equivalent to about 10 percent of the total employment in the country (MoLHR, 2019, cited in Zangpo, 2020).

Construction labour flows criss-cross the globe. Simplistically, this section reduces the labour mobility description along the developing-developed countries dichotomy. The reality is much more complex as some countries, such as the United Kingdom, Portugal, New Zealand, Malaysia and Thailand are simultaneously labour-sending as well as labour-receiving countries, although the magnitudes of the two segments may not even out. Complexity also manifests itself in the way migrants are mistreated. Nonetheless, it was possible to weave a tapestry of these mistreatments in the section below based on the fragmented patchwork of studies conducted at the global, regional, national and even individual levels.

Remittances and general agreement on trade in services

As highlighted in the previous section, the movement of construction workers from developing countries is a significant part of the narrative of global construction labour migration. Construction labour outflow from developing countries can be self-initiated; it can also be policy driven. This section focuses on the latter.

Several countries, mainly in Asia, promote labour migration to relieve unemployment pressures, thereby fulfilling the national objectives of poverty reduction, socio-economic development and foreign exchange earnings (Wickramasekera, 2002; Shah, 2018). Remittance flows to low- and middle-income countries worldwide are estimated at about USD 550 billion in 2019 (World Bank, 2021). These countries include the Philippines, Indonesia and Thailand. Nepal (Pant, 2011) and Bangladesh (Hassan & Shakur, 2017) also depend on remittances of their nationals working abroad. For Bangladesh, the total volume of remittances from workers in all sectors rose to over USD 15 billion in 2015 or about 8 percent of gross domestic product (GDP) and have become a major source of foreign exchange earnings, second only to ready-made garments (ADB & ILO, 2016). According to a 2017 Ukrainian official estimate, 39 percent of Ukrainian out-migrants worked in the construction sector (SSSU, 2017, cited in Pieńkowski, 2020). The Ukrainian economy benefited by way of remittances equivalent to 8 percent of GDP. Even North Korea exports its workers to 45 countries at one time or another (Connell, 2016). They have been detected mainly in China and Russia but also in Poland, Qatar, Kuwait, Nepal, Mongolia and Senegal (Connell, 2016; Lankov et al., 2020).

Labour-exporting countries such as Bangladesh, India, Indonesia and Sri Lanka have come up with legislation, policies, labour agreements and even dedicated ministries and labour attaches in host countries to streamline and regulate migration (Shah, 2018; Chanda & Gupta, 2018; Bosc, 2022). Some even provide pre-departure orientation sessions and vocational training. Inevitably, these countries face legal, diplomatic and institutional constraints to protect their citizens abroad from labour abuse (HRW, 2009; Malit, 2018).

Skills acquisition by workers while overseas is not a factor in the decision to promote the flow of national workers abroad. Trainee schemes used by the Republic of Korea (Abella et al., 1994) and Japan (Endoh, 2019) are really disguised cheap labour mechanisms. During the formulation of the General Agreement on Trade in Services (GATS) which was an integral part of the Uruguay Round of multilateral trade negotiations (1986–1993) under the auspices of the World Trade Organisation, labour-abundant countries called for the liberalisation of the regulations to enable free movement of across borders as quid pro quo for the liberalisation of the international movement of goods and services (Abdul-Aziz & Tan, 1997). Their attempts failed. To appease these labour-exporting countries, an annex was inserted in the GATS to permit further deliberations on the free movement of labour beyond the Uruguay round deadline, i.e., until June 1995 (Abdul-Aziz & Tan, 1998). Scrutinising the decisions that were made thereafter, mass movement of construction labour was again denied, but intra-corporate transfer was permitted. The argument put forward by the developed nations on behalf of their powerful multinational service companies was that without the movement of top executives, commercial presence would be difficult to be established. There are four modes of trade in services under GATS, the fourth which is of concern in the discussion in this chapter is the temporary movement of natural persons for the supply of services. 'Natural persons' refers to self-employed individuals or expatriates sent abroad to supply a service. Employment-seeking individuals do not qualify as natural persons.

Ill-treatment of in-migration construction workers

Hitherto, this chapter has mentioned the abuse that construction migrants face. This section provides full treatment to this topic. It is actually a prelude to the subsequent section which asks whether labour abuse can ever be resolved in the construction industry.

The construction labour markets can vary greatly between different places (Buckley et al., 2016). However, there are detectable similarities in unacceptable working conditions and employment relations experienced by cross-border construction migrant workers everywhere; only the intensity differs.

Exploitation wages

In-migrant construction workers all over the world, even those who enter the labour market legitimately, generally receive lower wages than the locals. An ILO survey before the

COVID-19 pandemic on wage differences between migrant workers and nationals for 49 countries found that wage inequalities between migrant workers and nationals were of very high levels in many countries and widening in some cases (Amo-Agyei, 2020). Migrants earned 12.6 percent less per hour than nationals in high-income countries and 17.3 percent per hour than nationals from low- and middle-income countries. The migrant workers also do not enjoy the same benefits as locals. For example, foreign site operatives in Japan receive wages that are, on average, a third of what their Japanese counterparts receive (BWI, 2018). Various construction industry studies specific to particular countries – Russia (HRW, 2009), Bahrain (HRW, 2012), Japan (Endoh, 2019), Maldives (Heslop & Jeffery, 2020), Bhutan (Wijunamai, 2020) and Kuwait (Bosc, 2022) – support this general observation. So too do various construction migration studies specific to nationality groups – North Koreans (Connell, 2016), Thais (Sakolnakorn, 2019) and Mexicans (Spindler-Ruiz, 2021).

Furthermore, migrant construction workers face the risk of late payments of wages, arbitrary deductions in wages or worst, non-payment (Wells, 2016), regardless of whether the workers are Indonesians in Malaysia (Wahyono, 2007), Kyrgyz in Russia (HRW, 2009), Indians in Bahrain (HRW, 2012) or Kenyans in Qatar (HRW, 2022). This is partly due to the multi-tier subcontracting system and the 'pay when paid' wage payment practice which has time ramifications for money to trickle down (Well, 2016; HRW, 2022). Migrant construction workers tend to be engaged by subcontractors and labour subcontractors at the bottom of the subcontracting hierarchy are paid last. Those with precarious legal status and with no formal access to workplace representation or recourse are the most vulnerable (Buckley, 2016). It has been suggested that withholding wages is a deliberate ploy to force companies into bankruptcy, at which point they have an excuse not to pay, at least not without protracted and expensive legal procedures that have no guarantee of success (Jureidini, 2018).

Not all the wages the migrants earn go to them. Out of the wages, migrants who make the journey by borrowing have to settle their debts which financed their migration (Buckley et al., 2016). Construction is one of three sectors where debt bondage is prominent (ILO, Walk Free & IOM, 2022). For example, Bangladeshi workers in the Maldives have to work for several years before paying off their brokers (Heslop & Jeffery, 2020). Most of the salaries of North Korean construction workers go directly to the state; to make money for themselves and their families, they moonlight at night and during weekends (Connell, 2016). Migrant workers also run the risk of wage theft, that is the siphoning of funds by their immediate employers (Buckley et al., 2016; Heslop & Jeffery, 2020). It is reported that money is extorted from construction migrants in Russia by police and other officials during spot document checks, and by border guards, customs officials and others during their travel to Russia (HRW, 2009). Indonesian construction workers in Malaysia have also complained of money being extorted from them by the police and immigration officials (Wahyono, 2007).

Insecure employment and forced labour

Foreign construction workers are vulnerable to job insecurity. They tend to be disproportionately represented in lay-offs and retrenchments during times of low activity in the industry (Krings et al., 2011; Buckley, 2012). Construction employers may maintain a small and diminishing number of 'core' workers employed on a permanent basis (if at all) and a much larger number of temporary workers, often migrants, who are employed on short-term contracts, are self-employed or work for labour-only subcontractors. Such job insecurity prevails even in the developed economies of Europe (Vogel, 2016).

Construction migrants are also made to work excessively long working hours without overtime payments as was detected in Russia (HRW, 2009), Qatar (HRW, 2022) and Bahrain (HRW, 2012). Construction is one of the five sectors which account for the majority of forced labour, i.e., 16 percent of the total, 22 percent for all men (ILO, Walk Free & IOM, 2022). The ILO Forced Labour Convention, 1930 (No. 29) defines forced labour as 'all work or service which is exacted from any person under the threat of a penalty and for which the person has not offered himself or herself voluntarily.' Forced labour in the construction industry has been reported in Malaysia (Augustin, 2021). The Department of Homeland Security acknowledged that forced labour also takes place in the US construction industry (DHL, 2022). In the EU, construction is the second highest on the list of economic sectors most prone to labour exploitation (FRA, 2015). Migrants come from within or outside the EU. Rampant subcontracting practice which obscures the legal situation of these migrants, and which also complicates monitoring is blamed for the exploitation of workers in construction. Simkunas and Thomsen (2018) provide narratives of migrant construction workers from Central and Eastern Europe in Denmark on how they cope with precarity and insecurity.

The *kafala* or sponsorship system practised by many countries in the Middle East such as Bahrain, UAE, Saudi Arabia, Oman, Qatar Jordan, Kuwait and Lebanon gives employers excessive power over the migrant workers, rendering the workers vulnerable to abuse and exploitative working conditions (Diop et al., 2018; Bosc et al., 2022). Foreign workers are tied for a fixed duration to employers with no option to switch employment, no matter how abusive the situation may be. Bahraini employers expect their migrant workers to pay for the passports to be returned (HRW, 2012). The *kafala* system can give rise to forced labour (ILO, 2017). Elsewhere, in places such as Russia (HRW, 2009) and Malaysia (Wahyono, 2007), employers habitually confiscate the migrants' passport as a weapon of coercion. Workers in Russia who complain about unpaid wages, or about poor work or living conditions face threats or violence (HRW, 2009). In the United States, other coercion strategies used by construction employers on their migrants include debt entrapment, withholding wages, restricted mobility and threatened abuse of legal processes (Halegua & Chin, 2021). The threat of deportation is also weaponised to dissuade the migrant workers from protesting (Erlich & Grabelsky, 2005).

Low safety standards

Construction work is intensive and temporary or seasonal, with significantly higher occupational hazards (Ujita et al., 2019). Migrants at the lower end of the construction wage ladder endure a disproportionately higher risk of workplace injuries and fatalities (Buckley, 2016). Construction migrants who work informally endure even worse safety standards (Ujita et al., 2019).

The outdoor nature of construction works exposes migrant construction workers to the elements. Heat-related hazards is the most important health issue in the countries in the Middle East such as the UAE (HRW, 2006). Mid-day work is officially banned by all GCC countries (ILO, 2019). However, temperatures are still extremely high outside of the hours covered by the ban, and limited labour inspection undermines the policy effectiveness. With global warming, heat stress will become an ever greater concern beyond the GCC countries.

Migrant workers may also suffer from mental distress. Every year, several Indian construction workers commit suicide in the GCC because of harsh working and living conditions, in particular, wage abuse (HRW, 2006; HRW, 2012; Chanda & Gupta, 2018).

Migrant workers are less eligible for social protection than their citizen counterparts (ILO, 2016). In the UAE (HRW, 2006) and Russia (HRW, 2009), construction migrant workers have to pay for medical treatment following their injuries. Illegal workers have the worst treatment. In the UAE, critically injured illegal migrant workers are 'dumped' at the hospitals by their employers without disclosing their identities to the hospital authorities (HRW, 2006).

Poor living conditions

If their employers can get away with it, in-migration workers are provided with the least expensive accommodation option. At worst, onsite accommodation is makeshift houses, windowless with bare furniture, as what migrant workers put up with in Bhutan (Wijunamai, 2020) and Malaysia (Abdul-Aziz, 1995). In the GCC countries such as Bahrain, Qatar and Kuwait, migrant construction workers, mainly from South Asia, are housed in dormitory-style accommodation that is cramped and dilapidated, with insufficient sanitation, running water or other basic amenities (HRW, 2012; HRW, 2022; Bosc et al., 2022). In Russia, migrants are housed in transport containers or trailers kept on worksites, buildings under construction or renovation and even tents with poor sanitary conditions (HRW, 2009). Substandard accommodation for construction migrants can also be found in many countries in the EU (FRA, 2015).

Poor living conditions can have an impact on the occupants' health. In Chiba City, Japan, crowded dormitories have been linked to comparatively higher tuberculosis transmission incidence among construction workers (Igari et al., 2009). Outbreaks of the zika virus, dengue and chikungunya in Singapore during the mid-2010s were suspected to stem from migrants living in dormitories (TWCT, 2016). Many factors can account for the higher risk of migrants to infectious diseases, one of which is their living conditions (Sadarangani et al., 2017). The Institute for Human Rights and Business, an independent think tank, highlighted poor, dense living conditions in dormitories in the GCC and Singapore as the source of the spread of COVID-19 among migrant workers (Subramaniam, 2020). It called for the provision of decent housing for workers.

Is fair, safe and decent employment a far-fetched idea?

Labour abuse in the construction industry is not a modern-day phenomenon. Its origins can be traced to decades, even centuries old practices. In Brazil, during the 'railway mania' of the nineteenth century, migrant workers were engaged by subcontractors, not by principal contractors (Lamounier, 2000). Therefore, subcontracting is not 'post-modern' as Wells (2007) suggests. The Canadian boom years from the late nineteenth century to the early twentieth century saw railways built across Canada (Lester, 2005). The lives of the railway workers, who were mainly immigrants, were nasty, brutish and often shortened due to accident and/or disease, the latter the result of unsanitary camp conditions. Workers who died were buried in the bush. The problems began at the start with the way the workers were recruited and continued with the form of transportation to and from the worksite, wages, housing and food, the provision of supplies and the medical treatments available to the sick and injured. Again, subcontracting was blamed for the callousness shown to the workers. The practice of bonded labour was evident more than two centuries ago in the United States when railroad companies needed unskilled labour from China (Sadowski-Smith, 2008). Chinese entrepreneurs already in the country took on the immigrants' passage debt, helped them find work, negotiated their terms of labour and ensured the repayment of their debts.

Structural changes to the construction industry in modern times in various countries further entrench undesirable working conditions and abuse perpetrated on migrant construction workers. Erlich and Grabelsky (2005) describe the dramatic transformation in the US construction industry since the 1970s due to a combination of interlinked factors that include the decline in union density, drop in construction wages, growth of anti-union forces, shift towards construction management and the emergence of an underground economy. Consequently, labour abuse has become widespread in the country. Furthermore, using selected countries in the Middle East, Bosc et al. (2022) highlight the intrinsically discriminatory feature of legislation and judicial systems of labour-receiving countries. For the first time ever in its post-war history, most likely in acknowledgement of its dwindling pool of domestic construction labour (one out of seven Japanese construction workers was older than 60), Japan introduced, in 2017, the Specific Skilled Workers (SSW) visas which allow construction immigrants to stay in the country indefinitely (Liu-Farrer, 2020). Other countries may have to follow suit to fill the gap which will be left by the departure from the industry of their ageing construction workers. However, lifting restrictive immigration policies may not mean automatic relief from labour abuse. Central to their poor treatment of the immigrant workers is the perception of these workers as no more than disposable construction resources.

External pressure may bring labour reforms, but only superficially. All the GCC countries except Bahrain introduced the Wage Protection System, but the online system fails to prevent wage violations (Jureidini, 2018; HRW, 2022). Because of unwanted negative publicity (the most recent is Qatar after winning the 2022 World Cup bid), a few of these GCC countries have also tried to reform the *kafala* system at different points in time. Only when labour maltreatment impacts the general public would the host country react. Acknowledging that confined space and poor hygiene could have led to the spread of COVID-19 among construction workers, the Malaysian government quickly revised the Workers' Minimum Standards of Housing and Amenities Act (Act 446) in 2019 with the aim of improving workers' housing standards (ADBI, ILO, & OECD, 2022). The Act was expanded to cover workers' accommodation provided by employers as well as centralised accommodation providers. The penalty was stiffened – errant employers or accommodation can be fined up to RM200,000 or jailed up to three years or both. In February 2021, the government published the Emergency (Employers' Minimum Standards of Housing, Accommodations and Amenities (Amendment) Ordinance. With the COVID-19 pandemic subsiding, it remains to be seen whether this regulation will be strictly enforced. Judging from past experience, it would not (Abdul-Aziz, 2001b).

Conclusion

Construction labour migration crisscrosses the globe. Therefore, the mistreatment of construction migrant workers is a global issue. Labour abuse of in-migrants is centuries old. As construction of megaliths in ancient times deployed armies of workers, labour abuse is likely as ancient as humankind. The movement of construction manual workers across borders will continue in years to come because of the features associated with the host and home countries. By extension, labour abuse will continue to prevail, not only because of the labour structure of the modern construction industry, but also, discriminatory legislation and judicial systems of host countries.

The ILO has been championing the rights of migrant workers to fair, safe and decent work in the construction sector for decades (Buckley et al., 2016). It concedes that this work has been arduous. Construction labour discrimination can only be overcome through a combination of short-, medium- and long-term policies (Bosc et al., 2022). All international organisations and NGOs including the UK Campaign Against Modern Slavery should orchestrate their efforts for greater impact. Their combined investigation of labour abuse in Qatar has shown the way; it culminated in the first inter-governmental agreement to address all aspects of international migration comprehensively, the Global Compact which was concluded in December 2018.

References

Abdul-Aziz, A.-R. 1995, *Foreign labour in the Malaysian construction industry*, International Labour Office, Geneva.

Abdul-Aziz, A.-R. 2001a, 'Bangladeshi Migrant Workers in Malaysia's Construction Sector', *Asia-Pacific Population Journal*, vol. 16, no. 1, pp. 3–22.

Abdul-Aziz, A.-R. & Tan, A.C.N. 1997, 'The General Agreement of Trade in Services (GATS) and its Implications for Non-Industrialised Contracting Countries', *Journal of Real Estate and Construction*, vol. 7, pp. 38–52.

Abdul-Aziz, A.-R. & Tan, A.C.N. 1998, 'GATT, GATS and the Global Construction Industry', *Engineering, Construction and Architectural Management*, vol. 5, no. 1, pp. 31–37.

Abdul-Aziz, A.-R., Olanrewaju, A. L. & Ahmed, A.U. 2018, 'South Asian Migrants and the Construction Sector of the Gulf', in M. Chowdhury and S. I. Rajan (eds.), *South Asian migration in the Gulf*, Palgrave Macmillan, Basingstoke, pp. 165–189.

Abdul-Aziz, A.R. 2001b, *Site operatives in Malaysia: examining the foreign-local asymmetry*, unpublished report for the Sectoral Activities Programme, ILO, Geneva.

Abella, M. I. 2018, 'The High Cost of Migrating for Work to the Gulf', in P. Fargues & N. M. Shah (eds.) *Migration to the Gulf. Policies in sending and receiving countries*, Gulf Research Centre, University of Cambridge, Cambridge, pp. 221–242.

Abella, M. I., Park, Y.-B., & Bohning, W. R. 1994, *Adjustments to labour shortages and foreign workers in the Republic of Korea*, International Migration Paper, International Labour Office, Geneva.

Allen, R. C. 1997, 'Agriculture and the Origins of the State in Ancient Egypt', *Explorations in Economic History*, vol. 34, pp. 135–154.

American New Economy. 2020, *Building America: immigrants in construction and infrastructure-related industries*, 3rd September. Available at: https://research.newamericaneconomy.org/report/covid19-immigrants-construction-infrastructure/ (Accessed on 10 October 2022).

Amo-Agyei, S. 2020, T*he migrant pay gap: understanding wage differences between migrants and nationals*, International Labour Organisation, Geneva.

Asian Development Bank & International Labour Organisation. 2016, 'Overseas Employment of Bangladeshi Workers: Trends, Prospects, and Challenges', *ADB Briefs*, No 63, August. Available from: https://www.adb.org/sites/default/files/publication/190600/overseas-employment-ban-workers.pdf (Accessed on 17 October 2022).

Asian Development Bank Institute, International Labour Organisation and Organisation for Economic Co-operation and Development. 2016, *Labor migration in Asia: building effective institutions*, Asian Development Bank Institute, Tokyo; International Labour Organisation, Bangkok; & Organisation for Economic Co-operation and Development, Paris.

Asian Development Bank Institute, International Labour Organisation and Organisation for Economic Co-operation and Development. 2022, *Labor migration in Asia covid-19 impacts, challenges, and policy responses*, Asian Development Bank Institute, Tokyo; International Labour Organisation, Bangkok; & Organisation for Economic Co-operation and Development, Paris.

Augustin, S. 2021, 'Complaints aplenty about Forced Labour in Construction Industry, says Suhakam', *Free Malaysia Today*, April 9, Kuala Lumpur Available at: https://www.freemalaysiatoday.com/category/nation/2021/04/09/complaints-aplenty-about-forced-labour-in-construction-industry-says-suhakam/ (Accessed on 24 October 2022).
Baker, J.O. & Bader, C.D. 2022, 'Xenophobia, Partisanship, and Support for Donald Trump and the Republican Party', *Race and Social Problems*, vol. 14, pp. 69–83.
Booth, C. 2015, *Lost voices of the Nile: everyday life in ancient Egypt*, Amberley Publishing Limited, Gloucestershire, England.
Bosc, I., Lerche, J., Shah, A., Fajerman, M. & Wadhawan, N. 2022, *Understanding patterns of structural discrimination against migrant and other workers in some countries of South and West Asia*, International Labour Organisation, Geneva.
Buckley, M. 2012, 'From Kerala to Dubai and Back Again: Construction Migrants and the Global Economic Crisis', *Geoforum*, vol. 43, no. 2, pp. 250–259.
Buckley, M., Zendel, A., Biggar, J., Frederiksen, L. & Wells, J. 2016, *Migrant work & employment in the construction sector*, International Labour Organisation, Geneva.
Buildforce. (2021, March 26) *Canada's construction industry rebounds post-pandemic, with more muted growth to come, 26th March*. Available at: https://www.buildforce.ca/en/node/12016 (Accessed on 3 October 2022).
Building and Wood Workers International, BWI. 2013, *Rights without frontiers: organising migrant workers in the global economy*, Geneva.
Building and Wood Workers International. 2018, *The dark side of the Tokyo 2020 Summer Olympics*, Geneva. Available at: https://www.bwint.org/web/content/cms.media/1542/datas/dark%20side%20report%20lores.pdf (Accessed on 13 October 2022).
Carreira, A. 1982, *The people of the Cape Verde Islands: exploitation and emigration*, C. Hurst & Co, London.
Castles, S. 2006. 'Guestworkers in Europe: A Resurrection?', *International Migration Review*, vol 40, no. 4, pp. 741–766.
Chanda, R. & Gupta, P. 2018, 'Indian Migration to the Gulf: Overview of Trends and Policy Initiatives by India', in P. Fargues & N. M. Shah (eds.) *Migration to the Gulf. Policies in sending and receiving countries*, Gulf Research Centre, University of Cambridge, Cambridge, pp. 179–197.
Chartered Institute of Building, CIOB. 2012, *The impact of the ageing population on the construction industry*, Berkshire, Ascot.
Connell, J. 2016. 'North Korea: Labour Migration from a Closed State', *Migration, Mobility, & Displacement*, vol. 2, no. 2, pp. 62–75.
Construction Industry Training Board, CITB. 2021, *Migration and UK construction*, CITB, Peterborough. Available at: https://www.citb.co.uk/media/cxxlqsps/citb-migration-and-uk-construction-report-2021.pdf (Accessed on 2 October 2022).
Cook, J. 2021, 'Immigrants 'Fundamental' to Tackling Canada's Shortage in Skilled Workers', *Building*, 18th August. Available at: https://building.ca/feature/immigrants-fundamental-to-tackling-canadas-shortage-in-skilled-trade-workers/ (Accessed on 4 October 2022).
Dearinger, R. 2015, *The filth of progress: Immigrants, Americans, and the building of canals and railroads in the west*, University of California Press, Oakland, Ca.
Department of Homeland Security, DHL. 2022, 'What is forced labour?', 22nd September 2022, Washington, DC. Available at: https://www.dhs.gov/blue-campaign/forced-labor (Accessed on 24 October 2022).
Diop, A., Johnston, T. & Le, K. T. 2018, 'Migration Policies across the GCC: Challenges in Reforming the Kafala', in P. Fargues & N. M. Shah (eds.) *Migration to the Gulf. Policies in sending and receiving countries*, Gulf Research Centre, University of Cambridge, Cambridge, pp. 33–60.
Endoh, T. 2019, 'The Politics of Japan's Immigration and Alien Residence Control', *Asian and Pacific Migration Journal*, vol. 28, no. 3, pp. 324–352.
Erlich, M. & Grabelsky, J. 2005, 'Standing at a Crossroads: The Building Trades in the Twenty-First Century', *Labor History*, vol. 46, no. 4, pp. 421–445.
Fundamental Rights Agency, FRA. 2015, *Severe labour exploitation: Workers moving within or into the European Union: States' obligations and victims' rights*, Luxembourg.
Halegua, A. & Chin, K. 2021, 'Forced Labor in the US Construction Industry', in Chisolm-Straker, M. and Chon, K. (eds.), *The historical roots of human trafficking*, Springer, Cham., pp. 85–99. 10.1007/978-3-030-70675-3_6

Hassan, G.M. & Shakur, S. 2017, 'Nonlinear Effects of Remittances on Per Capita GDP Growth in Bangladesh', *Economies*, vol. 5, no. 3, p. 25; 10.3390/economies5030025

Heslop, L. & Jeffery, L. 2020, 'Roadwork: Expertise at Work Building Roads in the Maldives', *Journal of the Royal Anthropological Institute*, vol. 26, no. 2, pp. 284–301.

Honsberg, B. 2004, 'Undocumented Migrants in the Construction Sector in Europe' in M. LeVoy, N. Verbruggen & J. Wets (eds.) *Undocumented migrant workers in Europe*, PICUM Brussels, pp. 47–52.

Human Rights Watch, HRW. 2006, *Building towers, cheating workers exploitation of migrant construction workers in the United Arab Emirates*, New York, NY. Available at: https://www.hrw.org/report/2006/11/11/building-towers-cheating-workers/exploitation-migrant-construction-workers-united (Accessed on 24 October 2022).

Human Rights Watch, HRW. 2009, *"Are you happy to cheat us?" Exploitation of migrant construction workers in Russia*, New York, NY.

Human Rights Watch, HRW. 2012, *For a better life: migrant worker abuse, Bahrain and government reform agenda*, Los Angeles. Available at: https://www.hrw.org/report/2012/09/30/better-life/migrant-worker-abuse-bahrain-and-government-reform-agenda (Accessed on 23 October 2022).

Human Rights Watch, HRW. 2022, *"How can we work without wages?" Salary abuses facing migrant workers ahead of Qatar's FIFA World Cup 2022*, August, New York, NY. Available at: https://www.hrw.org/report/2020/08/24/how-can-we-work-without-wages/salary-abuses-facing-migrant-workers-ahead-qatars (Accessed 23 October 2022).

Igari, H., Maebara, A., Suzuki, K. & Shimura, A. 2009. 'Tuberculosis among Construction Workers in Dormitory Housing in Chiba City', *Kekkaku:[Tuberculosis]*, vol. 84, no. 11, pp. 701–707.

International Labour Organisation, ILO. 2016, *Non-standard employment around the world: Understanding challenges*, shaping prospects, Geneva.

International Labour Organisation, ILO. 2017, *Employer-migrant worker relationships in the Middle East: Exploring scope for internal labour market mobility and fair migration*, White Paper, March, Geneva.

International Labour Organisation, ILO. 2019, *Working on a warmer planet: The impact of heat stress on labour productivity and decent work*, Geneva.

International Labour Organisation, Walk Free & International Organisation for Migration (ILO, WF & IOM). 2022, *Global estimates of modern slavery forced labour and forced marriage*, Geneva (ILO), Broadway, Nedlands (Walk Free) & London (IOM).

Immigration New Zealand, INZ. 2015, *Essential Skills instructions changes—Canterbury rebuild and new nationwide labour hire accreditation provisions*, 26th June. Available at: http://dol.govt.nz/immigration/knowledgebase/item/18949 (Assessed on 4 October 2022).

Jureidini, R. 2018, 'Wage Protection System and Programmes in the GCC' in P. Fargues & N. M. Shah (eds.), *Migration to the Gulf. Policies in sending and receiving countries*, Gulf Research Centre, University of Cambridge, Cambridge, pp. 9–32.

Kiganda, A. (2022, May 24) 'Africa's Skilled Labour: A Looming Supply Shortage', *Construction Review Online*, 24th May. Available at: https://constructionreviewonline.com/features/africas-skilled-labour-a-looming-supply-shortage/ (Accessed on 16 October 2022).

Kongchasing, N. & Sua-iam, G. 2021, 'The Main Issue Working with Migrant Construction Labour: A Case Study in Thailand', *Engineering, Construction and Architectural Management*, vol. 29, no. 4, pp. 1715–1730.

Krings, T., Bobek, A., Moriarty, E., Salamonska, J. & Wickham, J. 2011, 'From Boom to Bust: Migrant Labour and Employers in the Irish Construction Sector', *Economic and Industrial Democracy*, vol. 3, no. 3, pp. 459–476.

Lamounier, L. 2000, *The 'labour question'in nineteenth century Brazil: railways, export agriculture and labour scarcity*, Working Paper No 59/00, Department of Economic History, London School of Economics. Available from: http://eprints.lse.ac.uk/22379/1/wp59.pdf (Accessed on 16 October 2022).

Lankov, A., Ward, P. and Kim, J. 2020, 'The North Korean Workers in Russia: Problematising the "Forced Labor" Discourse', *Asian Perspective*, vol. 44, no. 1, pp. 31–53.

Lester, G. 2005, *Atlas of Alberta railway*, University of Alberta Press, Edmonton, AB. Available from: https://railways.library.ualberta.ca/Chapters-5-1/. (Accessed on 27 October 2022).

Lewis, F. D. & Robinson, D. R. 1984, 'The Timing of Railway Construction on the Canadian Prairies', *The Canadian Journal of Economics*, vol. 17, no. 20, pp. 340–352.

Liu-Farrer, G. 2020, 'Japan and Immigration: Looking Beyond the Tokyo Olympics', *The Asia-Pacific Journal*, vol. 18, no. 4, pp. 1–8.

Malit, Jr. F. 2018, 'Frontlines of Global Migration: Philippine State Bureaucrats' Role in Migration Diplomacy and Workers' Welfare in the Gulf Countries', in P. Fargues & N. M. Shah (eds.), *Migration to the Gulf. Policies in sending and receiving countries*, Gulf Research Centre, University of Cambridge, Cambridge, pp. 199–220.

Manzanilla, L. R. 2014, 'The basin of Mexico', in C. Renfrew & P. Bahn (eds.), *The Cambridge World Prehistory*, Cambridge University Press, Cambridge, pp. 986–1104.

Mendoza, C. 2000, 'African Employment in Iberian Construction: A Crossborder Analysis', *Journal of Ethnic and Migration Studies*, vol. 26, no. 4, pp. 609–634.

Ministry of Labour and Human Resources, MoLHR 2019, Labour Force Survey Report, 2018. Available from: http://www.nsb.gov.bt/publication/files/pub3td3256de.pdf (Accessed on 17 October 2020).

Moch, L. P. 2011, *Internal migration before and during the industrial revolution: the case of France and Germany*, European History Online. Available from: file:///C:/Users/User/Downloads/mochl-2011-en.pdf (Accessed on 18 October 2022).

Morris, M. 1994, 'Towards an Archaeology of Navvy Huts and Settlements of the Industrial Revolution', *Antiquity*, vol. 68, no. 260, pp. 573–584.

Moyo, I. 2018, 'Zimbabwean Dispensation, Special and Exemption Permits in South Africa: On Humanitarian Logic, Depoliticisation and Invisibilisation of Migrants', *Journal of Asian and African Studies*, vol. 53, no. 8, pp. 1141–1157.

Mukora, J. 2008, *Scarce and critical skills*, Research commissioned by Department of Labour South Africa, March. Available from: https://www.labour.gov.za/DocumentCenter/Research%20Documents/2001-2009/Artisans%20or%20Trades,%20Scarce%20and%20critical%20skills%20Research%20Project.pdf (Accessed on 18 October 2022).

Mulvey, M. 2018, 'Once Hard Men were Heroes': Masculinity, Cultural Heroism and Performative Irishness in the post-war British Construction Industry', in J. W. P. Campbell, N. Baker, A. Boyington, M. Driver, M. Heaton, Y. Pan, H. Schoenefeldt, M. Tutton & D. Yeomans (eds.), *Studies in the history of services and construction. The Proceedings of the Fifth Conference of the Construction History Society*, Queens' College, University of Cambridge, 6-8 April, The Construction History Society, pp. 443–450.

Office for National Statistics, ONS. 2018, *'Migrant labour force within the UK's construction industry'*, 23rd August. Available from: https://www.ons.gov.uk/peoplepopulationandcommunity/populationandmigration/internationalmigration/articles/migrantlabourforcewithintheconstructionindustry/august2018#considering-regional-and-the-uks-constituent-country-variation-of-migrant-labour-in-the-construction-industry. (Accessed on 19 October 2022).

Ofori-Kuragu, J. K. & Osei-Kyei, R. 2021, 'Mainstreaming Pre-Manufactured Offsite Processes in Construction–Are we nearly There?, *Construction Innovation*, vol. 21, no. 4, pp. 743–760.

Pant, B. 2011, 'Harnessing Remittances for Productive use in Nepal', *Nepal Rastra Bank Economic Review*, vol. 23, pp. 1–20.

Pearson, M.P., Pollard, J., Richards, C., Thomas, J., Welham, K., Albarella, U., Chan, B., Marshall, P. & Viner, S. 2011, 'Feeding Stonehenge: Feasting in late Neolithic Britain', in Aranda Jiménez, G., Montón-Subías, S. & Sánchez Romero, M., (eds.), *Guess who's coming to dinner. Feasting rituals in the prehistoric societies of Europe and the Near East*. Oxbow Books, Oxford, pp. 73–90.

Pereira, C., & Azevedo, J. (2019). 'The Fourth Wave of Portuguese Emigration: Austerity Policies, European Peripheries and Postcolonial Continuities', in C. Pereira & J. Azevedo (eds.), *New and old routes of Portuguese emigration*, Springer, Cham, Switzerland, pp. 1–26.

Pieńkowski, J. 2020, *The impact of labour migration on the Ukrainian economy*, European Economy Discussion Paper 123, April, Publications Office of the European Union, Luxembourg.

Reyneri, E. 2012, *Migrants' involvement in irregular employment in the Mediterreanean countries of the European Union*, International Migration Papers, International Labour Office, Geneva.

Royal Institution of Chartered Surveyors, RICS. 2022, *Q4 2021: global construction monitor*, London. Available from: https://www.rics.org/globalassets/rics-website/media/knowledge/research/market-surveys/construction-monitor/q4-2021-global-construction-monitor—headline-report.pdf (Assessed on 3 October 2022).

Sadarangani, S. P., Lim, P. L. & Vasoo, S. 2017, 'Infectious Diseases and Migrant Worker Health in Singapore: A Receiving Country's Perspective', *Journal of Travel Medicine*, vol. 24, no. 4.

Sadowski-Smith, C. 2008, 'Labor Migration and the Illegality Spiral: Chinese, European, and Mexican Indocumentados in the United States, 1882-2007', *American Quarterly*, vol. 60, no. 3, pp. 779–804.

Sakolnakorn, T.P.N. 2019, 'Problems, Obstacles, Challenges, and Government Policy Guidelines for Thai Migrant Workers in Singapore and Malaysia', *Kasetsart Journal of Social Sciences*, vol. 40, no. 1, pp. 98–104.

State Statistics Service of Ukraine, SSSU. 2017, Зовнішня Трудова Міграція Населення (за результатами модульного вибіркового обстеження), Kyiv. Available from: http://ukrstat.gov.ua/druk/publicat/kat_u/publ11_u.htm (Accessed on 17 October 2022).

Schiff, A. 2022, *Case study: construction industry and migration*, May, New Zealand Productivity Commission. Available from https://www.productivity.govt.nz/assets/Documents/Case-study_Construction-and-migration.pdf (Accessed on 8 October 2022).

Searle, W., McLeod, K. & Ellen-Eliza, N. 2015, *Vulnerable temporary migrant workers: Canterbury construction industry*, Ministry of Business, Innovation and Employment. Available from: http://www.mbie.govt.nz/publications-research/research/migrants—settlement/vulnerable-temporary-migrant-workers-canterbury-construction.pdf (Accessed on 13 October 2022).

Selby, J. 2012, 'One Step Forward: Alberta Workers 1885–1914', in A. Finkel (ed.) *Working people of Alberta: a history*, AU Press, Edmonton, Alberta, pp. 39–76.

Seol, D.-H. 2015, *Indicators used in determining admissions for foreign workers in Korea*. Presentation at the Regional Meeting to Validate the Guide on Measuring Migration Policy Impacts in ASEAN Member States, Bangkok, Thailand, 10–11 November.

Seol, D.-H. 2018, 'Population Aging and International Migration Policy in South Korea', *Journal of the Korean Welfare State and Social Policy*, vol. 2, no. 2, pp. 73–108.

Shah, N. M. 2018, 'Emigration Policies of Major Asian Countries Sending Temporary Labour Migrants to the Gulf', in P. Fargues & N. M. Shah (eds.) *Migration to the Gulf. Policies in sending and receiving countries*, Gulf Research Centre, University of Cambridge, Cambridge, pp. 127–149.

Simkunas, D. P. & Thomsen, T. L. 2018, 'Precarious Work? Migrants' Narratives of Coping with Working Conditions in the Danish Labour Market', *Central and Eastern European Migration Review*, vol. 7, no. 2, 35–51.

Sivam, A., Trasente, T., Karuppannan, S. & Chileshe, N. 2017, 'The Impact of an Ageing Workforce on the Construction Industry in Australia', In F. Emuze & J. Smallwood (eds.) *Valuing people in construction*, Routledge, Abigdon, Oxon and New York, NY, pp. 78–97.

Sokas, R.K., Dong, X.S. & Cain, C.T. 2019, 'Building a Sustainable Construction Workforce', *International Journal of Environmental Research and Public Health*, vol. 16, no. 21, 4202, 10.3390/ijerph16214202

Spindler-Ruiz, P. 2021, 'Mexican Niches in the US Construction Industry: 2009–2015', *Journal of International Migration and Integration*, vol. 22, no. 2, pp. 405–427.

Subramaniam, G. 2020, *Accommodating the spread – migrant worker accommodation driving COVID Infections (Pt 2)*, Institute for Human Rights and Business, East Sussex, UK. Available from: https://www.ihrb.org/focus-areas/covid-19/covid19-migrant-workers-accommodation (Accessed on 24 October 2022).

Svajlenka, N. P. 2021, *Undocumented immigrants in construction, 2nd February*, Center for American Progress. Available from: https://cdn.americanprogress.org/content/uploads/2021/01/01114810/EW-Construction-factsheet.pdf?_ga=2.235774981.508290061.1612545024-2015160577.1603194421 (Accessed on 10 October 2022).

Transient Workers Count Too. 2016, *Foreign workers hit badly by zika and dengue: better housing needed*, Singapore. Available from: https://twc2.org.sg/2016/09/04/foreign-workers-hit-badly-by-zika-and-dengue-better-housing-needed/ (Accessed on 24 October 2022).

Tshele, L. & Agumba, J. N. 2014, 'Investigating the Causes of Skills Shortages in South African Construction Industry: the Case of Artisans', *Proceeding of TG59 People in Construction*, Port Elizabeth. South Africa, 6–8.

Turnbull, C. M. 1970, 'Convicts in the Straits Settlements 1826-1867', *Journal of the Malaysian Branch of the Royal Asiatic Society*, vol. 43, pp. 1, pp. 87–103.

Ujita, Y., Douglas, P.J. & Adachi, M. 2019. 'Enhancing the Health and Safety of Migrant Workers', *Journal of Travel Medicine*, vol. 26, no. 2, pp. 1–3.

van Lottum, J. 2011, 'Labour Migration and Economic Performance: London and the Randstad, c. 1600–1800', *The Economic History Review*, vol. 64, no. 2, pp. 531–570.

Varvitsioti, E. & Kazmin, A. 2022, *Labour shortages in southern Europe threaten post-Covid building boom*, 14th July, Financial Times, London. Available from: https://www.ft.com/content/5a9919be-85ea-4c05-b978-65e91307771b (Accessed on October 4 2022).

Vogel, L. 2016, *Working conditions in construction: a paradoxical invisibility*, Special Report 3/33, Spring-summer, European Trade Union Institute, Brussels.

Wahyono, S. 2007, 'The problems of Indonesian Migrant Workers' Rights Protection in Malaysia', *Jurnal Kependudukan Indonesia*, vol. 2, no. 1, pp. 27–44.

Wee, K., Lam, T. & Yeoh, B. S. A. 2022, 'Migrant Construction Workers in Singapore: An Introduction', in Brenda S. A. K. Yeoh, Wee & T. Lam (eds.), *Migrant workers in Singapore*, World Scientific Publishing Co Pte Ltd, Singapore, pp. xiii–xlix.

Wells, J. 2007, 'Informality in the Construction Sector in Developing Countries', *Construction Management and Economics*, vol. 25, pp. 87–93.

Wells, J. 2016, *Protecting the wages of migrant construction workers*, Engineers Against Poverty, London.

Wickramasekera, P. 2002, *Asian labour migration: issues and challenges in an era of globalisation*, August, International Migration Papers, International Labour Office, Geneva.

Wijunamai, R. 2020, Migrant Construction Workers in Bhutan: Understanding Immigrant Flows and their Perceptions, September, Rig Tshoel – Research Journal of the Royal Thimphu College. Available from: SSRN: https://ssrn.com/abstract=3684293 (Accessed on 17 October 2020).

Woodhouse, M. 2015, *Immigration changes to support rebuild*, 13th May. Available from: https://www.beehive.govt.nz/release/immigration-changes-support-rebuild (Accessed on 4 October 2022).

World Bank. 2021, *Annual remittances data*, updated as of May 2021. Available from: https://www.worldbank.org/en/topic/migrationremittancesdiasporaissues/brief/migration-remittances-data (Accessed on 17 October 2022).

Yates, R. D. S. 2001, 'Cosmos, Central Authority and Communities in the Early Chinese Empire', in S.E. Alcock, T.N. D'Altroy, K.D. Morrison & C.M. Sinopoli, (eds.), *Empires. perspectives from archaeology and history*, Cambridge University Press, New York and Cambridge, pp. 351–368.

Zangpo, T. 2020, 'Local Firm Tries To Fill In The Gap For Construction Workers', 15th January, *Business Bhutan*. Available from: https://www.businessbhutan.bt/2020/01/15/local-firm-tries-to-fill-in-the-gap-for-construction-workers/ (Accessed on 17th October 2020).

13

THE SMART CITY AS THE FACTORY OF THE TWENTY-FIRST CENTURY?

How urban platforms reshape the nexus between the built environment, livelihoods, and labour

Jeroen Klink and Ângela Cristina Tepassê

FEDERAL UNIVERSITY OF THE METROPOLITAN REGION OF SAO PAULO (UFABC), BRAZIL

Introduction

The dramatic evolution in information and communication technologies (ICTs) has increased our capacity to collect, process, and transform big data. This is bound to influence the production of contemporary urban space and the quality of living of urban workers in ways we have only started to explore.

The city-technology nexus is not a new theme and has occupied urban scholars since the industrial revolution and the emergence of modern urban planning (Hall, 2014). To begin with, the roll-out of infrastructure networks, such as the telegraph, steam and railways, ports, roads, and, more recently, telecommunication and the internet, cannot be dissociated from the urbanisation-industrialisation nexus (Graham and Marvin, 2001). In his *Principles of Economics*, Marshall (1920) was one of the first to recognise the city as the privileged locus of the industrial revolution, considering it concentrated the economies of agglomeration in terms of the abundance and variety of labour and the dense networks of specialised suppliers. Moreover, successive cycles of regional learning and skills upgrading of the labour force (Storper and Salais, 1997), combined with the spatial density and proximity of consumers and producers, the associated circulation of information, and its effects on innovation have all continued to capture the imagination of urban scholars. For instance, contemporary economists such as Edward Glaeser (2012) analyse the "urban triumph" in terms of the economies of agglomeration that allow cities to trigger a virtuous trajectory driven by the regional learning and skills upgrading, innovation, and the generation of income, employment, and wealth.

However, despite the similarities, history in general, and urban history (Hall, 1998), seldom repeats itself. Particularly in the twenty-first century, the impressive increase in our capacity to collect, transform, and systematise big data and information systems has now disseminated debates on what is increasingly called the smart, intelligent city. Smart cities connect corporate providers of technological services and platforms such as Uber and

 DOI: 10.1201/9781003262671-13

Airbnb, local governments, and users-citizens to improve the planning, production, and maintenance of the built environment.

This debate has raised somewhat different questions. One of them relates to the potential of big data, increased computational capacity, and the dissemination of urban service provision platforms to effectively deliver better, that is, more inclusive and sustainable cities. Related to that, issues have emerged whether the technological development has effectively led to the emergence of platforms as a new, "smart" type of city builder and provider of urban infrastructure and services.

In this chapter we define platforms as socio-technological assemblages that perform as well as charge for strategic intermediation services of "ideas, knowledge, labor and use-rights for otherwise idle assets between geographically distributed but connected and interactive on-line communities" (Langley and Leyshon, 2017: 1925). In that sense, do we witness, as is claimed by some strands of the literature, the rise of a kind of platform urbanism, with yet unexplored consequences for the role of labour in the production, operating, and financing of cities and for its effects on urban livelihoods and the quality and affordability of housing for workers and families?

In this chapter we shall deal with two dimensions of the above-mentioned questions. First, we will highlight the broader socio-technical and economic restructuring that has taken place in the data and information-intensive industries and the impacts on corporate strategies, labour, and the production of urban space and livelihoods. This provides the broader picture of the drivers behind *the supply-side* of platform urbanism. Second, based on the rental housing sector, we provide an illustration of how this socio-technical and economic restructuring is likely to impact urban livelihoods and the cost of living of urban workers through a breakdown in traditional patterns of housing provision and finance. This is related to the *demand effects* of platforms on cities and labour.

Our methodology is based on a review of the associated literature on the platform economy and platform urbanism, with a more specific emphasis on their contradictory role in the finance and provision of affordable housing. We include a brief discussion of how countries such as Brazil, marked by a significant participation of informal settlements and slums, fit into this broader picture.

After this introduction, the chapter is organised around three complementary sections. The first provides a primer on the science and technology literature, platform capitalism, and the changing relations between corporate providers of information and data services, urban labour, and cities. The second, through a case study on rental housing, illustrates the effects of platform urbanism on the production and affordability of urban livelihoods. More specifically, through a detailed review of the literature, we flesh out the impacts of real estate platforms, both in terms of the expected cost of living and the type of product delivered in urban markets. This includes a brief illustration of the geographic and historical specificities of a country in the Global South such as Brazil. In the conclusion we wrap up the main arguments and suggest elements for a research agenda on platform urbanism, urban livelihoods, and labour along the lines discussed in this chapter.

From smart cities to platform capitalism and platform urbanism

Castell´s trilogy on the network society contributed to linking the science and technology literature with urban studies (Castells, 1999). According to him, the new ICTs performed a strategic role in the transformation of industrial capitalism, based on the generation of value added and wealth through manufacturing, into what he called informational or cognitive capitalism, driven by the nexus between data collection, information, knowledge, and innovation. Somewhat different from twentieth-century industrial capitalism, which had been marked by structural class conflict between capital and labour regarding the appropriation of value, the emerging divisions within informational capitalism depended on the access to the network and the associated capacity to extract value from it. The implication for labour was not only to accompany these transformations but also, to obtain new technological skills and competences to avoid becoming irrelevant.

Castells also argued that cities performed a strategic role in the new knowledge economy. While the global network society had generated a space of flows whereby information circulated in real time on a planetary scale, the space of places in cities, where the daily reproduction of life occurred, remained crucial. For Borja and Castells (1997), the contemporary city represented something of a *global* anchor, or hub, of the knowledge economy, and considering its economies of agglomeration were crucial for the capture and transformation of information into new knowledge and innovation. Cities emerged as new actors in the successful governance and transformation of urban and metropolitan industrial economies into dynamic learning and innovation systems.

The smart city literature has built upon this earlier work on the network society and the role of science and technology in urban planning and management. First, it has stressed the transformation of techno-bureaucratic, hierarchical planning into networked governance (Kjaer, 2012), which was also facilitated by the dissemination of new ICTs that could connect different stakeholders around a collaborate and communicative planning project (Healey, 2003). Moreover, smart cities mobilise non-state stakeholders, such as high-tech firms and service providers, in the design and implementation of new strategies for the delivery of specific urban services such as mobility and transportation, operation and maintenance of public rental housing, and the communication interface between local governments and the citizens, among other examples. The bottom line of the more optimistic strand of the literature has emphasised that smart cities contribute to more responsive, efficient, and transparent urban governance.

More recent research has started to investigate the underlying role of platforms in the post-industrial economy. The latter concentrates on the intermediation of "ideas, knowledge, labor and use-rights for otherwise idle assets between geographically distributed but connected and interactive on-line communities" (Langley and Leyshon, 2017: 1925). As argued by Caprotti et al. (2022), platforms, as "dense and socio-technical assemblages" of users, providers, and technology firms, have emerged as a phenomenon that goes beyond the scale of the smart city, which is associated with specific places and/or local administrations. Corporate operators such as Airbnb and Uber operate on a global level.

In a recent contribution, Nayaran (2022) provides a theoretical perspective on the rapid expansion of the platform economy that is based on three interrelated drivers. The first of them is the microeconomic underpinning of the increase in scale of networks. Platform companies are characterised by what economists call increasing returns to scale. The latter means that a proportionally small volume of additional investments in fixed capital and employees generates a more than proportional increase in output in terms of the number of connections and users. A quick look at the figures regarding the relationship between the number of users-clients and employees of the main players in the platform economy confirms this pattern. For instance, with only 4,422 employees the company Zoom can maintain a daily number of connections for 300 million meeting participants. Likewise, Facebook employs

71,000 people, while no less than approximately 3.52 billion people are using its products and services (Nayaran, 2022: 913). Moreover, networks have an in-built mechanism of "causal circulation" that reinforces growth, considering that additional users leverage the attractiveness and value of the system at relatively low marginal, that is, additional costs. For example, companies like Facebook generate a significant amount of revenue from third-party actors, that is, companies that are attracted by the sheer size and variety of network users that allow them to provide differential, complementary services.

A second, also microeconomic driver, is the entanglement between new network users and their role as implicit providers of big data as a new and distinct means of production in the corporate strategy of platforms. Companies like Netflix use algorithms to detect the behaviour of their large-scale consumer base, thereby strengthening their business model and reaping competitive advantages in their investment and expansion strategy. A key point here is that platforms don´t pay for this. As argued by Nayaran (2022: 914), "if labor is conceptualized as a value-generating activity, then it holds that 'data' in this context results from the unpaid and free labor of users". This data-driven expansion of networks, which is based on the low marginal costs associated with the utilisation of the existing computational infrastructure and user profiles, also sheds light on the surprisingly rapid growth of platforms in the last decade or so.

Finally, in addition to specific internal microeconomic drivers behind industry expansion, the platform economy has benefitted from a set of macro-institutional dimensions. A detailed analysis is outside the scope of this chapter, but it is nevertheless important to recognise their impact on platform expansion. First, successive rounds of regulatory roll-back and deregulation have facilitated the dissemination of platforms, for example, through lax zoning and land use regulation (e.g., AirBnb) and the lack of a solid legal framework and rules for enforcement to guide anti-trust policies and strategies (e.g., Facebook). Second, the abundant supply of speculative venture capital, characterised by an investment profile explicitly focused on growth industries, also allowed a rapid expansion of platforms that didn´t exclusively depend on the generation of internal revenue in the future. Instead, a relatively small amount of upfront cash that was provided by venture capital guaranteed the required financial and physical-technical leverage for the expansion of networks.

Considering the setting outlined above, the common narratives that surround platform capitalism (Langley and Leyshon, 2017) in terms of sharing, co-production, and collaborative governance should be put in its proper perspective. Instead, platforms are grounded in corporate strategies of data-driven network expansion, which is reinforced by a macro-institutional framework of deregulation and the abundant availability of speculative venture capital. Moreover, platform companies have specific mechanisms for pricing access and extracting rents from the network that require further research that is outside the scope of this chapter. In our discussion, however, two different mechanisms should be mentioned. First, companies such as AirBnB and Uber charge users to access their service provision, which can be considered a classical *direct* rent (Harvey, 2013). Moreover, even so-called open access sources, such as social media like FaceBook and Tweeter, obtain their main sources of revenue through *indirect* rents by charging their advertisers for the privileged access to data associated with the massive volume of global users of their network. In a way, platform capitalism performs the digital economy of intermediation through the valuation and capitalisation (anticipation) of specific streams of income associated with the provision and packaging of data-related services.

Cities concentrate the circulation of information and the dense agglomeration of service providers and high-tech firms, users, consumers, and labour that are all instrumental to the performance and dynamism of platform capitalism. The relevant literature on platform urbanism (Caprotti et al., 2022) considers cities as the backbone of the emerging platform economy in ways that go beyond the smart city paradigm that merely connects local government, users, and technology firms around the efficient planning and management of urban space.

At the same time, however, the rise of platform urbanism has raised several questions that require further research regarding its limits and potentials in generating more inclusive and sustainable cities. First, the increasing returns to scale that accompany the expansion of the platform economy are unlikely to lead to a significant generation of additional jobs. As mentioned, the business model of platforms is based on reaping the benefits of new users that connect to the network under conditions of low marginal costs, that is, dispensing with the need for a large volume of additional investments in fixed capital and employees. Moreover, without significant state-driven efforts for massive upscaling of the labour force, it is unlikely that much of the benefits of platform urbanism is going to "trickle down" to the unprepared industrial labour pool in metropolitan areas.

Second, platforms have become the object of an increasingly critical evaluation considering their role in the downgrading of urban labour. This is especially clear in the so-called "gig economy" whereby platforms intermediate the demand and supply of particular kinds of labour services. For example, Ebstebsari and Werna (2022) and Bauriedl and Strüver (2020) discuss the threat that is posed by platforms to the more vulnerable segments of the labour market such as migrants and women. The latter analyse the specific example of mobility and home-care applications.

Third, the platform economy is both *concentrated* in terms of ownership and *dispersed* in relation to its spatial patterns. These features are likely to challenge the constitution of a strong social dialogue between capital, labour, and (local) governments around standards for decent labour. Platform capitalism tends to become more *concentrated* in terms of ownership considering that a growing number of companies have changed from in-house strategies to subcontracting their demand for data and information processing to a few large cloud computing companies such as Amazon, Microsoft, and Google. At the same time, platform capitalism has created highly *dispersed,* trans-scalar, territorial networks that connect different cities, users, and urban workers. Concentration and fragmentation are likely to generate increasing challenges to constitute a transparent and equitable dialogue around decent labour standards in cities.

Finally, as is the object of the next session, in specific circumstances the pricing of user access to networks is likely to increase the cost of living in cities and exclude workers and families in ways that contradict the narrative of collaborative sharing and co-production of services. Moreover, it is even less clear how platform urbanism affects cities in the Global South, given the geographical and historical specificities of their urban development trajectories, where the relationship between property (of assets and platforms) and rents has evolved in very different and often contradictory circumstances marked by colonialism, slavery, and clientelist relations between the state and society. In the next section, we discuss a specific example of platform urbanism related to affordable housing in cities, including a brief discussion of the specific conditions under which real estate platforms are emerging in Brazil.

The impacts of real estate platforms as emerging producers and operators of urban livelihoods: An overview of the literature

This section provides an overview of the literature on real estate platforms. In the first part, we discuss the literature on how the subprime housing finance crisis of 2007–2008 triggered a shift from owner-occupied to rental housing, particularly through the emergence of institutional investors in the affordable housing sector, as well as the dissemination of built to rent housing projects. These trends somehow set the scene for a more active role of real estate and housing platforms in short term rentals (STRs) and single-family housing units (SFHs), which will be discussed in the second part of this section. The section is concluded with an illustration of how real estate platform emerge within the Brazilian scenario.

Setting the scene: Institutional investors discover affordable housing and Built to Rent in the post-subprime era

Affordable (rental) housing

Heeg (2013) discusses the commodification of state housing in Germany, particularly in the bigger cities such as Frankfurt, Berlin, Hamburg, Köln, and Munich. This process was triggered by a wave of acquisitions undertaken by private and institutional investors, which generated a significant escalation of real estate prices. According to this author, in the last 20 years the market for residential real estate has been going through a process of financialisation not unlike the trajectory of the commercial segment.

To be clear, the residential real estate investment trusts (REITs) were transformed into viable investment products through the regulatory roll-out that provided the instruments that enabled the private acquisition of housing units with support from state subsidies. This triggered a shift towards specialised, professional investment portfolios, whereby housing was treated as a tradeable, income-yielding asset. The author concludes that these institutional changes and the associated corporate search for high yields and shareholder value are likely to provoke gentrification and segregation in cities, providing challenges to socially vulnerable families and workers with deficient access to the housing market.

In the United Kingdom, Wijburg and Waldron (2020) also analyse how private equity, real estate investment funds (REIF) and other institutional investors entered the market for affordable housing in the context of the massive privatisation of public housing estates. Changes in housing policies, reinforced by a context of austerity politics, contributed to this scenario. The latter was marked by a significant reduction in the supply of social housing and increasing waiting lists, forcing low-income families into alternative housing rental arrangements (Wijburg and Waldron, 2020). As a result, the authors report that in 2017, the company *Legal & General* announced that it would build an annual stock of 3,000 new affordable rental housing units, to be sold either to residents associations or to rented out by the corporation itself. Along similar lines, BlackRock entered this market segment while Blackstone acquired Sage Housing, which was involved in the provision of 20,000 affordable housing units between 2018 and 2023 (Wijburg and Waldron, 2020). Capital investments undertaken by big players such as BlackRock potentially alleviate cash strapped housing associations. However, these partnerships tend to shift the emphasis to professional portfolio management aimed at the maximisation of financial rates of return through a more selective allocation of tenants to the housing stock. The authors conclude by arguing that we are witnessing a financialised privatisation, accompanied by the emergence of corporative actors in the domain of affordable rental housing within a broader process of restructuring of the relations between the state and capital.

Janoschka et al. (2020) identify similar investment strategies of Blackstone in Spain. They evaluate how the post-subprime political and economic restructuring in the country provided incentives for transnational real estate investments and the consequences for asset portfolios. To illustrate, between 2013 and 2018, Blackstone had acquired more than 120,000 assets, which practically consolidated the company as the main player in the Spanish real estate market (Janoschka et al., 2020). The first step to accomplish this hegemony was a highly polemic privatisation process that involved two public companies based in Madrid, which launched two competitive bidding exercises that allowed private investors to acquire 4,795 social housing units from the public stock. This process benefitted Goldman Sachs and Blackstone (Fidere) with 2,935 and 1,860 units, respectively (Janoschka et al., 2020). The privatisation process was widely contested and criticised by the regulatory authorities. As a result, Janoschka et al. (2020) identified six different strategies that allowed Fidere to maximise its rates of return and rental revenues while changing the socioeconomic composition of its tenants:

- Decrease in maintenance of the buildings and public spaces. Abandonment and ring fencing of commercial areas.
- Increase in the energy prices (through new contracts with suppliers). Local services started to be charged individually. Additional charges for rentals of garage and storage spaces were introduced.
- Lack of information regarding the implication of changing ownership structure of the rental stock. More particularly, the duration of contracts was reduced drastically. More than 40 percent of the existing tenants moved to contracts with a duration of less than one year.
- Intimidation of tenants who tried to mobilise residents in neighbourhood associations aimed at the collective negotiation of rental contracts. Several members of associations were not offered new contracts.
- Continuous search for new, more affluent tenants in line with the required corporate financial rates of return when contracts expired. As a rule, the periodic increases in rents followed the general market tendency, which led to the exclusion of the more vulnerable families and workers that could not afford to pay higher rents. For example, the authors′ research confirmed an average increase in rents for new locations of more than 50 percent in less than two years (Janoschka et al., 2020).

Built to Rent

Nethercote (2020) analyses the financialisation of housing through the emerging role of institutional investors in the development of Built to Rent (BtR) as a new type of asset class. The author compares the markets in Australia, the United Kingdom, Ireland, the United States, and Canada with the objective to generate a better understanding of the financialisation of rental real estate. The starting point of the analysis is the large-scale entrance of institutional investors such as private equity capital, hedge funds, REIFs, and listed real estate firms into the market for urban rentals.

All these actors have become involved in large-scale acquisition of the existing housing stock, or the development of new units for rental housing. In this scenario, they maintain a property portfolio with a large variety of diversified apartments as assets that enable a stable flow of income through rents and capital gains (Nethercote, 2020). This approach involves the large-scale construction of units, which is accompanied by professional management that provides amenities and complementary hotel-style services and more flexible and convenient locations, including long-term rentals and options to swap units (Nethercote, 2020). Moreover, the long-term investment profile has allowed the incorporation of short-term arrangements: the relatively stable, contractual flow of rental income facilitates the leverage of revenues through financial instruments, either through IPOs and the creation of REIFs, securitisation, or both. This process constitutes a broader change in urban housing through financialisation, exercising influence on non-financial domains such as social and cultural living conditions and individual subjectivities, delegating risk that was previously taken on by the state, and deepening inequalities. The author suggests a broader research agenda on housing organised around three main points: diversification beyond the build for sale arrangements, the evolution and increasing sophistication of the private rental market, and the changing entanglements between the labour and housing markets. Related to the latter, further research is required regarding how corporate strategies of developers and financial intermediaries threaten to hollow out the right to decent housing of urban labour.

Institutional investors and STR: Real estate platforms enter the scene

There is now a growing strand of work that investigates the effects of STRs and corporations such as Airbnb in tourist neighbourhoods in case study cities. Cocola-Gant et al. (2021) argue that the larger platforms that dominate the market have expanded after the subprime crisis. According to the authors, these firms have provided new real estate investment opportunities that have enabled the entrance of institutional actors and a professionalisation of the specialised suppliers of STR services. They have supported property owners in attending to a global demand and extract additional profits from the rental market in times of neo-liberalisation an financialisation.

The effective performance of real estate platforms does not match the officially disseminated narratives of the sharing economy and the efficient and equitable allocation of housing. Instead, platforms represent speculative investment and an instrument to increase middle-class incomes. Consequently, tenants lose their apartment, new forms of gentrification are set in motion, while the increasing scarcity of long-term rental housing contracts reduces the number of alternatives for workers while triggering higher prices.

To illustrate, Cocola-Gant and Gago (2021) analyse the Alfama neighbourhood in Lisbon, Portugal, during the period 2015–2017. Their objective was to identify to what extent the housing stock was being allocated to STR uses, the ownership structure in this segment, and the specific linkages between the STR market and the broader restructuring in the housing provision for workers and families in the city. Their findings indicate that the average size of portfolios of property managers in Lisbon was around 70–200 apartments, while there was no evidence of any sharing economies. Moreover, 78 percent of the property owners that had outstanding contracts with Airbnb were individual or corporate investors, indicating an acquisition process of housing units with the sole purpose of generating financial profits through rents, while displacing existing tenants with tourists. There were two reasons for this pattern. First, after the subprime crisis, Airbnb was actively involved in the transformation of rental housing into a new financial asset class. The suppliers-owners of housing units represented investors that used their housing as a tradeable income yielding asset. The short-term rental market offered the additional advantage of combining a regular stream of income with the flexibility to sell units without tenants at any moment. Second, the downside of this dramatic increase in flexibility was an increased insecurity for tenants associated with the perspective of relocation and deepening social injustice driven by gentrification and increasing rent levels.

Platforms perform the role of opening-up local residential real estate for investors at a truly global scale. This implies that ICTs and digital networks allow the reinforcement of globalisation and accelerate the spatial-temporal logics of real estate markets. The financial institutions and the super-rich are not the only actors in this process. The rise of professional property managers facilitates investments by owners of a second house and middle-class segments active in Airbnb. As a matter of fact, the proliferation of Airbnb and the increase in tourism were stimulated by the regulatory roll-out regarding the real estate sector, the opening-up of local markets to global capital and state strategies aimed at providing solutions to de-industrialisation and the subprime crisis of 2008. According to these authors, political solutions are required and should concentrate on the reversal of the insecurity that guides short-term rental contracts; a tax structure that provides incentives to long-term rental contracts; reform of the regulatory framework that provides tax incentives to foreign and institutional investors; reconsideration of the role of tourism as the main driver behind local economic development strategies; and a re-constitution of the role of housing as a fundamental right to the city and not a prime investment vehicle (Cocola-Gant e Gago, 2021).

Along similar lines, Hof (2021) analyses the effects of STRs in the tourist conurbation of the enclaves of Los Cristianos/Las Américas in Spain. Considering the post-subprime scenario whereby original residents had lost their job and were unable to pay the mortgage, holiday rentals increasingly came to dominate the local real estate market and provided an effective access barrier for low-income workers and families. The author concludes that, particularly from 2014 onwards, there is a significant increase in tenant eviction, indicating the emerging financialisation of rental housing and the associated removal of the original low-income workers from the social housing stock. The interviews undertaken by the author confirmed that corporate strategies organised around high yields and shareholder value led to escalated rent levels and were considered as one of the main drivers behind unvoluntary relocation (Hof, 2021).

In Toronto, Canada, skyscrapers in the central city area are increasingly being acquired by REITs and private equity companies that are in search of the upscaling (or gentrification) of entire apartment buildings to maximise shareholder value for investors (August e Walks, 2018 apud Grisdale, 2021). These "financialised" property owners are looking to reposition their buildings that are in privileged areas with a strong demand profile, transforming them into luxury STRs. In this setting, Grisdale (2021) investigates the data associated with Airbnb and others, with the objective to assess the role of STR in the gentrification process of Toronto. He finds that, while "corporate advertisement" represented something around 3 percent of the universe analysed, it generated approximately 15 percent of all the revenue that was generated by Airbnb in the city of Toronto. Moreover, in most of the neighbourhoods affected, the average property owner only required to rent out units for half of the year to obtain the equivalent yearly revenue associated with long-term contracts (Grisdale, 2021). The author concludes that the short-term segment drains a significant share of the long-term market for rental housing and thereby contributes to the process of gentrification (Grisdale, 2021).

In relation to the Irish scenario, Clancy (2020) analyses the impacts of STR offered by Airbnb in a setting whereby the city of Dublin was transformed into a target for the acquisition of bad assets by global financial players in the context of the subprime crisis. The analysis connects tourism and the housing crisis in Dublin. More specifically, rent levels in Dublin increased with approximately 100 percent since 2011, and while announcements of STRs by platforms such as Airbnb occurred in the city as a whole, the latter were nevertheless concentrated in two working class and predominantly ethnic neighbourhoods that were already facing rapid gentrification. There were evidently multiple drivers behind the housing crisis, starting with the structural disinvestment of the government from social housing and the associated reduction of the housing stock available for low-income workers and families. This was reinforced by the neoliberal Irish policy stance over the last 25 years, which was reflected in the regulatory roll-out and tax incentives aimed at attracting global finance capital, including, more recently, hedge funds, private equity, and pension funds. The latter aggressively stepped into the market for Irish residential real estate. For instance, according to the Irish Confederation for the Building Industry, 95 percent of the total volume of 3,644 final units that were delivered during 2019 were acquired by investment funds (Clancy, 2020).

Institutional investors and single-family rentals (SFR): Digital platform as emerging actors

Both Fields (2018) and Nethercote (2020) analyse the rise of SFR as a new asset class on the US financial market. These units represent a characteristic pattern of residential real estate in the country′s suburbs. While traditionally marked by its fragmentation, the authors analyse how this segment has become a viable investment option for the financial market and a new frontier for the large-scale extraction of rents.

The scenario of the post-global subprime crisis, marked by a huge supply of depreciated real estate, risk-averse and credit rationing banks, and an increased demand for rental housing, provided a window of opportunity for the acquisition of repossessed properties and their transformation in rental stock by institutional investors and private equity funds. This set the stage for the impressive growth trajectory of companies such as Blackstone. The SFR segment became a new driver behind the financialisation of US real estate after the subprime crisis of 2008. Fields (2018), for example, analyses the role of the state and capital markets in the resignification of mortgaged single-family units into rental property after 2008. She also discusses the calculative practices and the strategies adopted by issuers and rating agencies in the creation of a new asset class for institutional investors. To realise this, residential single-family units had to be disconnected from the 2008 subprime crisis and the previous massive dispossession, creating an environment favourable to reform and renovation, standardisation, institutionalisation, and commodification. The transformation of rental housing was organised as a manufacturing process, with the creation of a line of production whereby houses were acquired, reformed, renovated, and rented out. In the meanwhile, the State and financial intermediaries collaborated to guarantee the matching of several actors and their associated interests. Some examples to illustrate the argument:

- State regulatory roll-out to renew the role of housing in the circulation of capital in the context of increasing housing needs;
- Actors from the capital markets, including traders and rating agencies, banks with repossessed houses on the balance sheet, institutional investors looking for yields in an environment of low interest rates, and private equity funds that were keen to show that they could support this search for yields;

- Municipalities that suffered from a large stock of abandoned and vacant units, reinforced by growing levels of criminality;
- Property owners with negative net wealth due to repossessed properties in their neighbourhoods; and
- Former and prospective owners relegated to become tenants.

Fields and Rogers (2021) further develop this argument regarding "the making of a rental market" by finance, platforms, and the state. They provide a classification of real estate platforms according to their logics, the digital work, and the ongoing financialisation as perspectives to analyse the latter´s role in the housing sector. They suggest two related lines of investigation. The first is the role of platforms as facilitators in the circulation of capital and the absorption of overaccumulation in the economy, either through the allocation of financial resources in the development of platforms or by providing guidance to investment decisions in real estate as commodity. The bottom line here is that platforms perform a strategic role to coordinate and guarantee a quick payback period on specific investments. Second, platforms enable the constitution of real estate as a new asset class. In that sense, Fields (2022) argues that the favourable post-subprime conditions would have been insufficient to guarantee the success of a strategy based on the acquisition of repossessed single-family units and transform them into rent-backed tradeable securities. Digital innovation and big data proved to be instrumental in the automatisation of essential functions to efficiently manage large-scale diversified portfolios of geographically dispersed properties. These technologies have allowed the extraction of a continuous flow of rents and the transmission of a relative sense of security to capital markets. The financialisation of housing was redesigned by shifting a flow of funds backed by mortgages to the payments of rents as the basis of present and future value. But this only could have been realised through the development of big data and digital technologies which, in combination, reduced the costs associated with the management of large-scale, geographically dispersed housing. Fields (2022) provides several examples of how this system was set in motion:

- Data in the format of algorithms of subscription, including neighbourhood equipment, infrastructure and amenities, proximity to the employment centres and transportation corridors, construction standards and required capital investment, among others. This type of information, combined with the expected yields and related specifications from investors, generates a list of possible properties with their associated prices, which allows the relatively quick design of a portfolio with a mix of real estate in the right neighbourhoods, in addition to making acquisitions more competitive by reducing time-consuming and costly site visits;
- Technology to administer and monitor the supply of business services, which allows the management of large real estate portfolios by providing real time information regarding the "stage of specific assets", their geographical location, and the state of the art of ongoing business and transactions of the portfolio. Moreover, these technologies facilitate the flow of work by providing "big data" on a specific stock of assets, make operations more transparent and commensurable for rating agencies, banks, and financial institutions that design, coordinate, and issue operations of securitisation and attract investors;
- Platforms for rental housing and administration of real estate, which represent a response to the obstacle of distance and centralise large and geographically dispersed portfolios of property. They provide portals for the on-line search for specific real estate; rental

payments and demand for maintenance; and guidelines and orientation for the maintenance of housing units through blogs and videos. Other technologies also allow the presence and immediate interaction between landlord and tenant, reduce cost levels, and create efficiency for owners, such as key-less entrance systems that enable visits without the use of real estate brokers; request for maintenance through the use of photos, which dispenses with the sending of personnel to make a diagnosis or reduces the associated time of a diagnosis; scoring cards for tenants that can be used to allocate benefits; and smart housing services that are useful to manage vacant units during their rotation; and
- Platforms for operation and maintenance, which require developers to use smartphone applications that provide geolocations, optimise the routes to minimise trip time, require personnel to check in and out when coming to their workplace, and provide pictures that register the services that have been implemented (allowing quality checks and the identification of possible inefficiencies). These platforms are focused on "targeting bodies and subjectivities so as to secure the supply chain and render it transparent" (Fields, 2022: 12).

Fields concludes that the above-mentioned technologies constitute a key ingredient of the operations of financial capital and platforms in real estate markets. They contribute to rent extraction and the improvement of the logistics that allow to accelerate the flow of capital in the built environment, and the creation of what she calls an "automated landlordism" (Fields, 2022). Nevertheless, the scenario potentially aggravates the unequal power dynamics in the housing field. It allows the user to generate and process a large quantity of data ("big data") aimed at the "classification and allocation of people in categories" through algorithms that maximise profit rates and financial rates of return on investments. These categories carry economic sticks and carrots that contribute to socio-spatial stratification (apud Fields, 2022). As such, big data and digital technologies will become part of present and future housing policies and threaten to emerge as a driver behind the increasing socio-spatial disparities.

Recapping the common themes in the literature

There are three common aspects in the above-discussed literature. The first is related to role of the subprime crisis of 2008, which somehow triggered the rise of institutional investors in the market for rental housing (Heeg, 2013; Cocola-Gant and Gago, 2021; Cocola-Gant et al., 2021; Hof, 2021; Fields, 2022; Rolnik, 2021). The second is the role of the state in the design of successive cycles of regulatory roll-out aimed at providing incentives to this new market segment, through outright, polemic privatisation of public assets (Janoschka et al., 2020); the allocation of subsidies (Heeg, 2013); the declining role of social housing in overall housing policies (Wijburg and Waldron, 2020; Clancy, 2020); the deregulation of real estate markets; the opening-up of local markets to global investors, frequently grounded in policies aimed at promoting tourism as a sustainable solution to the problem of de-industrialisation (Cocola-Gant and Gago, 2021); or by the re-signifying of repossessed single-family units after the 2008 subprime crisis (Fields, 2022). The third dimension stressed in the literature is the social consequence of this kind of investment in rental housing. Wijburg and Waldron (2020) observe that the capital cost reduction and the higher expected rates of return for investors through the selective allocation of tenants are matched by the influences on the reproduction of daily life in cities and the delegation of risks previously assumed by the state, thereby intensifying inequalities (Nethercote, 2020). The bottom line here is the escalation of rent levels, upward filtering of the socioeconomic profile of tenants and gentrification, socio-spatial segregation, insecurity of rental contracts, and crowding out of traditional tenants by tourists (Janoschka et al., 2020; Heeg, 2013; Wijburg and Waldron, 2020; Nethercote, 2020; Cocola-Gant and Gago, 2021; Grisdale, 2021; Clancy, 2020; Cocola-Gant et al., 2021; Hof, 2021; Fields, 2022).

And what about cities in the Global South? A sample of the Brazilian debate

Rolnik et al. (2021) describe the emerging patterns of financialisation of rental housing in Brazil and Chile. Their hypothesis is that a more flexible regime of rental housing, which allows the transformation of housing into a service, represents a potential new frontier that connects the financial and real estate sectors with the informalities and peculiarities of low-income settlements (Rolnik et al., 2021: 21).

The connection between the financial sector and real estate in Brazil was initially established in the commercial sector in the 1990s. This was followed by the creation of the first REITs and REIFs, backed by corporate towers, shopping complexes, and industrial buildings, and finally moving into housing. But in many countries of the Global South, informality and the lack of predictability continued to be a bottleneck for the financialisation of low-income housing.

Nevertheless, based on Nethercote's (2020) analysis regarding the global, structural transformation towards rental housing in all income segments, the authors argue that this process has more recently also gradually advanced in Latin America. The highest growth rates in the region are seen in the apartment segment, which combines a relatively favourable mix of risk and return. It involves multi-storey buildings with rental units that are managed according to a centralised and standardised model, with a variety of services that are frequently linked to STR platforms such as Airbnb (Vannuchi, 2020 apud Rolnik et al., 2021). In general, owners are corporate players such as institutional or pension fund investors, real estate asset managers, family offices, and insurance firms. Frequently, the effective management is undertaken outside host countries.

The shift from owner-occupied to rental housing also occurs in the low-income, precarious housing areas, irrespective of the specific legal situation of occupations. For decades, the outskirts of many metropolitan areas in Latin American countries have witnessed the proliferation of slums and precarious housing units. There is a large literature on the urbanisation-industrialisation nexus that explains how wages have never fully incorporated the housing and living costs in cities. According to that perspective, slums and informal tenement housing represented a structural feature of the Brazilian developmental trajectory based on economic growth without sufficient redistribution of income (Oliveira, 2003). While since the mid-1980s successive upgrading strategies, coordinated by the state, have brought improvements such as basic sanitation, urban infrastructure, investments in physical housing, among others, slums and informal housing areas have become increasingly dense and are still marked by precariousness. The market pressure of more centrally located areas has now spread out to more distant slums and has contributed to an overall increase in land prices and, consequently, rent levels in general. The authors argue that we are witnessing a centralisation of capital and an advance of private management arrangements of space that gradually encapsulates the sphere of informality, "including business institutions, providers of urban services, and 'community' rental platforms, which operate in a blurry zone between contractual formality and urban and building informality" (Rolnik et al., 2021: 46). The latter can be explained considering that in Brazil, both owner-occupied regimes and rental housing are able to expand without full formalisation of contracts, or formal registers, or complete legalisation of units.

According to the authors, the gradual penetration of finance capital in precarious and informal settlements is due to two complementary, on-going transformations in the outskirts of Brazilian cities and metropolitan areas. The first is the restructuring of the nexus between wage labour/owner-occupied formal or informal housing, marked by the trajectory of occupation of plots with self-help building (either or not supported by state through site and services, credit, and complementary measures). The newly emerging arrangement is characterised by more flexible labour relations, hollowing out the stable, wage-based industrial model. This is being accompanied by a more prominent role for rental housing in slums, which have been consolidated through previous upgrading policies in a precarious and imperfect manner, with high demographic and constructive densities. The spreading of housing demand pressure in formal areas to the outskirts has now implied that these partially upgraded informal settlements also witness the proliferation of contractual sublease arrangements among residents.

The second is the entanglement between the state and financial capital through temporary, emergency schemes for rent payment in state-sponsored relocation. An example of the latter is the relocation of communities that are in areas of environmental risk or watershed protected areas. In the city of São Paulo, for instance, the local government has explored real estate platforms and investment funds associated with the payment of emergency rents. This scenario has consolidated what the authors label as a "state of permanent transience", meaning that housing problems are not solved, but that crises associated with emergency situations and environmental risk are dealt with through temporary measures through public-private partnerships (PPP) backed by a stream of payments of emergency rents financed by the local budget (Guerreiro, 2020).

Conclusion and suggestions for an emerging research agenda

This chapter has concentrated on the entanglements between cities, labour, and the platform economy as reflected in the increased capacity to collect, process, and transform data that circulate in and throughout urban spaces. We have described the emerging platform economy in terms of a rapid, self-reinforcing trajectory of data-driven network expansion marked by increasing returns to scale. The latter means that an exponential growth of user connections can be realised through a relatively low level of investments in additional infrastructure and employees. New connections to the network trigger successive cycles of increased corporate competitiveness through the utilisation and sale of information obtained from users. These microeconomic drivers behind the growth of corporate providers of big data have been complemented by successive rounds of deregulation and frequently lax enforcement of urban and anti-trust policies. Moreover, the venture capital industry and investment funds provided further support to aggressive corporate expansion.

Cities and metropolitan areas are the backbone of the emerging platform economy. They concentrate the economies of agglomeration, driven by the variety of skilled as well as unskilled labour supply, the presence of specialised supplier firms, and, crucially, the large-scale, rapid circulation of data and information, which represents the prime raw material for corporate platforms. The rise of platform urbanism goes beyond the smart city phenomenon.

The latter concerns a process whereby the efficiency of urban service delivery is improved through governance arrangements that connect local governments, infrastructure companies, and providers of technology services. Platform urbanism represents a trans-scalar, systemic transformation in the role of information technology in the planning, management, and finance of cities with yet unexplored consequences for the quality of the built environment, livelihoods, and urban labour.

We have provided a first problematisation of the more optimistic narratives of platform urbanism structured around the sharing economy, whereby a common infrastructure is being allocated efficiently and transparently over an increasing number of networker users. Instead, we have stressed dimensions on the supply and demand side of platform capitalism that will not necessarily benefit cities and urban workers and require further empirical research.

Starting with the supply-side, the competitive strength of the platform economy is based on its data-driven virtuous expansion that allows one "to do more with less". In other words, platforms are unlikely to generate a significant expansion of employment and require specific niches of highly skilled and qualified labour. Without specific policies aimed at building new skills and competences, large segments of the labour force are likely to become irrelevant. A related point is that the platform economy has gradually consolidated an industry structure characterised by extremely large oligopolistic corporations (e.g., Google, Microsoft, and Amazon) that provide customised information services to firms that since the 1980s have downsized and subsequently outsourced their information and technology departments. This implies that the platform economy is both centralised in terms of industry structure and spatially dispersed through the global dissemination of platform urbanism. The bottom line here is that supply-side centralisation (i.e., consolidation of corporate oligopolist structures) in combination with spatial fragmentation of platform urbanism tends to challenge traditional strategies of social tripartite dialogue among the state, unions, and businesses aimed at providing decent labour standards and working conditions. For instance, an increasing number of studies indicate the deterioration of labour standards for migrants, ethnic minorities, and women in the gig economy, Uber, and home-care applications.

Demand side impacts of platforms on cities and labour are related to the changes in the built environment and its effects on the cost of living in metropolitan areas. We have illustrated this dimension through a revision of the literature on the penetration of finance capital and real estate platforms in the transformation of the built environment through rental housing. This case suggests at least five themes for a broader research agenda.

First, most of the work has stressed the key role of the subprime crisis in the emergence of rental housing as a new asset class. In a way, the Global Financial Crisis (GFC) generated the ideal circumstances for the constitution of what Neil Smith (1990) labelled as a rent gap, that is, a differential between the existing and potential levels of land and housing rents. In other words, the depreciation of real estate prices and the drying up of the market for mortgages in a context of high levels of debt-to-income ratios of most families generated an enabling environment for the constitution of a new circuit of private accumulation. The latter was supported by successive rounds of regulatory roll-out by the state and increasing levels of rent. However, considering the lack of maturity of secondary markets for real estate backed mortgages, countries such as Brazil were less affected by the GFC (Klink and Denaldi, 2014). Theoretically, this means that the constitution of a new frontier for the financialisation of housing through rent gaps requires a different analytical-empirical mediation than is provided by the nexus between deflation-debt of the GFC. In such countries, we need alternative explanations for the existence and profitable utilisation of rent gaps as the driver behind the transformation of the built environment through rental housing and platforms. For example, in relation to the role of companies such as Airbnb in São Paulo, Rolnik et al. (2021) discuss factors such as the existing stock of physically deteriorated or abandoned housing units in central locations, which nevertheless offers good infrastructure, cultural and leisure amenities and, as such, a favourable environment for the profitable retrofit and development of investment opportunities in rental housing.

Which brings us to the second theme. At first sight, the dialogue between the literature from the North and countries such as Brazil seems more promising in relation to the consolidated, formal areas of cities and metropolitan areas, which are characterised by a reasonable supply of urban infrastructure and opportunities for employment and income generation. Both the central business district and older inner-city areas of Brazilian cities and metropolitan areas attract a significant share of the value added and employment, both in higher end segments such as professional producer services, real estate, insurance, and finance as well as sectors that aim to benefit from the economies of agglomeration concentrated in the centre. Nevertheless, as discussed previously, authors such as Rolnik et al. (2021) suggest the immanent penetration of finance capital and real estate platforms in the market for informal housing in slum settlements in the outskirts of Brazilian cities.

However, this thesis requires more in depth theoretical and empirical research which is beyond the scope of this chapter. For one, the transaction costs that are associated with the standardisation and circulation of adequate and reliable information regarding the scale, variety, and financial viability of the informal market in slums might well provide a formidable entrance barrier to consolidate this segment in the Brazilian setting. As a matter of fact, a new municipal law in São Paulo, which proposed the creation of a PPP for a popular real estate platform that would be responsible for the management of emergency rents in relocation projects and operation and maintenance services, was eventually not approved.

A third theme is the implicit role of investment elites in the literature. As mentioned, platforms have become attractive options for transnational investment elites (Fernandez et al., 2016). In countries of the Global South such as Brazil this should be complemented by an investigation regarding the role of *national* investment elites, considering the dramatic inequalities in income and wealth in these countries. As shown by authors such as Penha Filho (2020), despite the gradual entrance of international finance capital, institutional investors, and professional fund managers, traditional family-owned enterprises still perform a strategic role in Brazilian building and construction, including through the coordination of initial public offering of stock that allow them to maintain the majority ownership and the control over strategic corporate direction and management. As a matter of fact, the Brazilian rentier and patrimonial-driven economic trajectory has a long tradition, which suggests its investigation requires to be linked up with the contemporary literature on finance, platforms, and rental housing discussed in this chapter.

Fourth, the bulk of the literature has generated new knowledge on specific experiences and case studies. It should be recognised that doing similar, empirically based work on how real estate platforms and finance emerge as new builders and operators of cities in the Global South will be an indispensable element for a research agenda on the theme. At the same time, this empirical research requires to be articulated with new theories and analytical perspectives that do not only recognise the specificities of cities in countries such as Brazil (e.g., the degree of consolidation of secondary markets, informality and precarity, and role of local investment and building elites, etc.), but also how this knowledge regarding geographical and historical situatedness can contribute to innovative research on how the penetration of digital technologies and big data, in combination with finance, is able to transform the building and management of the built environment at a truly global scale.

Finally, the illustration of the impacts of real estate platforms on urban space and the affordability of housing represents only one of the possible threats posed by the platform economy to the right to the city. Further work remains to be done on how platforms advance in other circuits than housing – labour and the gig economy, transportation, logistics, consumption, urban infrastructure, intellectual property rights, among others – and change the building, operation, and maintenance of contemporary cities. In his provocative work on the rentier economy titled "Rentier capitalism: Who owns the economy, and who pays for it?", Christophers (2020) argues that platforms have emerged as the privileged device to extract rent from us all. According to the author, this involves a two-stage process whereby monopoly rents are initially constituted and charged to access platforms and subsequently circulate in non-competitive markets for a variety of goods and urban infrastructure services. This raises further questions whether platform urbanism tends to transform cities, as the privileged space for the reproduction of life and use value, into tradeable income yielding assets (Guironnet and Halbert, 2015).

Note

1 Or, in Marxian terms, the reproduction cost of urban labour.

References

Bauriedl, S., & Strüver, A. (2020). Platform urbanism: Technocapitalist production of private and public spaces. *Urban Planning*, *5*(4), 267–276.

Borja, J., & Castells, M. (1997). *Local y Global, La Gestión de las ciudades em la era de la información.* Madrid: Santillana de Ediciones, S.A.

Caprotti, F., Chang, I-C.C., & Joss, S. (2022). Beyond the smart city: A typology of platform urbanism. *Urban Transformations*, *4*, 4. Available at: 10.1186/s42854-022-00033-9. Accessed October 13, 2022.

Castells, M. (1999). *Sociedade em rede*. São Paulo: Paz e Terra.

Christophers, B. (2020). *Rentier capitalism. Who own the economy, and who pays for it?* London/New York: Verso.

Clancy, M. (2020). Tourism, financialization, and short-term rentals: The political economy of Dublin's housing crisis. *Current Issues in Tourism*, 1–18.

Cocola-Gant, A., & Gago, A. (2021). Airbnb, buy-to-let investment and tourism-driven displacement: A case study in Lisbon. *Environment and Planning A: Economy and Space*, *53*(7), 1671–1688.

Cocola-Gant, A., Hof, A., Smigiel, C., & Yrigoy, I. (2021). Short-term rentals as a new urban frontier–evidence from European cities. *Environment and Planning A: Economy and Space*, *53*(7), 1601–1608.

Ebstebsari, A., & Werna, E. (2022). *Smart cities, homes and the workforce: challenges and prospects.* Forthcoming.

Fernandez, R., Hofman, A., & Aalbers, M. B. (2016). London and New York as a safe deposit box for the transnational wealth elite. *Environment and Planning A: Economy and Space*, *48*(12), 2443–2461.

Fields, D. (2018). Constructing a new asset class: Property-led financial accumulation after the crisis. *Economic Geography*, *94*(2), 118–140.

Fields, D. (2022). Automated landlord: Digital technologies and post-crisis financial accumulation. *Environment and Planning A: Economy and Space*, *54*(1), 160–181.

Fields, D., & Rogers, D. (2021). Towards a critical housing studies research agenda on platform real estate. *Housing, Theory and Society*, *38*(1), 72–94.

Glaeser, E. (2012). *Triumph of the city. How our greatest invention makes us richer, smarter, greener, healthier, and happier*. London: The Pinguin Press.
Graham, S., & Marvin, S. (2001). *Splintering urbanism: Networked infrastructures, technological mobilities and the urban condition*. London: Routledge.
Grisdale, S. (2021). Displacement by disruption: Short-term rentals and the political economy of "belonging anywhere" in Toronto. *Urban Geography*, *42*(5), 654–680.
Guerreiro, I. D. A. (2020). O aluguel como gestão da insegurança habitacional: possibilidades de securitização do direito à moradia. *Cadernos Metrópole*, *22*, 729–756.
Guironnet, A., & Halbert, L. (2015). *Urban development projects, financial markets, and investors: A research note*. Chairville: École des Ponts Paritech. 30 p.
Hall, P. (1998). *Cities in civilization*. New York: Pantheon Books.
Hall, P. (2014). *Cities of tomorrow. An intellectual history of urban planning and design since 1880* (Fourth Edition). Main Street/MA (USA)/Oxford/UK/West Sussex (UK): Wiley-Blackwell.
Harvey, D. (2013). *Os limites do capital*. (tradução de Magda Lopes), São Paulo: Boitempo.
Healey, P. 2003). Collaborative planning in perspective. *Planning Theory*, *2*(2), 101–123.
Heeg, S. (2013). Wohnungen als Finanzanlage. Auswirkungen von Responsibilisierung und Finanzialisierung im Bereich des Wohnens. *Sub\urban. zeitschrift für kritische stadtforschung*, *1*(1), 75–99.
Hof, D. (2021). Vom Eigentum zur Miete. *Berichte Geographie und Landeskunde*, *94*(1), 41–63.
Janoschka, M., Alexandri, G., Ramos, H. O., & Vives-Miró, S. (2020). Tracing the socio-spatial logics of transnational landlords' real estate investment: Blackstone in Madrid. *European urban and regional studies*, *27*(2), 125–141.
Kjaer, A.M. (2012). Governance and the urban bureaucracy. In: Davies, J.S.., & Imbroscio, D. *Theories of* urban politics (Second Edition). Los Angeles/London/New Delhi/Singalpore/ Washington, DC: Sage, pp. 125–136.
Klink, J., & Denaldi, R. (2014). On financialization and state spatial fixes in Brazil. A geographical and historical interpretation of the housing program My House My Life. *Habitat International*, *44*, 220–226.
Langley, P., & Leyshon A. (2017). Capitalizing on the crowd: The monetary and financial ecologies of crowdfunding. *Environ Plan A.*, *49*(5), 1019–1039.
Marshall, A. (1920) *Principles of economics*. London: Macmillan.
Nayaran, D. (2022). Platform capitalism and cloud infrastructure: Theorizing a hyper-scalable computing regime. *Environment and Planning A. Economy and Space*, *54*(5), 911–929.
Nethercote, M. (2020). Build-to-Rent and the financialization of rental housing: Future research directions. *Housing Studies*, *35*(5), 839–874.
Oliveira, F. (2003). *Crítica à Razão dualista/O Ornitorrinco*. São Paulo: Boitempo.
Penha Filho, C. A. (2020). *Estratégias e dinâmica de acumulação das incorporadoras listadas (2010/ 2018)* (Doctoral dissertation, Tese de doutorado. Campinas: Universidade Estadual de Campinas).
Rolnik, R. (2021). Direito à moradia em rara semana vitoriosa. Disponível em: http://www.labcidade.fau.usp.br/direito-a-moradia-em-rara-semana-vitoriosa/. Acesso em: 12 Nov. 2021.
Rolnik, R., Guerreiro, I. D. A., & Marín-Toro, A. (2021). El arriendo-formal e informal-como nueva frontera de la financiarización de la vivienda en América Latina. *Revista INVI*, *36*(103), 19–53.
Smith, N. (1990). *Uneven development, nature, capital, and the production of space*. London: Basil Blackwell.
Storper, M., Salais, R. (1997). *Worlds of production. The action frameworks of the economy*. Cambridge (Massachusetts) & London (England): Harvard University Press.
Vannuchi, L. V. B. (2020). *O centro & os centros: produção e feituras da cidade em disputa*. Doctoral Thesis Faculty of Architecture and Urbanism. São Paulo: University of São Paulo.
Wijburg, G., & Waldron, R. (2020). Financialised privatisation, affordable housing and institutional investment: The case of England. *Critical Housing Analysis*, *7*(1), 114.

14

HOUSING MARKETS AND LABOUR MARKETS

Towards a new research agenda for the Global South

Edmundo Werna[1], *Ramin Keivani*[2], *and Youngha Cho*[2]

[1]London South Bank University, UK; [2]Oxford Brookes University, UK

Introduction

Housing and labour markets are strongly related. They are fundamental aspects of urbanisation and, together, they can make or break the development of a city. Housing market decisions have important effects on social well-being of residents, in addition to notable macroeconomic implications, such as the influence on labour mobility and work productivity. Many people cannot get proper employment because of limited accessibility to the housing market. They live far away from good job opportunities (see Cho et al., 2018, for an overview).

Labour decisions are directly related to permanent or long-term income that, in turn, influence household decisions about housing (Colom and Moles, 2013). This view is supported by Haas and Osland (2014) who argue that income and income prospects determine whether people can afford to buy an appropriate house in a suitable place. For example, many urban workers cannot access a proper housing unit in the city because they receive low wages that are not sufficient to enable them to save for buying a decent home, access a mortgage scheme or, in some cases, even afford rent for decent housing in appropriate locations. This is exacerbated by the fact that jobs are not secure, and many workers shift between periods of formal, informal or self-employment, sometimes with spells of lack of jobs altogether (see, for example, Bird, 2021). For a broad view, UN-Habitat publishes a World Cities Report every two years. Basically, all of these reports note that there are large numbers of urban poor people. The same set of publications also emphasises the difficulties that the poor have to obtain decent housing (see, for example, UN-Habitat, 2022).

At the same time, while houses constitute a major part of a household's wealth, earned income from employment is a major source of household revenue and is also an important determinant of house prices (Meen and Andrew, 1998; Muellbauer and Murphy, 2008). Finally, at the micro-level, poor quality of housing also affects people's health and physical well-being, which can, in turn, affect their joining the labour market and work (due to ill health).

As is presented later in the chapter, there is a wealth of research on housing markets and related policies. The situation is the same with respect to labour markets. Yet, the combined

DOI: 10.1201/9781003262671-14

literature on the connection between the two markets is still limited, both geographically and thematically. This lack of connection is particularly evident in the literature on the context of the Global South. Here, the works on both markets are highly segmented. Informal arrangements are the default; compounded by complex governance and management issues. This chapter argues that it is time to review such combined literature and to explain the missing links between the housing and labour markets. It argues that the literature which combines analysis of both markets, is preponderantly focused on developed countries (Global North), and at the same time focused on the formal housing market and its connection with wage work. It is still a rather Fordist-inspired vision of both markets. However, the complexity of both housing and labour markets in the developing world (Global South) makes such an analysis limited in its application. More so, both markets are getting more complex even in developed countries.

The chapter starts by highlighting main points of the two sets of literature, on labour markets and housing markets, noting their respective pluralistic nature. Next, it reviews the combined literature, highlighting the missing analytical linkages. It concludes with the proposal for a research agenda.

The labour context

Cities and towns will not be sustainable if the livelihoods of their inhabitants are not adequately addressed. This is particularly challenging in the Global South (World Bank, 2019). In reality, unemployment may not be as high as underemployment and inappropriate working conditions. Meaning that the poor are working one way or another, otherwise they cannot survive because of inadequate or non-existent social security (although some do rely on their family network). But this often means that they are working under improper conditions, which perpetuates urban poverty (for example, large numbers of people are living below $1.90 per day, do not have adequate protection against occupational safety and health hazards (Ferreira, Jolliffeespen, and Prydz, 2015)).

From this perspective, the Sustainable Urban Development (SDG11) can be only meaningfully aspired to if it is aligned with measures for achieving economic growth, Employment and Decent Work for All (SDG8). Indeed, the United Nations' New Urban Agenda launched during the Habitat III UN Summit in 2016 specifically introduced labour issues as part of urban policies (United Nations, 2017).

For this chapter, the task is not just about facilitating access to work per se. But the kind of work that sums up the aspirations of people in their working lives and, among other objectives, to have a fair income, security in the workplace and social protection for families (ILO, 1999).

Urban labour markets are, to a large extent, pluralistic; they encompass a wide range of employment forms, duration and status. Cases of a common and integrated cluster of supply and demand for labour in a given city are increasingly fewer and farther between (Buffie et al., 2022; Deakin, 2013; Leontaridi, 2002).

Throughout the twentieth century, until around the mid-1970s, trends in the labour process were heavily influenced by the overall strategy of industrial development and capital accumulation prevailing in most industrialised countries, termed Fordism. This entailed mass production of standardised goods, with production lines along a Taylorist model, as well as direct, long-term and secure employment. While the majority of developing countries were far from such a Fordist model, it was believed that this paradigm would propagate itself throughout the world. Although the labour market was still pluralistic to some extent, Fordism was seen as 'the future', and, as such, it was expected that there would be a cohesive urban labour market (Harvey, 1989; Werna, 2000; 2016).

From the mid-1970s the world experienced major shifts in the labour process, embedded in a new mode of industrial development and accumulation broadly called 'post-fordism' or 'flexible accumulation' (Harvey, 1989; Antunes, 2021), which still prevails. One of the curious features of post-Fordism is that in many ways it resembles pre-Fordism. It is pertinent to note that the concept of the informal sector was coined in the 1970s, which seems to suggest that people no longer held the belief that the Fordist model would be the solution for the developing countries and would open up a far-reaching school of thought on urban labour (i.e., informal employment) (Werna, 2016). It is now widely accepted that the urban economy has two concomitant labour markets: formal and informal. This has been an important step in the effort to disentangle the analysis. Yet, it is more complex than this.

There has been a considerable increase in the casualisation of labour, meaning large changes, on and off, in the number of people employed; or constant shifts from job to job, many times also coping with different part-time jobs in parallel. There are also considerable movements between formal and informal employment – those who lose their jobs, or are in-between jobs in the formal market, resort to informal work. Others actually find it profitable to move between the formal and informal markets, when they manage to avoid taxes in the latter.

A related trend is the zero-hour contract, where the number of hours an employee gets paid for every week solely depends on the demand of the employer for the worker's labour. The worker, similar to a subcontracted enterprise, is forced to have multiple employers at once. Actually, in many countries, more and more workers are turning into 'one-person enterprises'. The concept of 'one-person enterprise' in fact blends into one of 'self-employment'. Such type of employment is widely prevalent in many developing countries and is also spreading into the developed countries. The difficulties of – or the disbelief in – the possibility of finding jobs for everyone has led many institutions to promote how to 'start and improve your own business' (Antunes, 2021; Eliasson et al., 2003; Ferreira et al., 2015; ILO, 1999; Standing, 2011).

The universe of micro and small enterprises and complex chains of subcontracting add another layer of complexity to the notion of the urban labour market. One day, a person works on her/his own as a self-employed worker or as a 'one person enterprise', providing services directly to clients. On another day, the same person may be an employer, if s/he gets a contract that needs more than one worker. Subsequently, the same person can become an employee, then unemployed and so on.

All these trends illustrate the plurality and complexity of the urban labour market. As noted before, in a given urban area, there is at least a formal and an informal market. As noted before, many people also fluctuate between the two. In sum, the labour market in urban areas is increasingly detached from the Fordist/Taylorist predictability. There is seldom a common and integrated cluster of supply and demand for labour in a given city. The impact this has had on housing entails an increase in the transient use of space and enhanced mobility to look for new or multiple jobs and contracts.

The housing context

On a global level, the housing backlog is alarming, with around 59 percent (200 million) of the urban population in sub-Saharan Africa, 25 percent in India and 22 percent in Brazil living in slum areas (UN-Habitat, 2022; World Bank, 2019, Sengupta, 2019). This is an indication of inappropriate housing conditions. In addition to slums, many people live in other types of substandard housing, such as derelict buildings (sometimes termed 'squats'). There is also a large population of homeless urban residents.

Housing is a complex commodity. There is seldom a single mode of provision in a given country or even city. The pluralistic nature of housing provision has been analysed by a number of authors (Keivani and Werna, 2001a, 2001b; Werna et al., 2000; Wakely, 2018; Sengupta, 2019). The main forms of formal market delivery coincide with the stages of public housing policy shifts. These can be summarised in terms of a) direct public delivery, b) supported public delivery, c) private market-led delivery and d) public-led large-scale housing delivery in partnership with the private sector. In his recent book, Geoff Payne (2022) similarly identifies the complexity of land tenure arrangements in the Global South with indigenous and modern (often imposed from the colonial era) tenure systems coexisting at the same time and location leading to legal plurality that can undermine legal tenure regimes which are not in harmony with local contextual social and cultural values. He highlights the point that this plurality is recognised in the New Urban Agenda (UN, 2017) requiring a range of (housing) solutions based on the continuum of land and property rights and goes on to conclude:

> At a structural level of urban management, the broad aim of planning and urban management should be to create pluralistic systems of supply which can respond to variations in demand and needs (p. 190).

This has clear impacts on housing markets, making them equally pluralistic. A summary of types of delivery is next presented; each type of delivery comprises different variants (Keivani and Werna, 2001b; Wakely, 2018).

Direct public delivery

In the context of the Global South, the heyday of public housing delivery was in the late 1950s, 1960s and mid-1970s. However, with the exception of Singapore and Hong Kong which used housing as a means of nation building and a few socialist countries, these programmes largely failed on almost all relevant criteria. In terms of scope, they have largely contributed to around 10 percent of housing delivery (Keivani and Werna, 2001b; Wakely, 2018). They were often badly built with minimal internal services and located in peripheral areas with insufficient social and infrastructure facilities. They were unsuitable for living requirements of low-income populations who were relocated there. Many times, this led to their abandonment or ingenious and often extensive but illegal changes to the original buildings by the occupiers. Finally, many of these projects were prone to corruption and syphoning of public funds through collusion of the officials with contractors leading to even abandonment of projects or misallocation to middle-income people or government functionaries (Keivani and Werna, 2001b; Wakely, 2018).

Nevertheless, as stated above, some former socialist countries have had major successes in public housing delivery primarily for rental purposes. UN-Habitat (2022), for example, cites the Chinese Dibao social housing programme as a successful example of integrating a fragmented system to establish a security net for meeting the basic needs of all people with a view of instituting social justice and inclusive cities.

Indirect public delivery

This housing type refers to Sites and Services schemes and Settlement Upgrading. Considering the failure of direct public housing delivery and increasing housing shortages for the low-income populations these delivery modes came to the fore from the mid-1970s onwards. They are both inspired to various levels by the informal self-help housing delivery mode of the informal settlements (Keivani and Werna, 2001b and Wakely, 2018).

The former can be described as aided self-help and was particularly prominent in the 1980s. The main rationale here was to repeat the success of the incremental house building and improvement process of the informal settlements by the provision of subdivided and serviced sites to households who would then be primarily responsible for the building of their own housing units, at their own pace. An alternative version of this form is the provision of core units, that is, a minimum-shelter unit which can be occupied relatively quickly by the household and extended when the occupier can afford the time or money. The schemes were assumed to largely utilise self-help labour of the occupants, thus, reducing the cost of construction. In reality, however, most self-help projects involve a high degree of paid labour and contract building. In Matero in Lusaka, for example, 92 percent of households in the sites and services projects used hired labour (Keivani and Werna, 2001b).

The latter, in itself, does not necessarily constitute additions to housing stocks but as the name implies it leads to higher standards of housing particularly through infrastructure and service improvements although in some cases there have been small- or large-scale redevelopment of the housing units as well. As with Sites and Services, low-income neighbourhoods upgrading programmes have been usually on a project basis. However, in some cases, such as the slum upgrading programme of Karachi and Kampung Improvement Programme of Indonesia, it can become an essential part of the overall physical planning process of the urban areas of the country, affecting all informal settlements. Cost recovery for the projects occurs through household payments on the basis of the size of the land occupied and type of land use (Keivani and Werna, 2001b).

In both cases, the project-based nature of the schemes limited output, and it did not match expectations. For example, in the first decade, output was about 10 percent of the required volume. Bureaucratic costs consumed around 50 percent of the budgets (Keivani and Werna, 2001b). The sites and services schemes were particularly badly located in the peripheries and the residents did not receive appropriate technical support for self-build construction. Consequently, from the 1990s onwards, the policy domain shifted to enabling private markets through various mechanisms including government withdrawal from delivery, focusing on the provision of infrastructure and services, and supporting mortgage finance. At the same time while settlement upgrading continued in modified forms, sites and services schemes were largely abandoned. Nevertheless, Wakely (2018) argues that these schemes were judged too early and many of them have subsequently developed into vibrant communities.

Formal private delivery

In the Global South, this mode of delivery has mainly been for the middle- and high-income groups of the urban population. In most countries, it has contributed to about 20 percent of the housing stock in the more developed and higher-income countries this can rise to 60–70 percent (Keivani and Werna, 2001b). This mode can take many forms, from individual house building which is initiated and financed by the individual owner-occupier and built by a small contractor, to small- and large-scale speculative residential developments (being small contractors the vast majority). The determining factor in this regard is the level of development of the private-housing market and the presence of commercial housing developers whose function is the initiation of, and speculation in, housing development. It is worth noting that even in Metropolitan Sao Paulo, the industrial heart of Brazil and South America, formal developers only accounted for around 23 percent of housing construction during the 1990s (Torres et al., 2007). There are other variations of formal private delivery, but the two most important modes are public-private partnerships and joint ventures between small-scale landowners and developers and public and private sectors, as presented next.

Combined public-private

Public-private partnerships for housing delivery have existed in various forms for a long time. This is particularly useful when the public sector is able to leverage private-sector investment through offering various regulatory, fiscal, financial and land incentives. For example, in Iran, following the 1978 revolution and the introduction of the Land Ceiling Act, the government was able to amass large tracts of urban land and therefore introduced partnerships with private developers to enhance the supply of low-income housing (Keivani et al., 2008). Here, the main incentive was to offer public land to entice private finance for housing development shared between the two parties with the public share allocated to eligible households and the private share sold on the open market. While this was only partially successful, it did provide an additional tool for more affordable housing delivery in the country.

More recently, there has been a resurgence of large-scale government-led low-income housing delivery across a range of countries of the Global South. Exemplary programmes include Minha Casa Minha Vida (MCMV) in Brazil, launched in 2009 with the aim of delivering 3 million housing units; Pradhan Mantra Awas Yojana (PMAY) in India, launched in 2015 with the aim of delivering 20 million units by 2022; and the on-going Dibao Chinese public-housing programme with the aim of delivering 10 million housing units a year (Sengupta, 2019; Keivani et al., 2020). A key aspect of these programmes is their objective to enhance public-private partnership through various subsidies including access to cheap and subsidised loans, access to public land, and supporting effective demand by low-income potential homeowners through grants and subsidised loans. As such, Sengupta (2019: 511) argues that the re-emergence of state-led housing is, in fact, an additional means of supporting markets as they 'are creating and formalising new areas of market engagement, and are far less radical and transformative than is assumed'. In addition, in spite of some measure of success, there have also been major concerns about the quantity and quality of their outputs. For example, Klink and Denaldi (2014) note the contradictory application of MCMV (Minha Casa Minha Vida) including lack of coordination with the national plan for social housing, lack of sufficient attention to the needs of the lowest income groups (in Brazil it means those who earn up to three minimum salary), particularly isolated and peripheral locations, and lack of sufficient leverage on the real estate markets.

Joint venture between small-scale developers and landowners

This is a very common mode of housing delivery. Particularly in middle-income countries, it accounts for a substantial number of speculative private residential developments. It is, for example, very prominent in Turkey and Iran, and relies on small-scale developers who make a joint contract with landowners to develop their lands into apartment blocks in return for a share of the apartment units (Keivani and Werna, 2001a; 2001b). In the context of Turkey in the 1990s, Baharoglu (1996) notes that developers would usually have a small amount of working capital to start with, comprising about 26 percent of the total cost of the development. With this capital, developers would start the work to build a basement and ground floor flats which are then sold. With the money gained from this initial sale, work is restarted and the remaining flats are pre-sold before completion; that provides additional working capital. In this way, the small-scale developer normally secures about 60 percent of the total cost of the project, borrows another 14 percent from building materials producers and others as required, and finishes the project without any recourse to the formal financial institutions at all.

Co-operative housing

Housing co-operatives can be formed through different means, depending on the socio-economic organisation of a country. For example, in Zimbabwe and Iran, housing co-operatives were actively encouraged by the government through the workers in specific workplaces (Keivani and Werna, 2001a; 2001b; Vakil, 1996). As a result, many housing co-operatives have been formed, either through factories, offices, ministries, and so on, or through professional associations and trade organisations such as the teachers' associations or guilds of taxi drivers, tailors, shoemakers, and so on. In some other countries, such as Mexico, neighbourhood and social organisations and movements, sometimes in conjunction with political parties, also play a large role in the formation of housing co-operatives.

This system allows groups of people to organise themselves and pool their resources and efforts into a formal organisation, thereby enabling it to negotiate on behalf of its members for acquiring land from the government or the private market, applying for and receiving credit or mortgage loans from government and formal sector institutions, receiving building materials, and commissioning contractors for building the housing units. This form of housing delivery can be substantial in some countries. For example, in Turkey, in the 1980s, it contributed to around 13 percent of housing delivery (Okpala, 1992).

Formal private rental housing

This mode of housing is a major form of housing delivery to a wide range of socio-economic sectors, particularly lower middle-class and middle-class segments of the population. Some low-income households can also be covered through public rental provision that are usually small-scale but there are exceptions such as the aforementioned Chinese Dibao programme. In most countries, formal private rental provision is achieved through small-scale landlords who may develop their land into small apartment units and rent out all or part of it. This point is re-emphasised by UN-Habitat (2003) who note that small landlords provide the bulk of private rental housing across the globe including Europe and North America although in these places, private companies also own substantial stocks of rental housing.

The proportion of households in rental accommodation tends to vary among countries and cities. In some low-income cities such as Port Harcourt, Nigeria, in the 1980s, this was as high as 89 percent, although by 2006 it had fallen to about 72 percent (Ayotamuno and Owei, 2015). This reduction is indicative that the rate of tenancy reduces with economic development of the countries and cities concerned and improvements in the income of households. Similarly, the proportion of renters in urban centres of India fell from 53.7 percent in 1961 to 46.4 percent in 1981 and then to 37 percent in 1988. In Sao Paulo, on the other hand, rental housing fell from about 75 percent in 1940 to about 30 percent in 1970 (UN-Habitat, 2003). However, the trend seems to have slowed since the 1980s and in some cases reversed with increasing shares of rental housing (UN-Habitat, 2003). This may be due to increasing economic challenges and high inflation in certain countries that has led to rising house prices putting house ownership out of reach of many households and therefore increasing demand for renting. The share of rental housing in Iran, for example, has increased from 25 percent in 1986 to about 30 percent in 2020 (40 percent in Tehran). At the same time, both due to increased demand and worsening economic situation, particularly high inflation, we have massively increasing rents in the formal sector with a major squeeze on the middle-income renters. A case in point is the current situation of the rental market in Tehran where in July 2022 house rental prices increased by 46.5 percent year on year (Tehran Times, 2022). A situation that has led to many middle-income families moving out to cheaper suburbs and even more informal communities in the surrounding areas of Tehran which in turn transmits the inflationary rental and housing pressures onto the lower-income groups in these settlements.

Unconventional (informal) delivery modes

The informal mode of housing delivery primarily arises due to the inability of capitalist development to absorb large sections of the low-income urban population into the formal private employment market or to provide them with adequate wages to be able to participate in the formal housing market. Consequently, the low-income urban population have had to resort to informal means (outside planning and legal regulations) to provide housing for themselves. Overall, this mode is the main form of low-income housing delivery in cities of the Global South particularly in the lower-income countries. Globally, it is estimated that around 23 percent of the urban population (or about 1 billion people) live in informal settlements but 80 percent of them are in East and South Asia and sub-Saharan Africa (UN, 2018). Therefore, the contribution of this mode in some cities will be far higher, and in Nairobi, for example, it accounts for 80 percent. The exact form of such provision depends on the political, socio-economic and cultural conditions of the relevant countries and cities within each country. However, there are three main delivery modes comprising squatter occupation, informal subdivision and informal rental housing.

Squatter settlements

In the years between 1950 and the mid-1980s, squatter settlements were the most common form of housing provision in developing countries with an annual expansion rate of between 15 and 20 percent up to the mid-1970s (Keivani and Werna, 2001a; 2001b; Wakely, 2018). Squatter invasions occur through two methods: organised mass invasions; and gradual infiltration by individual families or small groups. However, on the whole, it can be stated that large-scale land invasions are relatively rare and have generally occurred at times of particular political situations when either the government has been too weak to enforce eviction or where the government has required political support, and therefore, accepted the invasions.

The actual settlement development and construction varies widely across cities, depending on the income level of squatters and the general socio-economic and political situation of the city concerned. In certain cases, very sophisticated and expensive methods are employed even at the earliest stage of land development. In one settlement in Lima, Peru, for example, a group of topographers were hired by the squatters to survey, subdivide, and mark out the land at a cost of US$1,000 in the mid-1960s (Turner, 1967, quoted in Drakakis-Smith, 1981). However, in the majority of cases, the priority for the settlers is to build on the land as soon as possible. Hence, rough subdivisions are undertaken by the squatters and very simple structures are built in the shortest possible time to gain de facto possession of the land at the initial stage. In most cases, such units are then improved upon by the owner-occupiers over a period of several years, depending on their income and the political situation of the country with regard to the acceptance of their settlements.

Informal subdivisions

During the 1980s, the increasing penetration of market relations in the informal housing provision system and the consolidation of state power in preventing land invasions led to the replacement of squatter housing by informal subdivision as the dominant mode of housing provision for low-income households in developing countries (Keivani and Werna, 2001a; 2001b; Wakely, 2018).

This mode primarily relies on informal developers who are generally well connected to the formal sector, particularly municipal and government officials who take over public land tracts, subdivide them, often providing basic layout, roads and infrastructure, and selling them to low-income households (Keivani and Werna 2001a; 2001b; Wakely, 2018). The actual characteristics of informal subdivisions, their integration into local housing and land markets, and the social position of developers and type of families who buy such land vary from city to city. However, the following general features can be identified:

1 Planned layout, where the developer hires an engineer or surveyor to subdivide the land into residential plots and, in some cases, even public utility spaces such as parks and schools and the road networks, and so on. These usually follow the pattern set by formal subdivision of lands.
2 Service standards can vary from non-existence to basic. However, usually, settlers are promised higher standards at later stages.
3 The receipt of sale documents by settlers which register the sale and the belief in the connection of the developers to authorities gives settlers a high degree of confidence in their security of tenure.
4 Similar to squatter settlements, there is a high degree of self-building in the housing development process of informal subdivisions. However, due to the commercial nature of land provision, wage labour and contractors are used more often and at earlier stages than in squatter settlements. In addition, there is also a relatively high degree of speculative house building by incipient contractors and building for rental purposes by household settlers and others from the beginning.

The beneficiary households of this mode of provision, in general, tend to be the higher-income groups and also the middle-income households who have been hit hardest by the withdrawal of government provision of direct housing and have saved up to gain access to land for housing (Keivani and Werna, 2001a; 2001b; Wakely, 2018).

Informal rental housing

Many sections of the low-income and first-time immigrants to urban areas, neither have the effective demand for an informal subdivision plot nor a firm foothold in the urban environment with the appropriate social connections to join a squatter settlement as owner-occupiers. Consequently, they find rented shelter in overcrowded central city slums, specialised rental slums in the periphery areas or in squatter settlements and informal subdivisions. Alternatively, low-income households may also build their own units on rented land. It should also be noted that there are special rental areas where major and minor landlords develop their land particularly for renting to low-income migrants. The best example of this type of housing provision is probably the Bustees of Calcutta, India.

Overall, Amis (1996) notes that the de facto recognition of informal settlements and tolerance of informal commercial activities has led to a significant increase in rental housing provision in informal settlements. However, as Scheba and Turok (2020) stated, the true extent of informal rental housing provision is difficult to estimate due to its nature, but they note that studies that suggest that 92 percent of surveyed residents in Kibera in Nairobi, Kenya, and 80 percent of informal settlements resident in the Philippines are, in fact, informal renters. They further suggest that informal rent forms the majority of all rental housing in many cities of the Global South with high levels of informal settlements.

In summary, the definition of a housing market entails the type, size, cost, and availability of housing in a particular area. In fact, there are many sub 'markets' in the same area operating in parallel, with some interfaces. The preceding discussions have provided some detail on the characteristics of different delivery modes that relate to different housing submarkets. However, as will be shown in the following section, their connection with labour markets is based on macroeconomic studies which do not fully capture the pluralism of the Global South and are limited to consideration of owner-occupation, renters, and in some cases, mortgage holders. There is, therefore, room for improvement, as will be shown in the Conclusion.

The connection between housing, labour and their markets

The links between the housing and labour markets have long been recognised, particularly in the Global North (Hanson and Pratt, 1988). In spite of this, the largest proportion of the research in the field has focused on tenure (owner-occupation and rental), mobility and other housing and labour market issues mainly in isolation (Allen and Hamnett, 1996; Jones and Hyclak, 1999; Eliasson et al., 2003; Wegener, 2004; Hincks and Wong, 2010). This is particularly so for empirical works in the field (Eliasson et al., 2003). The section below elaborates upon the connections which have not been outlined yet.

The Global North

Housing tenure and mobility

The impact of housing on labour mobility has received increasing attention since Oswald and colleagues seminal work suggested that there is a negative correlation between home ownership and labour mobility, leading to greater unemployment or lower wages (Oswald, 1997; Hildreth and Oswald, 1997; Blanchflower and Oswald, 2013). The basic point here was that home ownership makes workers immobile. If demand for labour falls in a region (even if it leads to job losses or lower wages), homeowners do not respond by migrating to other regions to find new jobs because the expected benefits from such a move do not necessarily outweigh the high transactions cost of moving from one owner-occupied house to another. This argument has been used in many studies by a variety of authors in different countries viewing longer residence spells not as stability, but as immobility, and therefore as a cost rather than a benefit since owners are not able to respond to local labour demand shocks and therefore undermine the economy as a whole. Therefore, there is a coordination failure that creates inefficiency.

Further work, has followed that, in the main, has been limited to macro and micro econometric analysis, with the former largely supporting Oswald's hypothesis and the latter largely rejecting it (Isebaert et al, 2015; Haas and Osland, 2014; Colom and Moles, 2013). Baert et al. (2014) provided a more refined analysis than the two groups of work using micro-level data and argued that, in Belgium, outright owners indeed follow Oswald's hypothesis but mortgage holders, in fact, exit unemployment first, followed by tenants.

These studies are methodologically and geographically limited. As already noted, they are all based on longitudinal econometric analysis and are almost exclusively focused on the Global North. An exception to this is Gangopadhyay and Shankar (2016) who looked at the situation in Bangladesh from a political economy perspective. They argued that employers in monopsonistic competition have the market power to set wages below the productivity of workers if they are immobile. Therefore, if the housing market does not allow mobility, it will have a negative impact on wages. Otherwise, such studies require the availability of large data sets that are often lacking in the context of developing countries, particularly, in informal housing and labour conditions. In addition, the studies are unable to account for the reasons underlying the behaviours noted. Oswald's hypothesis generally assumes that high transaction costs of buying and selling houses limit the scope for labour mobility. On the other hand, Baert et al (2014) argued that the key issue is overall housing costs which lead to induced reduction in consumption that for mortgagees and renters act as higher impetus for finding employment. However, the statistical data do not allow for definitive answers and these explanations can only remain conjectures.

In sum, the pro- and anti-Oswald debate is still inconclusive. One reason is that each study comes to a different conclusion in a different place and this supports exactly the point made in this chapter about pluralism and the one-size-does-not-fit-all policy. This is compounded by the lack of disaggregation of the analysis of the labour-housing market nexus, which, as noted before, tends to privilege only limited issues. As such, there are missing elements which could throw light on the debate on the work of Oswald and his colleagues.

Location

The impact of location in the labour-housing markets connection has also been the subject of attention, as spatial variation in housing prices and/or income prospects may determine commuting distance and may influence decisions about whether to move or stay in various places (see Vickerman, 1984; Van der Vlist et al., 2002; Hämäläinen and Böckerman, 2004; Saks, 2008; van Ewijk and van Leuvensteijn, 2009; Zabel, 2012).

This literature has, somehow, informed the Oswaldian debate (Oswald and colleagues, noted before) in one direction or another, but also has its own contribution due to the specific nature of the housing products (fixed in space) and its impact on locational decisions, possibilities or limitations of people to move between home and workplace. For example, how much people are prepared to pay to avoid long commuting, and sometimes the willingness of people to actually live far away from the workplace, and the effects on prices. This is different from the willingness to move from one housing unit to another.

One innovative element of this body of literature relates to teleworking. According to O'Connor and Healy (2001), teleworking and teleservices seem to be developing hand-in-hand with lower-density, less nodal urban forms and with travel behaviour that is more car-dependent than before. These technologies are helping to territorially expand and disperse the metropolitan area by enabling more complex linkages between jobs and homes; that is, the cities become less dense.

A question that this chapter poses is what would be the long-term impact of the 'tele-labour market' on the housing market, since the teleworker is liberated from the need to find a home in a specific location vis-à-vis her/his workplace. Their housing becomes their workplace. Yet, the need to eventually commute to a specific office for periodic meetings should continue to be taken on board. At any rate, while the literature on the housing-labour markets' nexus privileges the effects of commuting, the teleworking trend is one example of a reverse effect. This is an area of research that deserves to be further developed.

In addition, there is a growth of home-based enterprises. This is taking place in both the Global North and South. They are reinforced due to reduced, or lack of, alternatives in the urban economy. They also interface with the concept of the informal sector but are not exactly the same. In developed countries, they are part of the transition away from Fordism. This requires re-thinking housing not only as a place to live, but also as a place to work. Part-time work and unemployment may also entail spending more time at home, requiring further changes in the housing unit. The potential impact of home-based enterprises in the housing market may be analogous to that of teleworking. Specific points regarding the Global South will be presented afterwards in the chapter (Bonnet et al., 2021; Kellet and Tipple, 2011; Reuschke and Domecka, 2018; Sohane et al., 2021; Tipple, 2005).

Careers

An important derivative of the literature on the housing-labour nexus regards the reasons for housing choices through studies of careers. This area of research goes back to Pickwance's (1974) work on life cycle, housing tenure and residential mobility. Subsequently, the introduction of life course, as opposed to life cycle, has enabled a more dynamic understanding of housing choices and behaviour based on broader life events including changes in income and employment (Ozuekren and van Kempen, 2002; Abramsson, 2008). Many of these studies have relied on more in-depth cross-sectional surveys and/or qualitative interviews particularly focusing on lower-income and minority groups. However, again, with few exceptions, they are focused on the Global North or thematically limited to impact of life events on housing choices. The exceptions include Marias et al. (2018), who examined the housing careers of middle class black South Africans and Watt (2005) who studied housing histories and career progression/mobility of marginal professionals in London, UK.

Housing booms

Ernst and Saliba (2015) approached the housing-labour markets nexus from a different standpoint. They reviewed literature on housing production cycles (see, for example, Iacoviello, 2005; 2010; 2011; Iacoviello and Neri, 2007; 2010; Iacoviello and Pavan, 2013; Estevão and Tsounta, 2011; Askenazy, 2013) and argued that housing booms with rising house prices may have a positive effect on wages in the short-run but a negative one in the long-run. Their argument is that as house prices rise, wages tend to follow to make up for the loss in real disposable income, which limits employment creation. In addition, with rising house prices, the relative size of the construction sector, a low-productivity industry, tends to increase, lowering aggregate productivity growth, further dampening competitiveness. They conclude by recommending macro-prudential monetary policies to increase the resilience to housing market swings, and therefore avoid periods of boom.

One could also add that booms in housing construction (and also in other sub-sectors of construction) are usually followed by periods of bust. Bust periods may also have negative implications for the labour market, therefore there is a double rationale to avoid market swings. Macro-prudential policies seem to be a fair answer to this.

At the same time, the present chapter argues that it is important to avoid swings but without necessarily limiting housing supply. Housing is known to be fundamental for workers' physical and social well-being and productivity. That is, if they do not have a good home, they will not perform well at work. Therefore, it is important to think about policies that, while avoiding the possible negative effects of swings, guarantee a steady flow of housing production and supply, especially in developing countries, which still have large housing deficits. This deficit has been seen as having negative impacts on the ability of people to make ends meet. Although such an equation may be more complex than one of 'pure' un/employment, it needs to be taken into account in the analysis of whether a housing boom is indeed negative.

The specific focus of Ernst and Saliba (2015) and the literature they review is the set of developed countries. The same set analysed by the bulk of the authors who concentrate on tenure and location, as noted in previous sections. The set of developed countries is, to some extent, relatively similar in terms of housing and construction markets. When one considers emerging, middle-income, low-income and least-developed countries, the complexity increases. At the same time, these are the countries that may benefit most from policy advice, considering their challenges both to create decent jobs and to provide decent housing.

As mentioned in a previous section, the 'traditional' concept of employment (and unemployment) is changing, with large numbers of workers on casual, part-time, multiple jobs, plus micro (one person) enterprises and self-employed workers. The impact of booms in the housing market on workers in this growing range of employment types is not the same and deserves to be disaggregated and further investigated.

The connection between the housing market and home-based workers is also worth considering in this context. For this burgeoning number of workers, it is difficult to understand that a boom in the housing market would be detrimental to them, considering that they need housing in the first place in order to earn a living. The analysis of Ernst and Saliba (2015) and the authors referred to does not include such workers, as they are self-employed and as such they are not wage earners. This is another reason for the need to expand the analysis, as argued by the present chapter.

In a similar vein, the positive effect of housing construction on employment does not seem to have been taken into account either. Housing construction is usually accompanied by the construction of other elements of the built environment to complement the housing units (roads, water and sanitation systems, health and education facilities, and so on). Especially in times of boom, this can account for up to 10 to 15 percent of direct local employment, and there is also indirect and induced employment. While this only takes place during the time of the construction works themselves, there is a longer-term positive impact on local livelihoods.

The conclusion is that the analysis of the housing booms, while bringing value to the literature, is incomplete, and is undertaken, and seen, in isolation.

The Global South

In the context of the low-income settlements in the Global South, the relationship between housing and livelihoods has also long been recognised but from a different perspective. In the seminal work of Turner (1976), the choice of low-income settlements was, in part, seen to be driven by their role in supporting livelihoods that better suited the income and employment situation of residents. A range of work examining low-income housing provision in different geographical settings and tenures has confirmed this finding (Keivani and Werna (2001a); and Wakely (2016; 2018) provide comprehensive reviews).

First, rental housing has, perhaps, received the most attention in terms of creating livelihood opportunities particularly for subsistence landlords and building/shack owners. At the same time, it offers tenants affordable housing, residential choice, mobility, flexibility in managing budgets, and freedom from large financial commitments that are particularly important to vulnerable households and transient workers (Lonardoni and Bolay, 2016; Kumar, 2011; Gilbert, 2008; 2011). In a way, these authors support Oswald's hypothesis, and argue in favour of rental housing as opposed to owner-occupation, although theirs is a qualitative analysis and does not include the macroeconomics in which Oswaldism is grounded.

Second, the role of home-based enterprises in job creation and supporting livelihoods was highlighted by other authors (UNCHS-ILO, 1995; Gough et al, 2003; Gough 2010). In addition, there is a growth of home-based enterprises, especially in the Global South, not necessarily linked to information and communication technology or digitalisation, but catering for different sectors such as food and clothes production, repair and maintenance of mechanical instruments and personal services. At least part of this trend involves the continuation and expansion of traditional crafts and client-producer relations which never ceased to exist. They are now reinforced due to reduced or lack of alternatives in the urban economy. They also interface with the concept of the informal sector. In developed countries, they are part of the transition away from Fordism.

Third, attention was given to the operation of informal urban land markets, highlighting that it often outperforms the formal supply of land, services and housing (Napier et al, 2013).

Fourth, additional factors for housing choices in low-income communities have also been highlighted, including the gender dimension, asset accumulation strategies, kinship, ethnic and geographical linkages, sense of belonging and social networks as well as forced refugee migrations (Moser, 2009; Varley, 2015; Sanyal, 2017; Napier, 2007).

Low-income communities often rely on local networks and proximity to find jobs. Public programmes for relocation without considering livelihoods, even with improvements in the built milieu, often lead to resistance because of the loss of networks and reduced livelihood opportunities (Keivani and Werna, 2001a; Huchzermeyer, 2009; 2016). To this must be added negative sustainability and employment impacts of sprawl due to locations of public housing programmes (both directly provided or promoted through private sector) on more remote and cheaper land. It can also lead to the abandonment of the new housing estates or resettlement sites by some families and re-migration to other areas closer to employment locations. Among many examples, one can note the Paseos de San Juan development in Mexico City, Public housing estates in the Niger Delta in Nigeria, and Sapang Palay Settlement in Manila, the Philippines (Janoschka and Arreortua, 2017; Ihuah and Fortune, 2013; Downs, 1980).

However, apart from the broader anthropological and sociological insights that have already been noted and the aforementioned work of Gangopadhyay and Shankar (2016) that also employs a political economy perspective, there is a lack of work on the relationship between housing and labour markets and their impact on livelihoods for low-income households in the Global South. Considering this, some suggestions for research are made in the next section.

Towards a new research agenda

Pluralism in housing and pluralism in labour

Pluralism in housing and pluralism in labour, as mentioned before, is the overall approach which should be taken into consideration when analysing housing and labour markets in a given city. As noted in the section on labour markets, in a given urban area, there are often many people who fluctuate between the formal and informal markets. People earn a living in many different ways (casual work concomitantly or subsequently, self-employment, shifts between being an employer and a worker). All these entail different characteristics of demand for housing and many of them exist in parallel. They will affect the housing markets in different ways.

From the housing side, there are also many types, as explained before. Many of them exist in parallel, and the argument related to labour is mirrored: each will affect the labour markets in different ways. In a given urban area, application of the case study method rather than statistical analysis has the potential to provide a better understanding of these intricacies and inform policy making.

In addition to this overall approach, this section suggests three trends of research for policy making, related to: (a) livelihoods; (b) the economics of housing construction and home-based work; and (c) health and physical and social well-being.

Housing as an asset in the labour market

The urban livelihoods approach has been researched well, and so has the role of housing as an asset for livelihoods (Rakodi, 2001). An important point to be taken into account is that, due to the limitations of obtaining employment in the traditional sense, many people make ends meet partially or sometimes totally outside the traditional definition of a labour market. Housing often plays an important role as an asset for livelihoods.

This chapter has already mentioned home-based work, which is burgeoning, and provides an illustration of how housing is used as an asset for livelihoods. They are, on the one hand, caused by a combination of digital opportunities (working from a computer, cyber-taylorism) which can reduce office costs by working from home, and also allow people to be digitally mobile while being physically fixed. On the other hand, especially in developing countries, home-based work is a consequence of lack of employment opportunities combined with lack of possibility to afford a separate workplace. There are many people engaged in home-based work in catering, small manufacturing, textiles, repair and maintenance ofappliances, personal services, and so on.

Other equally relevant examples include housing as a long-term investment, as a place for rent (either the full housing unit or part of it), as a guarantee for borrowing, and even, where there is space for growing food, as a foundation for urban agriculture (for subsistence or/and commercial purposes).

Therefore, the growing importance of housing as an asset for livelihoods and partially or totally replacing wage labour brings a new dimension for the housing-labour market nexus, which needs to be better understood. That is, part of the housing market becomes the labour market or vice versa.

The economics of housing construction

Following the analysis of the previous section, housing construction has a positive impact on employment and wages that does not seem to have been taken into account in the housing-labour market nexus.

As noted before, housing construction is usually accompanied by construction of complementary elements of the built environment which provide further employment. This has a positive impact (even if limited) in the labour market, although it is often neglected in the aforementioned types of analyses.

Again, similarly, in the case of livelihoods, the case of employment in housing construction brings a new dimension to the housing-labour market nexus, which needs to be better understood. That is, it is another part of the housing market which becomes part of the labour market or vice versa.

The housing market and workers' health and well-being

As also noted before, housing is known to be fundamental for a worker's health and physical and social well-being and productivity. Therefore, an analysis of the housing-labour markets connection needs to take into account the quality of housing that is being offered and its impacts on different aspects of well-being. There is a burgeoning literature on the relationship between housing and health (WHO, 2018), yet the market angle needs to be better understood.

Conclusion

Why do policies fail? Macroeconomic analyses have presupposed the existence of one housing market (comprising owner-occupation and rental) and one labour market in a given urban area, and therefore how these markets interact. In fact, there are several of each of these markets in the same city, especially in developing countries. Following, conclusions come from a limited picture of what, in fact, is taking place in both markets. This is compounded by difficulties faced in attempts to access sound data that would be necessary to inform policy making.

To give one example, Adiaba et al. (2011) note that presently, sub-Saharan African urban real estate markets, namely housing, income-earning property and land markets are confronted with deficient information regimes resulting in their underperformance. It is estimated that between 97 and 99 percent of the market transactions fall outside the radar of formal land registration systems. With such high levels of information deficit, it is easy to conclude that policy making is ill-informed. In turn, the magnitude of informal, self-employed and/or casual workers and those migrating among different categories of employment is usually also not captured in statistics.

Comprehensive discussion on how to address the current limitations of data collection and analysis is beyond the scope of the present chapter, although it can be suggested that alternatives such as community enumeration and other bottom-up approaches could be explored.

While the respective literature on housing markets and labour markets are extensive, the one which links both is somehow incipient and, at the same time, oriented to the Global North. This chapter has highlighted key aspects of such combined literature and pointed out room for improvement. This chapter adds value and innovation by analysing the housing-labour markets nexus through the lens of their respective complexities, and by bringing developing countries into the forefront. The chapter aimed to present important strands of literature that account for interactions that are foundational for studying this topic and for policy making. The chapter also acknowledges that the relationship between work and housing is much more than a relationship between labour markets and housing markets. Yet, it is beyond the scope of the chapter to articulate every possible connection between such broad themes as labour and housing. At the same time, the relationships between the two sets of markets are fundamental and have interlinkages with other issues, such as the ones mentioned in the research agenda.

References

Abramsson, M. (2008). Housing careers in a changing welfare state – A Swedish cohort study. *Housing, Theory and Society*. 25(4), pp231–253.

Adiaba, S., Hammond, F., Proverbs, D., Lamond, J. and Booth, C. (2011). Sources of deficient information regime in urban real estate markets in sub-Saharan African countries. In Laryea, S., Leiringer, R. and Hughes, W. (eds.). Proceedings of the West Africa Built Environment Research (WABER) Conference 2011 (19–21 July 2011, Accra, Ghana).University of Reading (UK).

Allen, J. and Hamnett, C. (1996). *Housing and Labour Markets: Building the Connections*. Routledge: London.

Amis, P. (1996). Long-run trends in Nairobi's informal housing market. *Third World Planning Review*. 18(3), pp271–285.

Antunes, R (2021). *Farewell to work? Essays on the world of work's metamorphoses and centrality*. Brill Publishers: Leiden, Netherlands.

Askenazy, P. (2013). *Capital Prices and Eurozone Competitiveness Differentials*. IZA Discussion Paper 7219. Institute for the Study of Labor: Bonn.

Ayotamuno, A. and Owei, O.B. (2015) Housing in Port Harcourt, Nigeria: The modified building approval process. *Environmental Management and Sustainable Development*. 4(1), pp16–28, file:///C:/Users/p0074006/Downloads/6824-24245-1-PB.pdf. Accessed 25 February 2023.

Baert, S., Heylen, F. and Isebaert, D. (2014). Does homeownership lead to longer unemployment spells? The role of mortgage payments. *De Economist*. 162(3), pp263–286.

Baharoglu, D. (1996). Housing supply under different economic development strategies and the forms of state intervention: The experience of Turkey. *Habitat International*. 20(1), pp43–60.

Bird, J. (2021). Low wages are the bedrock of our housing woes. The Big Issue. 01 February 2021. Acesssed 13 April 2023 https://www.bigissue.com/opinion/low-wages-the-bedrock-of-our-housing-woes/

Blanchflower D. and Oswald, A. (2013). *Does High-income Ownership Impair the Labour Market?* Working Paper 19079. Washington, D.C.: National Bureau of Economic Research. NBER.

Bonnet, F., Carré, F., Chen, M. and Vanek, J. (2021). *Home-Based Workers in the World: A Statistical Profile*. Statistical Brief No 27. January. WIEGO (Women in the Informal Economy Governing and Organizing).

Buffie, E., Adam, C., Zanna, L.F., Balma, L., Tessema, D. and Kpodar, K. (2022). Public investment and human capital with segmented labour markets. *Oxford Economic Papers*. gpac051, https://doi.org/10.1093/oep/gpac051

Cho, Y.; Keivani, R. & Werna, E. (2018). Room for improvement: How to (better) integrate housing and labour markets. *URBANET*. Accessed 13 April 2023 https://www.urbanet.info/housing-labour-global-south/

Colom, M. and Moles, C. (2013). Housing and labor decisions of households. *Review of Economics of the Household*. 11, pp55–82.

Deakin, S. (2013). *Addressing Labour Market Segmentation: The Role of* Labour Law. Governance and Tripartism Department, ILO. Working Paper No. 52. ILO: Geneva.

Downs, B.E. (1980). Housing the urban poor in developing countries. *Georgia Journal of International & Comparative Law*. 10(3), pp527–578.

Drakakis-Smith, D. (1981). *Housing and the Urban Development Process*. Croom Helm: London.

Eliasson, K., Lindgren, U and Westerlund, O. (2003). Geographical labour mobility: Migration or commuting? *Regional Studies*. 37, pp827–837

Ernst, E. and Saliba, F. (2015). "Are house prices responsible for unemployment persistence?" Paper presented at the 7th ReCapNet Conference (Leibniz Network on Real Estate Markets and Capital Markets (ReCapNet). ZEW: Mannheim. November 6 and 7.

Estevão, M. and Tsounta, E. (2011). *Has the Great Recession raised U.S. Structural Unemployment?*, Working Paper 11/105, IMF: Washington, D.C.

Ferreira, F., Jolliffe, D.M. and Prydz, E.B. (2015). The international poverty line has just been raised to $1.90 a day, but global poverty is basically unchanged. How is that even possible? World Bank Blogs. Retrieved 1 March 2023 https://blogs.worldbank.org/developmenttalk/international-poverty-line-has-just-been-raised-190-day-global-poverty-basically-unchanged-how-even

Gangopadhyay, P. and Shankar, S. (2016). Labour (im)mobility and monopsonistic exploitation of workers in the urban informal sector: Lessons from a field study. *Urban Studies*. 53(5), pp1042–1060.

Gilbert, A. (2008). Slums, tenants and home-ownership: On blindness to the obvious. *International Development Planning Review*. 30(2), pp1–10.

Gilbert, A (2011). *A Policy Guide to Rental Housing in Developing Countries*. UN-Habitat: Nairobi.

Gough, K. (2010). Continuity and adaptability of home-based enterprises: A longitudinal study from Accra, Ghana. *International Development Planning Review*. 32(1), pp45–70.

Gough, K.V., Tipple, G.A. and Napier, M. (2003). Making a living in African Cities: The role of home-based enterprises in Accra and Pretoria. *International Planning studies*. 8(4), pp253–277.

Haas, A. and Osland, L. (2014). Commuting, migration, housing and labour markets: Complex interactions. *Urban Studies*. 51(3), pp463–476.

Hämäläinen, K. and Böckerman, P. (2004). Regional labor market dynamics, housing, and migration. *Journal of Regional Science*. 44(3), pp543–568.

Hanson, S. and Pratt, G. (1988). Reconceptualizing the links between home and work in urban Geography. *Economic Geography*. 64(4), pp299–321.

Harvey, D. (1989). *The Condition of Postmodernity: An Enquiry into the Origins of Cultural Change*. Blackwell; Hoboken, NJ, USA.

Hildreth, A. & Oswald, A. (1997). Rent-sharing and wages: Evidence from company and establishment panels. *Journal of Labor Economics*. 15(2), pp318–337.

Hincks, S. and Wong, C. (2010). The spatial interaction of housing and labour markets: Commuting flow analysis of North West England. *Urban Studies*. 47(3), pp620–649.

Huchzermeyer, M. (2009). The struggle for in situ upgrading of informal settlements: A reflection on cases in Gauteng. *Development Southern Africa*. 26(1), pp59–73.

Huchzermeyer, M. (2016) Informal settlements at the intersection between urban planning and rights: advances through judicialisation in the South African case. In Agnès Deboulet (ed.)*Rethinking Precarious Neighborhoods*, pp195–210. Agence Francaise de Developpement (AFD): Paris.

Iacoviello, M. (2005). House prices, borrowing constraints and monetary policy in the business cycle. *American Economic Review*. 95(3), pp739–764.

Iacoviello, M. (2010). Housing in DSGE models: Findings and new directions. In O.d. Bandt, T. Knetsch, J. Peñalosa, F. Zollino (eds.), *Housing Markets in Europe: A Macroeconomic Perspective*, pp3–16. Springer; Bonn, Germany.

Iacoviello, M. (2011). *Housing Wealth and Consumption*. International Finance Discussion Papers 1027. Federal Reserve Board: Washington, D.C.

Iacoviello, M. and Neri, S. (2007). *The Role of Housing Collateral in an Estimated Two-sector Model of the US Economy, Working Papers in Economics 412*. Boston College Economics Department: Boston, MA.

Iacoviello, M. and Neri, S. (2010). Housing market spillovers: Evidence from an estimated DSGE model *American Economic Journal: Macroeconomics*. 2, pp125–164.

Iacoviello, M. and Pavan, M. (2013). Housing and debt over the life cycle and over the business cycle. *Journal of Monetary Economics*. 60, pp221–238.

Ihuah, P.W. and Fortune C.J. (2013). Toward a framework for the sustainable management of social (public) housing estates in Nigeria. *Journal of US-China Public Administration*. 10(9), pp901–913.

ILO. (1999). Decent work. International Labour Conference, 87th Session, Report of the Director-General, International Labour Office, Geneva.

Isebaert, D.; Heylen, F.; & Smolders, C. (2015) Houses and/or jobs: Ownership and the labour market in Belgian districts. *Regional Studies*. 49 (8) pp1387–1406.

Janoschka, M and Arreortua, L.S. (2017). Peripheral urbanisation in Mexico City. A comparative analysis of uneven social and material geographies in low-income housing estates. *Habitat International*. 70, pp43–49.

Jones, G. and Hyclak, T. (1999). House prices and regional labor markets. *The Annals of Regional Science*. 33, pp33–49.

Keivani, R. and Werna, E. (2001a). Modes of housing provision in developing countries. *Progress in Planning*. 55(2), pp. 66–118.

Keivani, R. and Werna, E. (2001b). Refocusing the housing debate in developing countries from a pluralist perspective. *Habitat International*. 25(2), pp. 191–208.

Keivani, R., Mattingly, M. and Majedi, H. (2008). Public management of urban land, enabling markets and low income housing provision: The overlooked experience of Iran, *Urban Studies*. 45(9), pp1825–1854.

Keivani, R., Omena de Melo, E. and Brownill, S. (2020) Durable inequality and the scope for pro-poor development in a globalising world – Lessons from Rio de Janeiro. *CITY*. 24(3–4), pp530–551.

Kellet, P. and Tipple, G. (2011) The home as workplace: A study of income-generating activities within the domestic setting. *Environment and Urbanisation*. 24(1), pp. 203–213.

Klink, J. and Denaldi, R. (2014). On financialization and state spatial fixes in Brazil. A geographical and historical interpretation of the housing program My House My Life. *Habitat International*. 44, pp220–226.

Kumar, S. (2011). The research–policy dialectic: A critical reflection on the virility of landlord–tenant research and the impotence of rental housing policy formulation in the urban Global South. *Cities*. 15(6), pp662–673.

Leontaridi, M. (2002). Segmented labour markets: Theory and evidence. *Journal of Economic Surveys*. 12(1), pp103–109.

Lonardoni, F. and Bolay, J.C. (2016). Rental housing and the urban poor: Understanding the growth and production of rental housing in Brazilian favelas. *International Journal of Urban Sustainable Development*. 8(1), pp10–24.

Marias, L. Hoekstra, J., Napier, M., Cloete, J. and Lenka, M. (2018). The housing careers of black middle-class residents in a South African metropolitan area. *Journal of Housing and the Built Environment*. 33, pp843–860.

Meen, G. and Andrew, M. (1998). On the aggregate housing market implications of labour market change. *Scottish Journal of Political Economy*. 45(4), pp393–419.

Moser, C. (2009). *Ordinary Families, Extraordinary Lives: Assets and Poverty Reduction in Guayaquil, 1978–2004*. Brookings Press: Washington, D.C.

Muellbauer, J. and Murphy, A. (2008). Housing markets and the economy: The assessment. *Oxford Review of Economic Policy*. 24(1) pp1–33.

Napier, M. (2007). *Informal Settlement Integration, the Environment and Sustainable Livelihoods in Sub-Saharan Africa*. Council for Scientific & Industrial Research in South Africa.

Napier, M., Berrisford, S., Wanjiku Kihato, C., McGaffin, R. and Royston, L. (2013). *Trading Places: Accessing Land in African Cities*. African Minds for Urban LandMark: Somerset West.
Okpala, D.C.I. (1992). Housing production systems and technologies in developing countries: A review of the experiences and possible future trends/prospects. *Habitat International*. 16(3), pp9–32.
Oswald, A. (1997). Thoughts on NAIRU. *Journal of Economic Perspectives*. 11, pp227–228.
O'Connor, K. and Healy, E. (2001). The links between housing markets and labour markets: Positioning paper prepared by the Australian Housing and Urban Research Institute Swinburne-Monash Research Centre.
Payne, G. (2022). *Somewhere to Live: Rising to the Global Urban Land and Housing Challenge*. Practical Action Publishing: Rugby, UK.
Pickwance, C. (1974). On a materialist critique of urban sociology. *The Sociological Review*. 22(2), pp203–211.
Rakodi, C. (2001) Urban governance and poverty-addressing needs, asserting claims: An editorial introduction. *International Planning Studies*. 6(4), pp343–356.
Reuschke, D. and M. Domecka (2018). Policy brief on homebased businesses. In *OECD SME and Entrepreneurship Papers, No. 11*. Paris: OECD Publishing.
Saks, R.E. (2008). Job creation and housing construction: Constraints on metropolitan area employment growth. *Journal of Urban Economics*. 64, pp178–195.
Sanyal, R. (2017). A no-camp policy: Interrogating informal settlements in Lebanon. *Geoforum*. 84, pp117–125.
Scheba, A. and Turok, I. (2020) Informal rental housing in the South: Dynamic but neglected. *Environment and Urbanization*. 32(1), pp109–132, https://journals.sagepub.com/doi/full/10.1177/0956247819895958. Accessed 25/02/2023.
Sengupta, U. (2019) State-led housing development in Brazil and India: A machinery for enabling strategy? *International Journal of Housing Policy*. 19(1), pp1–27.
Sohane, N., Lall, R., Chandran, A., Lala, R., Kapoor, N. and Gajjar, H. (2021). *Home as Workplace: A Spatial Reading of Workhomes*. Indian Institute for Human Settlements (IIHS) and Women in Informal Employment Globalising and Organising (WIEGO).
Standing, G. (2011). *The Precariat: The New Dangerous Class*. Bloomsbury: London.
Sule Ozuekren, A. and van Kempen, R. (2002) Housing careers of minority ethnic groups: Experiences, explanations and prospects. *Housing Studies*. 17(3), pp365–379.
Tehran Times. (2022) https://www.tehrantimes.com/news/474309/Housing-rental-rises-46-5-in-Tehran-city-in-a-month-on-year. Accessed 25 February 2023.
Tipple, G. (2005). The place of home-based enterprises in the informal sector: Evidence from Cochabamba, New Delhi, Surabaya and Pretoria. *Urban Studies*. 42(4), pp611–632.
Torres, H., Alves, H. and De Oliveira, M.A. (2007). São Paulo peri-urban dynamics: Some social causes and environmental consequences. *Environment and Urbanization*. 19(1), pp207–223.
Turner, J. (1976). *Housing by People: Towards Autonomy in Building Environments, Ideas in Progress*. Marion: London.
UNCHS-ILO (UN Centre for Human Settlements (Habitat) and International Labour. Office(1995). Shelter provision and employment generation. Nairobi and Geneva: UNHCR and ILO.
UN (United Nations). (2017). The New Urban Agenda. Report of Habitat III – The UN Summit on Housing and Sustainable Urban Development. UN: New York.
UN (United Nations). (2018) SDG Indicators – The United Nations, https://unstats.un.org/sdgs/report/2019/goal-11/. Accessed 25 February 2023.
UN-Habitat. (2003). Rental Housing – An essential option for the urban poor in developing countries. https://www.humanitarianlibrary.org/resource/rental-housing-essential-option-urban-poor-developing-countries-2. Accessed 25 February 2023.
UN-Habitat. (2022). *World Cities Report 2022: Envisaging the Future of Cities*. UN-Habitat: Nairobi.
Vakil, A.C. (1996). Understanding housing CBOs. *Third World Planning Review*. 18(3), pp325–347.
Van Ewijk, C. and Van Leuvensteijn, M. (eds). (2009). *Homeownership and the Labour Market in Europe*. Oxford University Press: Oxford.
Van der Vlist, A.; Gorter, C. and Rietveld, P. (2002). Residential mobility and local housing-market differences. *Environment and Planning A: Economy and Space*. 34(7), pp 1147–1164.
Varley, A. (2015). Home and belonging: Reflections from urban Mexico. In C. Klaufus and A. Ouweneel (eds.), *Housing and Belonging in Latin America*, pp275–293. Berghahn Books: New York.

Vickerman, R. W. (1984). *Urban Economies: Analysis and Policy*. Humanities Press: Atlantic Highlands, NJ, USA.
Wakely, P. (2016). Reflections on urban public housing paradigms, policies, programmes and projects in developing countries. *IJUSD*. 8(1), pp10–24.
Wakely, P. (2018). *Housing in Developing Cities –Experience and Lessons*. Routledge: London.
Watt, P. (2005). Housing histories and fragmented middle-class careers: The case of marginal professionals in London Council Housing. *Housing Studies*. 20(3), pp359–381.
Wegener, M. (2004). Overview of land use transport models. *Handbook of Transport Geography and Spatial Systems*. Vol. 5 pp127–146. Emerald Group Publishing Limited: Bingley, UK.
Werna, E. (2000). *Combating Urban Inequalities – Challenges for Managing Cities in the Developing World*. Aldershot: Edward Elgar Publishers.
Werna, E. (2016), From Blade Runner to Habitat IV – How livelihoods can make or break the City of Tomorrow – Part I. URBANET (News and Debates on Municipal and Local Governance, Sustainable Urban Development and Decentralization). http://www.urbanet.info/blade-runner-to-habitat-iv-city-of-tomorrow/ downloaded 10 February 2023.
Werna, E., Abiko, A.K., Coelho, L.O., Simas, R., Keivani, R., Hamburger, D. and Almeida, M. A. (2000). *Pluralism in Housing Provision: Lessons from Brasil*. Nova Science: New York.
WHO (World Health Organization). (2018). *Housing and Health Guidelines*. WHO: Geneva.
World Bank (2019). *World Development Report: The Changing Nature of Work*. World Bank: Washington, D.C.
Zabel, J. (2012). Migration, housing market, and labor market responses to employment shocks. *Journal of Urban Economics*. 72(2), pp267–284.

15

SOCIAL INNOVATION OF WORKPLACES IN THE BUILT ENVIRONMENT

How public spaces have become central workplaces – Lessons from Kampala City, Uganda

Andrew Gilbert Were[1], *Stephen Mukiibi*[1], *Michael Majale*[2], *and Barnabas Nawangwe*[1]

[1]Makerere University, Uganda; [2]Independent Consultant/Researcher

Introduction

This chapter considers the socio-economic factors, conditions in the built environment, and policy and regulatory frameworks that influence the generation of workplaces in cities, with particular reference to Kampala, the capital city of Uganda. It provides an introduction and background to the market concept, and how it influences labour and market behaviour of workers in the urban informal sector, which is a major source of livelihood in cities throughout the Global South. It provides an account of the historical context of informality in Kampala, discusses the status of workplaces in the city, considers future prospects for workplaces, and offers recommendations. It discusses why and how homes, streets, and public spaces have become important workplaces.

Public space – which includes streets, pavements, plazas, squares and parks – is a productive asset for the livelihoods of many urban informal workers, notably, street vendors. It is for this reason that regulated access to public space was recognized as a key dimension of formalization of the informal economy by the International Labour Organization (ILO) Recommendation 204 on the Formalization of the Informal Economy, as well as in the New Urban Agenda (NUA) adopted by the world's nations at Habitat III, the United Nations Conference on Housing and Sustainable Urban Development, in 2016. Yet, cities around the world, which have the mandate to manage and regulate public space, do not typically recognize the need, and much less, the right, of urban informal workers to use public space to pursue their livelihoods (Chen et al., 2018). Thus, city authorities should recognize the importance of public spaces in the built environment and enable enterprises to thrive in such spaces as well, as a way of sustaining the city micro-economy.

The built environment and decisions related to the management of public spaces enable, and can also disable, people's workplaces which they need to pursue livelihoods and employment in various ways. In free market economies, city governments across the world

DOI: 10.1201/9781003262671-15

tend to prioritize the allocation of built environment spaces to businesses and enterprises that generate sufficient revenue and pay taxes to contribute to the funds they need for running the cities. Less attention is paid to providing workplaces for small-, micro-, and medium-sized enterprises (SMMEs). Because workers with SMMEs have low capital bases, smaller and unpredictable working places, and tend to be unable to regularize their enterprises, their working spaces are either not recognized, are considered illegal, or are reallocated to more significant projects with larger investments. Similarly, reorganization of places in the built environment tends to affect the public space workers more than those in formal establishments.

For example, the spatial location of land uses or zoning of activity areas such as 'commercial', 'industrial', 'educational', or 'health institutions' enables people to locate their enterprises in public spaces around the facilities built on them. This is due to their ability to draw large populations that form a market for their enterprises. The chapter focuses on public spaces such as streets and road reserves, markets, railway reserves, public leisure, and recreational spaces.

The market concept and the place for workers in public spaces

The term 'market' has various perceptions, connotations, and interpretations. However, there are two theories that can be used to explain the evolution of market institutions in Uganda. The first is based on 'the individual's propensity to barter, perhaps involving silent barter; deduces from this the necessity for local exchange, the division of labour and local markets; and infers, finally, the necessity for long-distance or at least external exchange or trade' (Hodder, 1965, p. 97).

The second theory argues that:

> Markets are not the starting point but rather the result of long-distance trading, itself the result of division of labour and the variable geographical location of goods.
>
> *(Hodder, 1965, p.98)*

The above theories on the origin of market institutions are mirrored in present-day physical markets, including the presence of large markets, as well as of smaller markets in close proximity to the large markets, from which they obtain goods for distribution and trade. The larger markets tend to be located near high concentrations of populations, which is the real and perceived 'market'.

The principle provides that anything that has a price tag can also have a market. This means that transactions involving buying and selling of goods and services can take place without necessarily having a physical location (Uzoigwe, 1972). However, the traditional African market was perceived to involve the presence of a physical space in a specific location. The physical site, whether open or enclosed, would facilitate face-to-face transactions between buyers and sellers and would be regulated by a traditional authority (Uzoigwe, 1972). In Kampala City, there are official physical markets in Nakasero, St. Balikuddembe, Kibuye, Nakawa, Busega, and in major shopping arcades. These markets are supported by public space workers such as street vendors who expand the distribution, through their convenient location, and their ability to trigger impulse buying.

Uzoigwe (1972) argues that 'small and unauthorized gatherings of hawkers and vendors do not constitute markets' (p.423). However, when hawkers and vendors are seen on streets and other public spaces, they are perceived to constitute a market. This is so because globally, even with the emergence of online marketing and shopping, physical interactions between humans as social beings remain widely preferred. That is why vending in public spaces still takes place worldwide, even in the Global North, under strict regulation, as there is an opportunity to 'feel' the goods and services offered. The perception of a 'market' as a physical space in a specific location could explain why public spaces as a confluence of people have continued to be attractive to sellers.

There seems to be no significant difference in the conceptualization of 'the market' between Kampala Capital City Authority (KCCA), which is responsible for the administration and management of the city, and the residents pursuing informal street-based livelihoods – and the contestation between the two parties regarding the use of the street space. The dominant alternative options to working in public spaces proposed by both parties are availing space in other locations or allocating time for particular street-based activities in specific locations. This could be because alternatives to public spaces and physical markets have not yet been developed or tailored to socio-economic and cultural structures related to the shopping experience. Uganda is still largely a country of people who have strong social ties that require human interaction, and markets are still associated with physical engagement.

In Uganda, the physical location of markets remains an important pull factor in the development of major trading centres, which often expand, grow, and become upgraded into towns and cities. Trading activities often attract related support facilities, services, and amenities. These, in turn, catalyse the construction of physical infrastructure and utilities in locations where markets are situated to serve users. The construction of physical markets is often used by governments as an indicator for the development of particular regions and urban centres. The construction and upgrading of markets and supporting infrastructure in major cities and towns by the Government of Uganda using World Bank funding under the Uganda Support to Municipal Infrastructure Development (USMID) programme attests to this.

Most of the built-up markets in Uganda tend to be spatially located where open-air markets previously existed. However, when the open-air markets are 'upgraded' into built-up markets, their location and capacity usually do not meet the diverse needs of workers. Usually, the built-up markets are inadequate in terms of workers' space requirements. The cost of new stalls puts them beyond the reach of all but a few. The built-up markets also tend to be insufficient with respect to the availability of support facilities and services. Moreover, they do not take into account equity and inclusivity, considering people in the diverse segments of society. They do not provide for flexibility to reallocate or reorganize the spaces within the built-up markets. Thus, the markets are unable to accommodate emerging needs and changing trends. They also fail to take into account convenience of the shoppers. Consequently, workers dissatisfied with built-up market spaces, or unable to gain access to them, or to afford the costs, often appropriate public spaces as workplaces.

For example, following the upgrading of Wandegeya Market in Kampala City into a multi-storey building with stalls, large numbers of workers continued selling their goods and providing their services on the roads adjacent to the market, while others relocated from the market. Similar relocations by market vendors happened when parts of St. Balikuddembe Market and Nakawa Market were upgraded. Ironically, the 'upgrading' of these markets has contributed to the increase in public space workers. Consequently, the view that street vending is taking place outside 'the physical market' or 'authorized market' is a contested issue between the authorities and the street vendors.

Efforts to resolve the stalemate should therefore address the conceptualization of 'market' in the Ugandan context, as a market is still understood to be spatially defined as being in a physical location while the streets are perceived by the workers to offer the ideal market for goods and services due to the (real and potential) traffic they generate. They are also perceived as such by much of the city population who patronize their 'establishments' owing to the convenience they offer.

The informal economy and street vending

The informal economy, as defined by the ILO, includes 'all economic activities that are, in law or practice, not covered or insufficiently covered by formal arrangements' (ILO, 2007, p. 3). The informal economy encompasses wage workers and own-account workers, contributing household members and those moving from one situation to another. It also includes those engaged in new flexible work arrangements and who are at the periphery of the core enterprise or at the lowest end of the production chain (ILO, 2007).

It is important to retrace the origin and evolution of informal workers in public spaces who largely fall into the category of informal sector workers in Uganda. The origin of informal workers in public spaces in Kampala City, and in Uganda as a whole, can be traced back to the colonial planning regime and the political economy of Uganda during the country's history as a protectorate of the British Empire from 1894 to 1962 (Wereet al., 2022a). The coming of the colonizers saw the society structured into formal and informal sectors, with the former being businesses registered by the state system (Kabanda, 2017). Consequently, Uganda's economy developed within the loosely structured framework of an officially undefined dual economy between 1919 and 1921 (Taylor, 1978). Thus, the significance of having both formal and informal sector activities as complements of each other was recognized in the city even in the colonial era (Were et al., 2022b).

The emergence of public space workers can also be traced back to the policies introduced in the late 1980s and early 1990s in the form of Structural Adjustment Programmes (SAPs). The SAPs reduced the state's role in the economy, cut services provided by the government to the people (which had a significantly adverse effect on the poor), and imposed a prescribed governance system. The liberalization of the economy in 1987 facilitated the import and export trade, creating employment opportunities in the distribution value chain which largely employed street vendors either retailing the goods as owners of their own enterprises or as agents of large formal sector businesses (see, for example, Bromley, 2000; Chen, 2008). The diminishing role of the state and cooperatives that marketed agricultural products because of SAPs conditions, compelled people to find alternative ways to access products, of which supply in public spaces was one. Kampala City, with the largest confluence of people, offered reliable markets for goods and services in public spaces which set the stage for contestation for working spaces. These conditions have been perceived as the real stimulants for the appropriation of public space for private enterprises.

Informal public space workers complement formal sector workers due to their role in the distribution value chain for goods and services in the economy. Currently, the informal sector in Uganda, of which the workers operating from public spaces in the cities are an important segment, employs over 70 percent of the population and contributes over 52 percent of the national GDP (UBOS, 2020). In Kampala City alone, over 70 percent of the 3.5 million people who work there daily are engaged in informal sector activities, from which they earn a modest income (Gumisiriza, 2021). The scarcity of job opportunities in

Uganda has also led to a large number of unemployed or underemployed youth, many of whom are forced to take up work in the informal sector (SIHA, 2018). Workers who operate from public spaces are enabled by the political economy of the state. Uganda is a neo-liberal economy, where the free market system is dominant. A system of urban space governance and management in a neo-liberal political economy prioritizes space allocation to firms with large returns to investment. We therefore briefly discuss how neoliberalism and how it influences public space and public space workers.

Neo-liberal urbanism and public space workers

The philosophy of neoliberal urbanism and the privatization or financialization of public urban spaces that have influenced the form, function, and scope of entrepreneurial urban governance (Tasan-Kok, 2009) present a challenge as the social functions of the city that allowed common use of public spaces are no longer given due consideration. Similarly, 'current urban revitalization strategies and privatized service delivery exclude the poor and obliterate the potential for alternative development models to emerge' (Brown and Mackie, 2018).

Thus, entrepreneurial urban governance promotes efficient and effective implementation of strategic projects without taking their social impacts into account, resulting in limited social gains compared to the economic ones (Tasan-Kok, 2009). Thus, common land or spaces for all are another set of significant factors given due consideration in public space allocation. Cities have consequently become 'hostile' to the welfare of low-income and vulnerable groups that are unable to integrate into formal market systems owing to neo-liberal policies. Using public spaces as working spaces is partly a consequence of neoliberal policies that undermined social welfare in favour of a capitalist market economy.

Despite being neglected by neo-liberal practices of space allocation, it is these public space workers who occupy and appropriate public spaces in sometimes unexpected ways, bringing new meanings and unforeseen functions to these spaces (Hou, 2020). Therefore, activities of workers in public spaces may seem a challenge to urban planning, urban design, urban policy and urban management. However, as disorderly as it may look, the workers present opportunities for a new epistemology of city and city making; contestation, collaboration and composition; and rethinking of democracy, justice and resilience (Hou, 2013).

The context of Kampala City and the persistence of public space workers

Kampala City is governed by the Kampala Capital City Act 2010, which provides for the Kampala Capital City Authority (KCCA), comprising both a technical wing and a political wing, to administer the city. The latter is elected by universal adult suffrage and is headed by the Lord Mayor, while the former is headed by the Executive Director appointed by the President of Uganda. With the Executive Director directly accountable to the executive and the Lord Mayor directly accountable to the electorate, the two have often clashed on matters of policy in Kampala City. As a result, matters including the management of public space and the workers in such spaces have not been successfully addressed because of a lack of consensus between the two wings of city governance and management. Enforcement efforts have also not been effective owing to a lack of support from the technocrats and politicians. The politicians protect public space workers against harassment as they, and the citizens who buy their goods or patronize their services, form a significant portion of the electorate.

Another contributing factor to the use of public spaces is the spatial layout of Kampala City. Spatially, the city layout is a product of colonial physical planning that segregated British colonialists, Asians, and native Africans (cf. Mirams, 1930). The colonialists occupied the hills and the prime land located around that area, where they built churches, educational institutions, health centres, and markets. The Asians were mainly engaged in manufacturing and business activities and hence purchased most of the prime land along major city streets. There were no specific planning provisions for native Africans in all aspects of life, except perhaps the Naguru and Nakawa housing schemes which were built for African low-skilled workers. Consequently, native Africans were relegated to the status of consumers of the goods and services provided by the Asians. They also provided labour to the small-scale cottage industries and farms set up by the Asians. Indeed, most of the buildings in Kampala City, especially in the Central Division, are still owned by the Asians. Most of the rent, which, in some cases, has to be paid for in US dollars, is unaffordable for local business people. This has pushed low-income workers to consider public spaces as viable options for vending goods and services. For example, there are about 10,000 hawkers in Kampala City, and street vending in Kampala has emerged as an alternative source of employment for individuals who cannot afford the cost of paying regular rent in buildings or market stalls (Nakazi, 2019).

At the same time, the location of shopping malls and open-air markets in Kampala City is often inconvenient for daily shoppers. For example, in Central Division, they are located in Nakasero Market, St. Balikuddembe Market, and around public transport terminals. In other Divisions of Kampala, large supermarkets are located where they are most accessible using private transport. The location of upgraded markets is not suitable for the majority of shoppers, due to the unregulated housing and settlement patterns in Kampala City. In addition, the larger shopping malls are perceived as high-income shopping centres and the goods they trade in are unaffordable to the low-income earners. Furthermore, some shopping malls, such as Garden City, are not easily accessible via mass public transport, and are, therefore, inconvenient or expensive for 'pick and board/load' shoppers.

Although working in public spaces without a license is illegal, it is still practiced around Kampala City, but the pattern and intensity of vending varies. Some vendors work the whole day, while others work for specific durations during the day and night, depending on the goods or services they sell or offer, and the time it takes to sell them. The intensity of vending from public spaces also depends on the level of surveillance by law enforcement officers. There seems to be more of such surveillance in the areas near offices than in the downtown, where traffic terminals are located. Working space is generally acquired through social capital and by what could be referred to as 'spontaneous location'. Meanwhile, a number of products and services that are sold by public space workers are produced or processed by home-based enterprises (HBEs) in settlements within the city or in the city suburbs because public spaces are not always adequate or appropriate for both production and marketing.

The connection between home-based enterprises and workers in public spaces

HBEs can be described as entities involved in processes that involve the production of goods and services, using spaces in homes, dwellings, and obtaining most of their resources in the immediate surroundings. HBEs can involve the use of all or part of the internal or external home space. They sometimes also involve the appropriation of road reserves that are adjacent to the home to expand their workplaces.

Most public space workers in Kampala City are slum dwellers. Each of the five city divisions has several slums which are predominantly occupied by low-income earners. Many slum households are engaged in various HBEs, as artisans and craft-persons, retailers, wholesalers, and distributors of various goods and services. Because of a lack of markets in the slums owing to the low purchasing power of the dwellers, products from the slums are sold on streets in various parts of the city.

The connection between street or public spaces and HBEs is enabled by the marketing or exchange of goods and services. The internal spaces in dwellings are usually rearranged for the production of HBEs and temporary storage of the products before they are distributed. Some HBEs sell their goods and services at retail and wholesale prices to small-scale distributors throughout the city. For example, it is estimated that about 70 percent of food items (such as fries, corn, peanuts, and assorted fruits and vegetables) sold on streets and in public spaces are prepared in slums. Other goods include textiles, footwear, kitchenware, jewelry, and art and craft pieces. Some of the items are obtained from formal markets and reprocessed before they are retailed around the city using streets and other public spaces. Services commonly offered on streets and other public spaces include shoe shining and repair, mobile money services, the repair of watches and other electronic goods, tailoring services, nail polishing, transport services, and advertizing of goods and services.

Although all city divisions have workers who utilize public spaces, Central Division has the highest number of workers in public spaces, owing to the concentration of people in the central business district (CBD) and the presence of transport terminals. Thus, streets bordering the Old and New Taxi Parks, the Bus Parks, traffic lights, the City Square, and markets are hotspots for public space workers because of the volume of human traffic in these places. Goods sold by the vendors in public spaces are mainly obtained from the open-air markets, HBEs, and rural producers, and are processed and repackaged both to meet the expectations of the customers and make them suitable for carrying by the vendors. The goods sold by public space workers include items and other industrial products sourced from wholesale outlets in formal business establishments.

The above discussion shows that the sustainability of workers in public spaces is based on their capacity to utilize their homes as production and processing spaces, and their ability to connect their enterprises to city markets. HBEs also enable the development of various skills and innovations, influenced by the needs of the market and the application or enforcement of regulations on the use of public spaces.

The case of public space workers in the Central Division of Kampala City

KCCA has attempted to manage public space workers in the Central Division of Kampala City through various ways such as licensing, taxation, eviction, confiscation of property, warnings, fines and arrests. It has also allocated working spaces outside the CBD which the workers have rejected, arguing that they are inaccessible to their customers (Mitullah, 2003). Nonetheless, KCCA still considers vending from public spaces an illegal activity that undermines the proper functioning of the formal economy. This perception has resulted in continued conflicts between KCCA and public space workers over licensing, taxation, location of their operations, sanitation facilities and working conditions (Mitullah, 2003). Despite the clashes, it has been difficult to evict already entrenched workers from public spaces as they are pursuing livelihoods and economic opportunities that can help them escape extreme poverty (Ghida, 2020).

Working from public spaces has negative implications as well. These include congestion, littering, evasion of taxes, sale of substandard products, petty theft, and reduced sales for shopkeepers. For example, streets with the highest concentration of vendors are near traffic and transport terminals. This results in conflicts over space allocation, space use, and space interpretation by street space users. Consequently, public space and traffic management in Kampala Central Division is a major challenge. This has compelled city authorities to outlaw working from public spaces such as streets. The Kampala City Street Traders Byelaws (p. 2) states that 'No street trader shall carry on business without a permit from the town clerk'. However, due to recognition of the importance of the trade, there have been attempts to allow the practice by amending the current bye-laws by KCCA, such that licensed vending in public spaces is enabled.

The use of public spaces as workplaces has been affected by the political, socio-economic and physical dynamics leading to fluctuating numbers of public space workers. These include influence from politicians; lukewarm support and opposition from licensed traders; uncertain income streams owing to fluctuating demand for their goods and services; the need to provide employment opportunities; and the fight for the limited street spaces between competing uses. Thus, regulating workers in public spaces and incorporating their activities into mainstream urban planning, design and management in the face of the above issues is a challenge for KCCA.

Public spaces are apportioned among workers based on a number of factors such as the type of goods and services sold, and the capacity of the vendor to negotiate for working space. The spaces can be obtained through purchase, free transfer from one vendor to another, sharing of space by the vendors by working in shifts, official permission/allocation by KCCA upon payment for a trading license and obtaining an appropriate space after satisfying all requirements for street trading as specified in the street trading bye-laws. Inheritance or transfer of trade from one family member to another is common, especially for the trading of food items that have been carried out on the streets for generations. Workplace-related conflicts among the workers are resolved informally by elderly vendors and established social relations and hierarchies.

Public spaces, for example street spaces, are often repartitioned during peak hours to allow both vending of goods and services to take place alongside pedestrian and vehicular movement. In addition, spaces such as streets are used for various purposes during different times of the day. However, it is the partitioning of street space use that creates conflict among street users, street managers and politicians. Whereas the city authority is focused on cleaning up the city, to developing a smart city, creating orderliness in trading and driving out street vendors, street vending has persisted within the city.

Buildings serve as storage for goods, facilities and materials used for various street enterprises. Sometimes, the public locations or buildings adjacent to them have storage facilities where workers and street-based entrepreneurs pay for storage facilities on a monthly, weekly or daily basis, in cash or in kind. Storage charges vary according to available storage, the types of goods and duration of storage, and the distance from the storage location to the workplace. Storage in close proximity to the workplace is necessary because some of the street vendors are agents of formal businesses, and hence return unsold stock to the stores from which they were obtained.

Projected trend in the use of public spaces for private enterprises

The use of public spaces, especially streets, for private enterprises will likely continue, especially in parts of cities where human traffic is expected, and for cities in the developing world, especially where human settlement patterns are not related to the location of shopping centres. The convenience provided by vendors and hawkers to clients in accessing goods and services in places and at the time they are required is a main driver for public space workers and their patrons.

Support in terms of patronage of public spaces workers is still strong in certain cultures. Moreover, it takes time, generations, changes in technology, changes to the political economy and economic changes to break norms and traditions such as vending from streets and other public spaces. The public space workers will continue to trigger impulse buying from the public. In addition, street-based vendors and hawkers will continue being an important block in the distribution value chain for goods and services. Due to the nature of their operations, where they do not pay direct taxes and have no permanent fixed stalls in public spaces, they will likely continue to sell at cheaper prices in comparison with their counterparts in formal workplaces.

Due to the convenient location of malls in the suburbs, more shopping for low- and middle-order goods and services will continue in the suburbs than in the city. Conversely, shopping malls in the city will likely continue to thrive in the sale of higher order goods, but will decline in lower order goods. Some stores located in the city centre will likely close, due to the low client base for low-order goods and as a result of competition from suburban shopping as well as traffic congestion that deter suburban shoppers to purchase items they need from inner city malls. Vending will continue to be practiced in public spaces such as streets, railways stations, walkways and alleys, edges or walls. The walls provide an important platform for hoisting and displaying various goods and services to pedestrians and people in the vehicular traffic. The walls have an added advantage for public space workers, because instead of displaying goods on tables, or on their bodies, or on the streets, they instead hang them on the walls, thereby minimizing interference with the movement of traffic.

Meanwhile, subsistence consumption will still dominate the volume of purchases from public space workers. This is because industrial production is still low in developing countries. Considering that workers in public spaces predominantly sell goods and services for daily use, their activities will continue to influence shopping patterns as well as attract the attention of formal businesses and city authorities.

Therefore, the (re)valorization and social innovation of public spaces as workplaces will continue, as spaces are utilized in ways not conceived by city planners, designers, and managers. This will include the reorganization of dwellings to accommodate HBEs, the appropriation of public spaces in proximity to dwellings and finally, gradual withdrawal from the inner-city spaces as the flow of human traffic shifts to self-contained suburbs. Again, a significant proportion of workers will hold their workspaces in the inner city, owing to the continued social acceptance of public space workers, lukewarm management from KCCA, proximity of their dwellings within walkable distances to the city, the permanent presence of transport terminals that draw people, and commuter traffic which attracts hawkers, vendors, and their enterprises.

Conclusions and recommendations

In many cities in the Global South, the available formal sector jobs are not enough for the large urban populations. Streets and other public spaces offer opportunities as workplaces and can be utilized to provide sustainable livelihoods for a significant number of people who live in the cities and are unable to find jobs in the formal sector.

Public spaces have been used as workplaces over a long period of time; this practice has its roots in the pre-colonial and colonial periods and has continued into the post-colonial periods, up to today. The markets emerge in places where traffic is anticipated, or which attract high vehicular and human traffic.

Markets for goods and services are perceived to be in spatial form by a large majority of people, especially workers in the SMMEs who ply various trades. Online marketing and e-commerce are poorly understood, especially by low-income earners. Many middle-income earners and business entrepreneurs have also not embraced the technological advances in online workplace opportunities. This is also reflected in the general public and the clientele of the enterprises in the public spaces. Thus, street trading and public markets are due to continue.

Therefore, sensitization and investment in electronic commerce and online workplaces should be included in analyses of socio-economic, political and cultural change. This involves a paradigm shift in the governance and administration of cities and values that the law enforcement authorities uphold, which, in turn, determines how they relate with workers in public spaces.

Recommendations on policy and planning guidelines for the development of public spaces

When city planners, managers, and policymakers find themselves in situations where they have to address a large informal economy, high unemployment rates, the presence of a significant population of street vendors and hawkers working in public spaces, unplanned settlements unrelated to shopping complexes, inner city traffic congestion, large commuter populations, and availability of public spaces, may need to rethink the conception and use of public spaces as workplaces, and develop inclusive urban policies, plans, and designs which reflect both the temporal and spatial realities of informal livelihoods in the cities of the Global South.

The planning and design of public spaces, especially in the CBD, should be done with the participation of the people that frequently use the streets and the public spaces. For example, public space workers should work together with developers and shopkeepers to ensure that the design or rearrangement of workplaces set aside sufficient and appropriate space for other public space users. The views of pedestrians, cyclists, motorists and other street/road users should also be sought.

Consideration should be given to the development of appropriate policy approaches for workers in public spaces. Experience in Uganda and other developing countries has shown that it is difficult and expensive to eradicate working from public spaces especially in cities of the Global South. Hence, there is a need for a policy shift to enable the formalization of the use of public spaces as working spaces. In addition, taxation mechanisms could be introduced as a way of regulating public space workers as well as generating revenue for city governments.

With respect to the management of public spaces, there is a need for private/community public space management committees to work with city authorities, to manage public spaces. In addition, reorganization of public spaces as social spaces should be encouraged. For example, communities that live and work from certain public spaces should work with planners to design social streets which make this possible. For example, owing to nature of the current use of streets and social spaces, there is a need for the provision and routine management of facilities such as bathrooms, toilets, waste collection arrangements, lighting, and drainage to support the use of public spaces in the variety of ways that is currently observable, and the diversity of which is likely to increase.

References

Bromley, R. (2000). Street Vending and Public Policy: A Global Review. *International Journal of Sociology and Social Policy*, 20 (1\2). Emerald Group Publishing Ltd, Bingley, UK.

Brown, A. and Mackie, P. (2018). Politics and Street Trading in Africa: Developing a Comparative Frame. *Journal of Urban Research*. 17–18.

Chen, M., Harvey, J., Wanjiku Kihato, C. and Skinner, C. (2008). Informality and Social Protection: Theories and Realities. *Institute of Development Studies Bulletin*. 39(2), pp. 19–26.

Chen, M. et al. (2018). Inclusive Public Spaces for Informal Livelihoods: A Discussion Paper for Urban Planners and Policy Makers. *Prepared by WIEGO for the Cities Alliance Joint Work Programme for Equitable Economic Growth in Cities*. WIEGO, Manchester, UK.

Ghida, I. (2020). Time to Rethink Street Vending: Lessons from Uganda. Economic Transformation, Globalisation, Health. Accessed at https://globaldev.blog/blog/time-rethink-street-vending-lessons-Uganda.

Gumisiriza, P. (2021). Street Vending in Kampala: From Corruption to Crisis. Center African Studies Quarterly. 20(1). University of Florida, USA.

Hodder, B. (1965). Some Comments on the Origins of Traditional Markets in Africa South of the Sahara. *The Royal Geographical Society*.

Hou, J. (2013). *Transcultural Cities: Border-Crossing and Place making*. Routledge, London.

Hou, J. (2020). Guerilla Urbanism: Urban Design and Practices of Resistance. *Urban Design International*, Springer.

ILO. (2007). *The Informal Economy: Enabling Transition to Formalization*. Tripartite Interregional Symposium on the Informal Economy: Enabling Transition to Formalization, Geneva.

Kabanda, U. (2017). Evolution of the Ugandan Government, Its Regulatory Role to the Formal and Informal Sector in Managing Trade and Production. *Social Science Review*, 3(2), ISSN 2518-6825.

Mirams, E. A. (1930). *Report on the Town Planning and Development of Kampala*. Vol. I. *Government Printer,* Entebbe, Uganda.

Mitullah, V. W. (2003). Street Vending in African Cities. A Synthesis of Empirical Findings from Kenya, Cote d'Ivoire, Ghana, Zimbabwe, Uganda and South Africa. IDS, Background Paper for the 2005. *World Development Report*. World Bank

Nakazi Florence. (2019). The Plight of Kampala Street Vendors. Retrieved from https://eprcug.org/the-plight-of-kampala-street-vendors/

SIHA. (2018). The Invisible Labourers of Kampala. A research paper on women street vendors in Kampala. Siha Network.

Tasan-Kok, T. (2009). Entrepreneurial Governance: Challenges of Large-Scale Property-Led Urban Regeneration Projects. Tijdschrift voor Economische en Sociale Geografie–2010. 101(2), pp. 126–149.

Taylor, T. F. (1978). The Struggle for Economic Control of Uganda, 1919-1922: Formulation of an Economic Policy. The International Journal of African Historical Studies. 11(1), pp. 1–31. Boston University African Studies Center, Boston, USA.

UBOS. (2020). Background to the Budget. Retrieved from https://www.finance.go.ug/sites/default/files/Publications/BACK%20GROUND%20TO%20BUDGET%202020.pdf

Uzoigwe, N. G. (1972). Precolonial Markets in Bunyoro-Kitara. *Comparative Studies in Society and History*. 14(4), pp. 422–455. Cambridge University Press.

Were, A. G., Stephen, M., Barnabas, N., Bridget, N., Juliana, N., Isolo, M. and Daniel, K. (2022a). A Spontaneous Location Theory and How Street Vendors Acquire Spaces. The Case Study of Kampala City – Uganda. *International journal of Architecture and Urban Design*. 10.30495/IJAUD.2021.58555.1510.

Were, A. G., Stephen, M., Barnabas, N., Bridget, N., Juliana, N., Isolo, M. and Daniel, K. (2022b). A Moralist Theory and Persistence of Street Vending. The Case Study of Kampala city – Uganda. Published in *Proceedings Book of the 7. International Istanbul Scientific Research Congress via this link*: https://drive.google.com/file/d/1AuvVosVdpExCQNo1-JbTjU6V8-AkpzhO/view?usp=sharing

16

BUILDING RESILIENT WORKPLACES

Prioritizing safety and disaster risk reduction for the global workforce

Jane Katz[1], *Emma Harwood*[2], *and Olivia Nielsen*[2]

[1]Global Urban Development, USA and Urban Representative to the United Nations Disaster Risk Reduction Stakeholder Engagement Advisory Group SEM; [2]Miyamoto International, USA

Introduction: The urgent call to address workplace safety in the face of the increasing frequency of disasters

The devastating effects of climate change on communities, economies, and livelihoods have foregrounded the urgency of integrating resilience into the built environment. Today, 56 percent – or 4.4 billion – of the world's population work and live in cities and is expected to double by 2050. Yet, when it comes to disaster preparedness, the adaptive measures needed to guide the sustainable growth of cities and mitigate risk lack consistency in implementation are absent altogether. The resilience of buildings in the face of different disasters, including those linked to climate change, is crucial to ensure the safety of people and assets, and to minimize the economic impact of such events.

Cities play an increasingly important role in tackling climate change because their exposure to climate and disaster risk increases as they grow in size and density. Uncontrolled urban sprawl and the proliferation of informal settlements reduce the quality of the building stock and place strain on public infrastructure, services, and resources. Heightened demand for affordable housing can lead to hastily built structures that forego safety and quality in the interest of providing rapid housing solutions at reduced construction costs. In the absence of adaptive actions to enhance resilience and preparedness, high concentrations of people, infrastructure, and economic activities in these areas can magnify the extent of damage, economic breakdown, and loss of life in the event of a disaster.

Urban workers are subject to the risks of the built environment in which they operate, which is especially acute in emerging markets. Across developing countries, the existing building stock is often not built to withstand extreme weather events; rendering workers, businesses, and households vulnerable to the increasing impacts of climate change. The lack of building codes and standards for urban land consumption, planning, and construction is a major factor contributing to the vulnerability of workplaces to disasters. Equally pressing is the enforcement of codes and standards; lessons from the recent Turkey–Syria Earthquake

DOI: 10.1201/9781003262671-16

have demonstrated the importance of quality control and monitoring of code compliance. While building codes and standards are designed to ensure that buildings are safe, durable, and resilient to hazards, the absence of proper enforcement and stringent regulation to ensure their implementation in practice can render them obsolete.

The image of a traditional worker is often associated with working in a physical building, such as an office or a factory. However, with the advancement of technology and the increasing trend towards remote work following the outbreak of the COVID-19 pandemic, the image of a traditional worker is rapidly evolving. More and more people are working from home and other non-traditional workspaces, challenging the conventional notion of what it means to be a modern worker. The changing nature of working environments has highlighted the importance of resilient buildings and homes that can adapt to these new circumstances. As of 2021, Women in the Informal Economy Governing and Organizing (WIEGO) reported that 260 million workers operated from home. This number does not take into account the informal labor of women who often are in charge of the 'unseen' work behind a household, including childcare.

Developing countries account for 86 percent of home-based workers and upwards of 90 percent of the informal labor force, where employers are not subject to strict regulations or required to provide other forms of protection. A number of factors are believed to have contributed to the rise of informal employment, including the rapid pace of urbanization, lax labor market regulations, and the expansion of the gig economy. In addition, the COVID-19 pandemic has likely had a significant impact on the size and composition of the informal sector, as many formal workers have been laid off or furloughed, leading workers to gravitate towards more informal and precarious work.

Resilient buildings and homes are crucial not only for protecting lives, but also for maintaining economic stability in the face of disasters. As the world continues to grapple with the impacts of climate change, the safety of the built environment is becoming an increasingly urgent concern. From rising sea levels to extreme weather events, the effects of the climate crisis are putting our infrastructure and the people who rely on it at risk. The question is: how can we ensure the safety of our built environment in the face of these mounting challenges?

Fortunately, many structural and financial tools, as well as preparedness solutions, exist to strengthen the resilience and agency of communities and local economies in the face of disaster. For example, disaster insurance can provide small businesses with financial assistance to cover expenses such as repairs, replacements, and business interruption losses. National programs aimed at retrofitting buildings and the promotion of comprehensive building codes can reduce the risk of loss of life and property, while developing continuity plans can help small businesses withstand and recover from disruption faster after a disaster inevitably strikes.

While climate change is an issue that has been recognized by the scientific community for decades, developed and developing countries alike have failed to heed to the warnings of experts to urgently confront harmful practices and implement sustainable solutions to guide urban and economic development. The increasing frequency and intensity of disasters and their well-documented disruptions to daily life have demonstrated that climate change is not just an environmental issue, but one that has significant social, economic, and geopolitical implications. By investing in disaster preparedness and response, communities can better protect themselves from the devastating impacts of natural disasters, which can save lives and reduce collateral economic losses.

The root causes of systematic failures in disaster protection

Disasters do not discriminate and all countries must be prepared to face catastrophic events linked to climate change. However, developed countries typically have fewer disaster-related deaths and injuries due to their capacity to prepare for and quickly respond to disasters. Advanced early warning systems, evacuation plans, and emergency response teams readily available to mitigate the impact of a disaster are common features of the disaster risk reduction (DRR) landscape. Furthermore, strict building codes and safety regulations to ensure structures are built to withstand disasters reduce the visible impact of disasters on the built environment, while protecting the lives and assets of those they house. According to the World Bank Group (2016), within the last 10 years, high-income countries with more advanced building code systems experienced 47 percent of disasters globally yet accounted for only 7 percent of disaster fatalities.

At the granular level, the effects of climate change and natural disasters can be felt in the day-to-day lives of individuals and communities. The loss of homes, businesses, and infrastructure due to floods, cyclones, and other hazards can disrupt livelihoods and lead to long-term displacement and impoverishment. At a broader level, the effects of climate change can have far-reaching implications for societies, economies, and ecosystems. The loss of fertile land, dwindling freshwater resources, and more frequent natural disasters can lead to food and water insecurity, increased poverty and inequality, and the displacement of people. These effects can exacerbate social tensions, undermine economic growth, and damage biodiversity and ecosystem services.

The effects of climate change and natural disasters are not evenly distributed and can exacerbate existing inequalities and vulnerabilities. Women, children, older persons, and marginalized groups are disproportionately affected by these hazards due to their lower socio-economic status, limited access to resources and services, and cultural and social barriers. While certain disasters can be unpredictable and unavoidable, systemic failures in preparedness and response can exacerbate their impact, particularly for these vulnerable populations. From inadequate building codes and lack of infrastructure to poor communication and insufficient support for affected businesses, there are many factors that contribute to systemic failures in disaster protection. Understanding these failures and addressing their root causes is critical to ensuring that workers and businesses are better prepared and protected in the face of future disasters.

Vulnerability is a complex and multidimensional concept influenced by drivers of risk and inequality that originate in spatial, socio-economic, demographic, cultural, and institutional systems (Kuhlicke et al., 2011). Addressing these underlying drivers requires a comprehensive approach that considers both the physical and social dimensions of vulnerability, as well as the political and economic structures that shape risk and disaster resilience. Poor urban development practices, such as the construction of buildings in hazardous areas, can increase exposure to hazards and result in significant losses during disasters. This is compounded by a lack of appropriate standards to guide the integration of safety and resilience into the construction of buildings, as well as by weak enforcement by national authorities and regulatory bodies. The pressures of the global economy can contribute to the lack of appropriate regulations and enforcement in several ways. In order to remain competitive and attract investment, governments may prioritize attracting businesses and creating a favorable business environment over protecting workers' rights and safety. Further, businesses may be able to relocate to countries with weaker regulations and lower labor costs, creating a race to the bottom in terms of labor standards and worker protections.

Inadequate regulation can disproportionately affect marginalized communities and workers, enhancing vulnerability in the aftermath of disaster. Workers in low-wage and informal jobs are more likely to work in hazardous conditions and are less likely to have access to legal protections and resources to improve their working conditions (ILO, 2018). In a competitive global economy, it can be difficult to incentivize employers to prioritize worker safety and well-being, especially if they perceive safety measures as a barrier to profitability. However, when national and local governments and employers fail to address disaster risk, risks accumulate and can undermine current and projected social and economic development (Mead, 2022). Conversely, investing in DRR can lead to long-term reductions in potential losses and free up resources to support development objectives.

Many different types of hazards endanger workers around the world, of which a few are discussed in the following sections.

Factory fires: A historical hazard which remains a concern today

Factory fires have been a significant safety concern for workers and businesses, particularly in the Global South, where safety standards are often inadequate, and enforcement is weak. Preventable fires have claimed countless lives and caused significant economic damage. While measures have been taken to improve fire safety in factories, such as installing fire extinguishers and smoke detectors, many factories continue to lack basic safety equipment and procedures. The tragic consequences of these fires underscore the urgent need for stronger safety regulations and more effective enforcement measures to safeguard workers and businesses.

It is imperative that governments, corporations, and other stakeholders take concrete steps to ensure the safety and well-being of workers in factories, including investing in fire prevention and safety training, improving building codes and safety standards, and holding negligent employers accountable for their actions.

Failure to protect workers in factory fires

In developed and developing countries alike, factory fires have resulted in significant loss of life and property damage, highlighting the need for stronger safety regulations and enforcement measures to protect workers and prevent such disasters from occurring in the future. High-income countries typically respond to disaster by implementing a series of reforms or codes to enhance safety in the workplace. In fact, the 1911 Triangle Shirtwaist Factory fire, which claimed the lives of 146 garment workers in New York City, led to critical reforms in workplace safety and labor laws, including the creation of the Occupational Safety and Health Administration (OSHA) in the United States.

Before the fire, the working conditions in the garment industry were abysmal, and the factories were often overcrowded and lacked proper safety measures. The Triangle Shirtwaist Factory was no exception, with its doors being locked to prevent theft and unauthorized breaks, and its workers being unable to escape when the fire broke out. Following the disaster, public outcry demanded that the government take action to prevent similar tragedies from occurring in the future. The New York State Legislature created the Factory Investigating Commission (FIC) to investigate the conditions of the factories, and their findings resulted in the enactment of new laws and regulations for workplace safety.

The FIC findings led to the implementation of laws that mandated workplace safety measures such as fire exits and sprinklers, regular safety inspections, and the prohibition of locking exits during working hours. Prior to the Triangle Shirtwaist Factory fire, many factories lacked these essential safety features. This was particularly true for sweatshops that operated in cramped, multi-story buildings. The new laws mandated that all workplaces, particularly those in high-risk industries such as textiles and manufacturing, must have fire exits and sprinklers installed to help prevent fires and ensure workers could safely evacuate in the event of an emergency. Regular safety inspections were introduced to ensure that workplaces were up to code and complied with safety regulations. This practice helps to proactively identify and address potential safety hazards and hold employers accountable implementing corrective actions to ensure worker safety.

The FIC findings drastically changed the landscape of actors and institutions concerned with the enforcement of workplace safety standards in the United States. One such example was the creation of the OSHA, a highly influential regulatory body responsible for establishing safety norms, laws, and protocols in the United States. The Triangle Shirtwaist Factory fire also had a profound impact on community-led movements for workers' rights, particularly among young, female immigrant workers. The International Ladies' Garment Workers' Union (ILGWU) was formed in response to the tragedy, and it advocated for better wages, safer working conditions, and better treatment of workers. The union's efforts eventually led to the passage of laws protecting workers' rights to organize and form unions. The tragic consequences of a failure to adopt preemptive safety measures galvanized the labor movement in the United States and established important protective regulations and ideas that remain in effect today.

Global supply chains and the outsourcing of production to developing countries has pressured employers to cut costs and prioritize production efficiency over worker safety. This can create a race to the bottom, with companies seeking out countries with weaker labor laws and regulations in order to reduce costs and increase profits. Competitive markets, combined with weak or corrupt regulatory institutions that are not able to effectively enforce safety regulations, drastically elevate the vulnerability and risk of workers in developing countries.

While the World Health Organization (WHO) and International Labor Organization (ILO) (2018) report a steady decline in the proportion of occupational deaths attributable to factory fires, workers continue to be at risk as employers prioritize profit over safety investments. In many developing countries, safety regulations exist on paper but are poorly enforced, allowing businesses to cut corners on safety measures to save costs. In Bangladesh, many factories lack adequate fire safety measures, such as fire exits, alarms, and extinguishers, despite being mandated by official regulations. The high number of deaths and injuries in factory fires in developing countries can be attributed, in part, to the failure to enforce safety regulations. For example, the 2012 Tazreen Fashions factory fire in Bangladesh, which killed 112 workers, was largely caused by the lack of fire exits and inadequate firefighting equipment (Bangladesh Center for Workers Solidarity, 2012).

One of the key systemic failures in protecting workers from factory fires has been the lack of safety training and education. Many workers in developing countries lack basic safety training and are not aware of the hazards they face, putting them at greater risk of injury or death in the event of a fire. According to a study by the ILO (2013), only 14 percent of workers in Bangladesh, a major garment-producing country, had received safety training. Furthermore, many workers in developing countries work in informal settings, where safety standards are often ignored, increasing their risk of injury or death in the event of a fire.

Failure to protect businesses in factory fires

Inadequate insurance coverage is a significant obstacle to safeguarding businesses from fire hazards, especially in developing countries. Without sufficient coverage, businesses face considerable financial losses that can be challenging to recover from in the aftermath of a fire. According to a study by the International Finance Corporation (IFC) (2017), only 5 percent of small- and medium-sized enterprises (SMEs) in low-income countries have insurance coverage. Many businesses are deterred from accessing disaster insurance due to high premiums or a lack of insurance providers in their area, thus making it difficult to obtain the financing and credit needed to grow their business and invest in safety measures. In the 2013 Rena Plaza factory collapse in Bangladesh, which killed over 1,100 people and caused $1.2 billion in economic loss, many of the affected businesses did not have insurance coverage and were unable to recover (Kumar, 2014).

A prioritization of profit over safety too often leads to a reluctance to invest in safety measures that do not generate an immediate return on investment. Furthermore, the lack of enforcement of safety regulations means that businesses can forego safety measures without facing consequences, further reducing their incentives to invest in safety. This failure to incentivize safety investment has contributed to the high number of factory fires in developing countries. In the aftermath of the Rena Plaza disaster, many factory owners in Bangladesh resisted efforts to improve safety standards, arguing that it would increase their costs and make them less competitive in the global market (Kumar, 2014).

In the United States, OSHA encourages voluntary programs that extend beyond its regulatory requirements for employers. These programs are designed to incentivize employers to implement, practice, and improve upon health and safety management systems in their workplace. One of the voluntary programs offered by OSHA is the Voluntary Protection Program (VPP). VPP recognizes employers who have implemented effective safety and health management systems and have maintained injury and illness rates below their industry average. VPP participants are exempt from OSHA-programmed inspections, meaning that OSHA will not conduct a routine inspection of the workplace as long as the employer maintains its VPP status. VPP participants also have access to a variety of other benefits such as training, technical assistance, and networking opportunities.

Another voluntary program offered by OSHA is the Safety and Health Achievement Recognition Program (SHARP). SHARP is designed for small businesses with fewer than 250 employees who have demonstrated a strong commitment to workplace safety and health. Employers who participate in SHARP receive an exemption from OSHA-programmed inspections for a period of one year, which may be extended if the employer continues to maintain its safety and health program.

The systemic failures to protect workers and businesses from factory fires are a complex issue that requires significant changes in policy and practice. To protect workers, there needs to be improved safety training and education, stricter safety regulations, and better enforcement of safety standards. To protect businesses, there needs to be increased insurance coverage, better incentives for safety investment, and greater access to financing and credit. Addressing these systemic failures will require a concerted effort from governments, businesses, and workers to create a more resilient and equitable economy that can withstand future crises.

Seismic hazards: A major cause of building collapse and workplace vulnerability

Earthquakes are a major cause of destruction to all forms of real estate, including office buildings and retail spaces putting the lives of workers at risk. One especially dangerous type of structure are soft story buildings, which suffer a disproportionate amount of damage during earthquakes. Soft story buildings are commonly used by storekeepers in emerging markets and are a multi-story structure with a shop on the bottom and apartments above. Because shops often wish to have a vast space empty of columns, modifications are often made which create a weak or flexible ground floor that is unable to support the weight of the floors above it. The collapse of soft story buildings poses a significant risk to workers worldwide, as these buildings are often used for commercial or industrial purposes.

In the absence of retrofits or structural reinforcements, the structural deficiencies intrinsic to the design of soft story buildings can exacerbate the vulnerability of workers and other occupants of the building during earthquakes. Falling debris and other structural elements resulting in the partial or total collapse of these buildings may crush or trap workers inside. Flammable materials in commercial buildings may ignite and cause a fire, which can spread rapidly and pose a significant risk to the safety of occupants and property in the building. Limited means of egress prevent workers from evacuating quickly and safely, increasing the risk of injury or death.

The rapid, unplanned growth of cities and urban areas in earthquake-prone regions increases the number of people and buildings which can make the impact of disasters more severe. Some studies even suggest that climate change may be contributing to an increase in micro-seismic earthquakes as melting glaciers, fluctuating sea levels, and extended periods of drought affect the pressure on faults and trigger seismic activity. Earthquakes are a global phenomenon that affect developing and developed countries alike. Global examples of soft story building collapses can both identify patterns across countries in sources of vulnerability and adaptive solutions to safeguard assets and lives.

Global examples of earthquake responses

Addressing the vulnerabilities of these buildings requires effective building codes and standards, as well as retrofitting of existing structures to make them more resilient to earthquakes and other natural disasters. Additionally, it is important for businesses and employers to prioritize the safety of their workers and take steps to minimize risks in soft story buildings.

1989 LOMA PRIETA EARTHQUAKE

During the 1989 Loma Prieta earthquake in California, the lack of structural support in buildings caused many of them to collapse, resulting in a large number of fatalities. In fact, a study by the Earthquake Engineering Research Institute found that over 40 percent of the earthquake-related fatalities were attributed to the collapse of soft story buildings (Yashinsky and Kwok, 2017). Following the Loma Prieta earthquake, the city of San Francisco enacted mandatory retrofitting ordinances to ensure the safety of soft story buildings, which has since been adopted by other earthquake-prone cities. The retrofitting measures include adding structural support such as steel frames, bracing, or shear walls, making these buildings safer for occupants during seismic events.

1995 KOBE EARTHQUAKE

The 1995 Great Hanshin-Awaji Earthquake, commonly referred to as the Kobe Earthquake, was a catastrophic seismic event that struck the Hyogo Prefecture in Japan on January 17, 1995. With a magnitude of 6.9, the quake unleashed widespread destruction, particularly in the city of Kobe and its surrounding areas. The disaster claimed the lives of around 6,400 people, left over 40,000 injured, and caused immense infrastructural damage. Notably, the collapse of older buildings, including soft story structures, accentuated the devastation. The Kobe Earthquake served as a tragic wake-up call, leading to extensive reevaluation of Japan's earthquake preparedness, building codes, and disaster response mechanisms. This transformative event ushered in significant advancements in seismic engineering, disaster management, and community resilience to minimize the impact of future earthquakes.

The Kobe earthquake also highlighted the importance of retrofitting soft story buildings to make them more resilient to earthquakes. Following the earthquake, the Japanese government introduced stricter building codes and regulations, including guidelines for the seismic retrofitting of soft story buildings. The government also provided financial incentives and assistance to encourage building owners to retrofit their structures. In addition to government action, there has been a growing awareness among building owners and residents of the risks associated with soft story buildings. Many building owners have voluntarily undertaken seismic retrofitting of their structures, recognizing the importance of making their buildings more resilient to earthquakes and other disasters. There has also been a significant investment in research and development to improve the seismic performance of buildings. Japanese researchers have been at the forefront of developing innovative seismic-resistant technologies and materials, such as base isolation systems, energy-absorbing dampers, and seismic-resistant steel frames.

Japan's approach to addressing soft story buildings has been multi-faceted, combining government regulations and incentives with voluntary retrofitting efforts and investments in research and development. While these efforts are certainly a step in the right direction to reduce the risk of soft story building collapse during earthquakes and other disasters, many soft story buildings still exist in Japan and pose a risk to public safety. A study by the Tokyo Metropolitan Government found that there were approximately 10,000 soft story buildings in Tokyo alone, representing a significant risk to the city's residents. As an earthquake-prone country, failure to address the vulnerability of Japan's built environment can worsen the devastation and economic breakdown when future disasters strike.

2010 HAITI EARTHQUAKE

The 2010 earthquake in Haiti was one of the deadliest natural disasters in history, causing widespread destruction and claiming an estimated 200,000 lives. Soft story buildings, which are common in Haiti and other developing countries, were particularly vulnerable to the earthquake's seismic forces. The lack of structural support in these buildings made them more susceptible to collapse, leading to significant loss of life and property damage. A post-earthquake analysis led by the United Nations Development Program (UNDP) found that the vulnerability of soft story buildings was a major factor in the extent of damage caused by the earthquake. The UNDP study also highlighted the need for better building codes and regulations, as well as the retrofitting of existing structures to make them more resilient to seismic activity. The devastating impact of the earthquake in Haiti underscores the urgent need for proactive measures to improve building safety and protect vulnerable populations from the devastating effects of natural disasters.

ONGOING CONCERNS TODAY

The collapse of non-governmental organization (NGO) offices during the recent earthquakes in Turkey and Syria has brought international attention to the issue of worker vulnerability in structurally deficient buildings. In fact, many of these organizations with office locations in Syria, Turkey, and other disaster-prone countries have begun undertaking seismic assessments and implementing retrofits to both ensure workers can remain confident in the ability of their workplace to withstand future shocks and that humanitarian assistance can remain uninterrupted following a disaster. Nonetheless, the global risk caused by earthquakes continues to challenge local and international organizations that are placing greater emphasis on worker safety – which inadvertently establishes an important precedent for other employers to replicate.

Uncovering the vulnerabilities of home-based workers

The COVID-19 pandemic has had a significant impact on the way we live and work, with many businesses and organizations forced to adopt remote work arrangements to prevent the spread of the virus. With 18 percent of the global labor force working from home, buildings and homes must be designed to withstand a wide range of disasters, including pandemics, hazards, and other unexpected events (Bick et al., 2020).

Even prior to the pandemic, home-based work was the norm for many households across developing countries, and especially for women. In 2019, the ILO estimated that there were 260 million workers, accounting for 7.9 percent of the employed population, in the broader category of "home-based work", which also includes work carried out by independent, self-employed workers in their home (or adjacent grounds or premises) (ILO, 2019a). Of this amount, 147 million were women, amounting to 11.5 percent of women's employment share (5.6 percent of employment for men).

With less than 40 percent of women participating in the labor force globally – and 21.7 percent of women providing full-time, unpaid care to family members – women and children are most often the victims of disaster-related fatalities simply because they spend more time at home (CEPR, 2020). In the 2004 Indian Ocean Earthquake, for instance, women accounted for upwards of 70 percent of the total fatalities. Tsunami waves flattened coastal communities and killed many of the women and children, most of whom were traditionally at home on a Sunday morning, while men working away from the home experienced higher rates of survival.

Low-income status, informal employment arrangements, and the location of homes in hazard-prone areas each contribute to and exacerbate the vulnerability of home-based workers. Rapid urbanization has witnessed the proliferation of informal settlements, which importantly house a large share of the informal economy. In India, many workers of the informal sector were found to reside in housing in low-lying river areas or in areas with poor drainage systems that are highly vulnerable to flood risk. Limited resources and access to credit are significant barriers to coping with the impacts of flooding for home-based workers.

Due to their informal employment status, these workers are typically excluded from traditional safety nets, such as insurance, pensions, or sick pay. Furthermore, home-based workers often lack savings and financial resources to invest in protective measures, such as flood-resistant equipment, quality construction materials, or insurance coverage.

With the number of home-based workers expected to rise as workplaces rapidly transition to remote employment, there is a need for policies and programs that can support home-based workers to strengthen resilience and disaster preparedness in homes. These policies could include measures to improve the structural integrity of the housing stock, through incentives and access to credit, as well as post-disaster measures as the creation of emergency funds or the provision of financial and material assistance during disasters. Such policies can help to reduce the vulnerability of home-based workers in the informal sector and ensure that they are better equipped to cope with the impacts of natural disasters as they grow in intensity and frequency.

The impact of flooding and heat waves on workers

The accelerating effects of climate change will have numerous impacts on workers and economies worldwide. As floods, heat waves and high winds increase in frequency, losses will continue to be pronounced without significant interventions in adaptation.

Floods

With over 5,000 lives lost annually, flash floods are one of the deadliest natural disasters worldwide, causing significant social, economic, and environmental impacts.[1] Flash floods are characterized by a rapid and violent flow of water through a normally dry area, such as a desert or arid region, or a typically calm watercourse, such as a river or stream. Flash floods can be caused by heavy rainfall, snowmelt, dam failures, or other factors that can cause large volumes of water to accumulate quickly and overwhelm drainage systems or natural watercourses. Flash floods make up approximately 85 percent of all flooding cases and have the highest mortality rate among different types of flooding, including riverine and coastal.

The monsoon season is a critical time for countries in South Asia, where heavy rainfall is known to cause widespread flooding and destruction. Pakistan is no exception, and an unprecedented monsoon season in 2021 had a devastating impact on the country. with torrential rainfall that is 10 times heavier than usual, according to the Pakistan Meteorological Department. The heavy downpours have caused the Indus River to overflow, creating a massive lake tens of kilometers wide. As of August 30, the southern provinces of Sindh and Balochistan witnessed rainfall that is 500 percent above average, resulting in entire villages and farmland being engulfed, buildings being razed, and crops being wiped out. Despite being responsible for less than 1 percent of the world's greenhouse gas emissions, Pakistan is the eighth most vulnerable nation to the climate crisis, according to the Global Climate Risk Index. The country has experienced extreme weather events, including record heat waves and destructive floods, as the climate crisis worsens.

Since 2021, nearly 2.2 million homes have been destroyed by floods in Pakistan alone, leaving 33 million people affected and without access to basic necessities (United Nations, 2023). This has led to a significant strain on the country's already overburdened resources and infrastructure, with many struggling to access safe shelter and clean water. The widespread destruction of homes and electricity outages had a significant impact on the economy, with many businesses and markets shutting down as a result of the flooding. The heavy rainfall has also rendered numerous roads impassable, making it challenging for emergency services and aid workers to access affected areas.

Some 4.3 million workers in Pakistan have been affected by the disruptions and job losses due to flash flooding. This is particularly true for the agriculture sector, a key driver of development and economic growth in Pakistan. Currently, agriculture employs about 48 percent of the country's labor force, making it the main source of income for a significant portion of the population. Agriculture plays a major role in providing food to a large portion of the country's population and serves as a primary source of raw materials for the industrial sector. In fact, agriculture contributes around 23 percent to the national GDP, which is higher than any other sector.[2]

Despite comprising 65 percent of employment in agriculture, female workers in the sector are frequently deprived of labor rights and protections and often paid less than their male counterparts.[3] While floodwaters have receded in the majority of the affected regions, the United Nations Office for the Coordination of Humanitarian Affairs (OCHA) reports that approximately 1.8 million individuals reside in close proximity to stagnant and contaminated water, resulting in women losing their agricultural livelihoods for two consecutive crop seasons (OCHA, 2023). Thousands of female agricultural workers face significant debt as they attempt to rebuild their homes and businesses, with limited support from national and international programs to recover from the climate disaster.

Heat waves

Heat waves do not have a high-profile impact on industry and the economy in the same way as catastrophic floods or storms. Nonetheless, 1.4 percent of total working hours were lost worldwide in 1995 as a result of high heat levels – the equivalent of around 35 million full-time jobs (ILO, 2019b). By 2030, according to UNDRR (2022), this share will rise to 2.2 percent – a productivity loss equivalent to 80 million full-time jobs. The loss in monetary terms is then expected to total US$2,400 billion Purchasing Power Parity (PPP). Lower-middle- and low-income countries would be the worst affected, losing 4 and 1.5 percent of their GDP in 2030, respectively. By 2030, heat-related stress in the workplace is projected to result in US $4.2 trillion in productivity losses.

Climate change and wealth disparities each shape the everyday lives of workers. Basra – an Iraqi city at the forefront of the adverse effects of climate change – experienced a record high heat index of up to 120 degrees Fahrenheit in 2022, the highest temperatures have been in the country since the 1970s. While elevated temperatures are not uncommon in this region of the world, Persian Gulf countries have warmed almost twice as fast as the global average (Klein, 2021). Though global warming disproportionately affects different countries and regions, such as in the case of the Persian Gulf, climate change remains a global phenomenon; researchers at Harvard University warn that by 2050, nearly half of the world's population are expected to live in areas that experience dangerous levels of heat for at least a month.

The extent of wealth inequality and the nature of jobs that individuals hold as a result are significant factors in determining the working conditions they experience, especially in developing countries. High temperatures are linked to increased worker disorientation and a high incidence of work accidents and injuries. Sustained levels of heat exposure can increase the risk of heatstroke, kidney failure, and dehydration, as well as exacerbating long-term cognitive deficits among individuals with pre-existing health conditions. The avoidance of heat during peak sunlight can lead to truncated workdays that reduce productivity and earnings for workers. While in-office workers may have access to cooler, indoor working environments with air temperature control, workers engaged in construction and other outdoor professions typically do not enjoy the same employer-provided services.

The consequences of extreme heat are already staggering. Each year, the economic toll of lost work due to heat-related illness and injuries amounts to hundreds of billions of dollars worldwide. This includes not only direct costs such as medical treatment but also indirect costs such as lost productivity and reduced economic output. The global warming crisis is further compounded by an expected proliferation in climate migrants. As climate refugees are often from marginalized communities with limited resources, the lack of proper documentation leaves them vulnerable and marginalized in their new location. Without the necessary papers to work legally, they may turn to the informal economy, where they take on low-wage jobs that exclude workers from job security or protection, such as worker's compensation, healthcare, and retirement plans.

What can global examples teach us about disaster risk management?

What can international examples of systemic failures to protect workers and households from disaster teach us? First and foremost, they highlight the importance of strong and effective regulatory frameworks to protect public safety. In many cases, the lack of enforcement of safety regulations has contributed to the high number of deaths and injuries in disasters, particularly in developing countries.

Cases of systemic failures underscore the need for proactive measures to retrofit buildings and homes to make them more resilient to disasters. This is particularly important in regions with high levels of seismic activity, where soft story buildings and other vulnerable structures pose a significant risk to public safety. The importance of providing financial assistance and incentives to encourage businesses and households to invest in disaster preparedness measures cannot be overstated. This can include everything from developing business continuity plans to retrofit homes and buildings to be more disaster resilient.

International examples highlight the need for a coordinated and collaborative approach to disaster preparedness and response. This can involve working with local communities, businesses, and government agencies to develop comprehensive disaster management plans that prioritize public safety and resilience. In light of these threats, different countries and cities are adopting programs to support their companies, households, and communities in the face of disasters.

Successful structural retrofit programs

Structural retrofits are a critical component of DRR, particularly in developing countries where safety standards are often inadequate, and enforcement is weak. One such initiative is the United Nations Development Program's (UNDP) Global Program for Safer Schools, which aims to promote the safety and resilience of schools and educational institutions in disaster-prone areas. The World Bank has implemented several programs aimed at promoting structural retrofits to improve building safety in developing countries. For example, the World Bank's Global Facility for Disaster Reduction and

Recovery has funded projects in disaster-prone countries such as Bangladesh and Nepal to retrofit vulnerable buildings, including schools, hospitals, and government buildings, to make them more resilient to disasters.

Many international donors and key actors have provided governments with the necessary technical expertise to develop national programs that aim to promote structural retrofits of the built environment. The following sections discuss various national programs that enhance disaster preparedness and risk reduction through structural retrofits.

Japan

Japan is known for its seismic retrofitting programs that have made the country a leader in earthquake resilience. In 1995, the Great Hanshin-Awaji earthquake, also known as the Kobe earthquake, caused significant damage to buildings and infrastructure in the city of Kobe, resulting in the loss of over 6,000 lives (Holzer, 1995). Since then, Japan has implemented several programs aimed at retrofitting buildings to make them more resilient to earthquakes.

One of the most successful programs in Japan is the "Housing Quality Improvement Loan Program" that provides low-interest loans to homeowners for retrofitting their homes to improve earthquake resistance. According to the Ministry of Land, Infrastructure, Transport and Tourism (MLIT), from 1996 to 2018, over 2.2 million households participated in the program, resulting in the retrofitting of over 3.3 million homes (MLIT, 2019). The program has been effective in increasing earthquake resilience among Japanese households and has contributed to a significant reduction in earthquake-related fatalities and injuries.

In addition to the Housing Quality Improvement Loan Program, Japan has building codes that require new buildings to meet earthquake resistance standards. The Japanese building code, which was revised after the Kobe earthquake, sets strict seismic design standards for buildings, including requirements for reinforced concrete and steel frames (Holzer, 1995). The code also includes provisions for retrofits of existing buildings, including soft story buildings, which are particularly vulnerable to earthquakes. The building codes have been successful in reducing the vulnerability of buildings to earthquakes, and Japan's experience provides a replicable model for other countries facing earthquake risks.

Overall, Japan's seismic retrofitting programs and building codes have been effective in reducing the vulnerability of buildings to earthquakes, protecting workers and households from the devastating effects of seismic events. These programs demonstrate the importance of government support for retrofitting programs and the value of building codes in promoting resilience.

The United States

The retrofitting programs in the United States have been successful in improving the safety of buildings and homes. As of 2022, the Soft-Story Retrofit Program in San Francisco has resulted in the retrofitting of over 3,000 buildings, which represents approximately 80 percent of the city's potentially vulnerable buildings (SFDBI, 2022). The program has also contributed to a reduction in the number of soft story buildings in San Francisco by approximately 70 percent since the program began in 2013 (SFDBI, 2022).

The Stronger Homes, Safer Communities program by the US Department of Housing and Urban Development (HUD) has also been successful in improving the resilience of homes to natural disasters. Since the program began in 2015, over 5,000 homes have been retrofitted to improve safety and resilience (HUD, 2021). The program has also provided training to over 7,000 individuals on retrofitting techniques and best practices (HUD, 2021).

Research has shown that retrofitting programs can have significant benefits in terms of reducing the impact of disasters on buildings and communities. For example, a study by the National Institute of Standards and Technology (NIST) found that the cost of retrofitting a building is typically much lower than the cost of repairing or replacing a building after a disaster (Fung et al., 2021). The study also found that retrofitting can reduce the number of injuries and fatalities in the event of a disaster, as well as the time and cost required for recovery efforts (NIST, 2018).

Nepal

The "Safer, Stronger Construction" program in Nepal was launched in 2017, two years after the devastating 2015 earthquake that killed over 8,000 people and caused significant damage to buildings and infrastructure. Under the auspices of the World Bank, the Asian Development Bank, and UNDP, the Government of Nepal promoted the retrofit of public buildings, schools, hospitals, and other critical infrastructure to improve their resilience and performance during future seismic events.

As part of the program, over 40,000 buildings were retrofitted or reconstructed with improved seismic design and materials. This included retrofitting public buildings, such as government offices, hospitals, and schools, as well as retrofitting private homes. The program also provided training and technical assistance to local engineers and builders to improve their knowledge of seismic-resistant design and construction practices.

The "Safer, Stronger Construction" program in Nepal is an example of a successful national retrofit program that aimed to improve the seismic resilience of buildings and infrastructure in a developing country. While there were challenges in implementation, such as lack of skilled labor and resources, the program made significant progress in improving the safety and resilience of buildings in Nepal and set an important precedent for future national programs to replicate.

New technologies such as drones and AI are being deployed to quickly identify buildings at high risk of collapse and enable authorities to intervene before a disaster strikes. The World Bank Global Program for Resilient Housing is leveraging these new technologies to enable governments to prioritize vulnerable households and businesses in an ever-growing building stock.

Overall, the promotion of structural retrofits is an essential strategy to improve worker and occupational safety and prevent the loss of life and property damage during disasters. These international programs and initiatives demonstrate the importance of investing in building safety and resilience to mitigate the impact of disasters on communities and economies.

Building resilience: Business continuity planning

Business continuity planning enables businesses to assess their risk, identify weaknesses, and implement solutions before, during, and after a disaster strikes. While large businesses often have well-structured business continuity plans, smaller ones do not.

Small businesses are essential contributors to the economy, accounting for a significant portion of employment and economic activity worldwide. However, these businesses are often at greater risk of experiencing disruptions and failures, including those related to worker safety, compared to larger companies. According to a report by the ILO, small businesses are more vulnerable to occupational safety and health (OSH) risks due to their limited financial resources, lack of access to OSH information, and inadequate training (ILO, 2017).

Small businesses may also lack the resources and expertise to develop and implement effective business continuity plans, which can help ensure their survival and recovery from disruptions. A study by the US Federal Emergency Management Agency found that up to 40 percent of small businesses do not reopen after a natural disaster, in part due to a lack of preparedness and planning (Ready Business, 2020). This vulnerability is especially pronounced in developing countries, where small businesses make up a significant portion of the economy and may face additional challenges related to infrastructure, access to capital, and regulatory compliance. Moreover, small businesses in developing countries are often subject to weaker regulations and enforcement, making them even more vulnerable to occupational safety and health risks. In such contexts, small businesses may also face challenges in accessing the necessary resources and support to improve worker safety, including retrofitting or upgrading buildings and facilities to meet safety standards. Addressing these unique challenges is critical to improving worker safety and promoting economic resilience in the face of disasters.

In response, various governments and organizations have implemented programs and initiatives aimed at helping small businesses develop effective business continuity plans. For example, in the United States, the Small Business Administration (SBA) provides guidance and resources to small businesses to help them develop business continuity plans. The SBA's Business Continuity Planning Suite includes tools and templates that small businesses can use to develop plans that address potential hazards, such as natural disasters, cyberattacks, and pandemics.

The SBA's tools and templates are designed to be accessible and easy to use, even for businesses with limited resources or expertise in business continuity planning. The suite includes a Business Continuity Plan Template, which provides a framework for developing a comprehensive plan that covers all aspects of the business, including operations, financial management, and employee safety. The SBA also provides a Disaster Planning and Recovery Guide, which outlines steps that small businesses can take to prepare for and recover from disasters.

In addition to the SBA, other organizations have also implemented programs and initiatives aimed at helping small businesses develop effective business continuity plans. For example, the International Finance Corporation (IFC) provides guidance and resources to SMEs in developing countries to help them prepare for and respond to disruptions. The IFC's Business Continuity Management Handbook provides a step-by-step guide for SMEs to develop effective business continuity plans that can help them survive and recover from disruptions.

Small businesses are particularly vulnerable to disruptions related to worker safety, and their survival is crucial for the overall economic stability of a region. However, governments and organizations have implemented programs and initiatives aimed at helping small businesses develop effective business continuity plans to address this vulnerability. The SBA in the United States provides guidance and resources to small businesses to help them develop plans that address potential hazards, while the IFC provides similar guidance to SMEs in developing countries. These programs and initiatives can help small businesses prepare for, respond to, and recover from disruptions, improving their resilience and contributing to the overall economic stability of a region.

Indonesia

The USAID Advancing Private Sector Engagement in Disaster Preparedness & Response in Indonesia (ADVANCE Indonesia) is a program aimed at improving the private sector's role in disaster preparedness and response in Indonesia. The ADVANCE Indonesia program is part of a broader United States Agency for International Development (USAID) initiative to promote disaster resilience in Indonesia, which is one of the world's most disaster-prone countries. Through the program, USAID aims to strengthen the private sector's role in disaster preparedness and response, and ultimately contribute to the country's overall disaster resilience. The program is funded by the USAID and implemented by the American Indonesian Chamber of Commerce (AMCHAM Indonesia) in partnership with the Indonesian Chamber of Commerce and Industry (KADIN).

The program is focused on strengthening the capacity of private sector companies to better prepare for and respond to disasters, with the aim of reducing the impact of disasters on businesses, employees, and communities. Specifically, the program aims to increase private sector engagement in DRR, improve disaster preparedness capacity, and enhance coordination between the private sector and government agencies. To achieve these objectives, the program provides technical assistance, training, and resources to private sector companies in Indonesia, including SMEs. This includes support for the development of business continuity plans, emergency response procedures, and DRR strategies. The program also facilitates networking and collaboration between private sector companies, government agencies, and civil society organizations to enhance disaster preparedness and response efforts.

Since 2020, Miyamoto International, a global DRR and structural engineering firm, has provided structural and non-structural mitigation and resiliency technical assistance to a representative sample size of businesses in Jakarta. Miyamoto executed earthquake disaster scenario planning exercises to enable private sector participants to identify, understand, and use current and future earthquake risk scenarios, including threat and risk analysis, understanding of hazard, exposure, and vulnerability levels with respect to earthquake risks, and financial and legal implications.

Private sector engagement is a crucial aspect of disaster risk management (DRM) as the private sector often plays a significant role in disaster preparedness, response, and recovery efforts. The private sector can bring expertise, resources, and innovation to the table, which can be leveraged to enhance DRM efforts. However, effective private sector engagement requires collaboration and coordination between the private sector, government, and other DRM actors.

The facilitation of private sector stakeholder consultations and performance of DRM knowledge-sharing and coordination activities through private sector engagement coordination platforms can help achieve this collaboration and coordination. These platforms provide a space for private sector stakeholders to engage with government and other DRM actors to share knowledge, coordinate efforts, and identify opportunities for collaboration. By promoting the sustainability of these platforms beyond the life of the program, the ADVANCE Indonesia program was able to ensure that private sector engagement in DRM efforts continued even after the program ended. This is important because DRM is an ongoing process that requires continuous engagement and effort to maintain and improve preparedness, response, and recovery efforts.

By strengthening the capacity of private sector companies to prepare for and respond to disasters, the ADVANCE Indonesia program aims to mitigate the impact of disasters on businesses and their employees, thereby promoting economic resilience and stability. The improvement of disaster preparedness and response capacity among private sector companies, such as emergency evacuation plans, training on disaster response, and the promotion of financial tools to protect businesses, helps to ensure that workplaces are stable and structurally safe.

Innovative adaptation strategies for heat waves, flooding, and other disasters

Nature-based solutions are strategies that use natural resources to address environmental challenges in cities. These solutions include floodplain restoration, which involves allowing rivers to flood into areas that have been cut off by levees and embankments, and green infrastructure, which uses natural features like trees and wetlands to absorb excess water during floods. Mangrove restoration can also protect coastlines from storm surges and flooding. Rainwater harvesting reduces flooding and provides a source of water during droughts, while planting urban forests can reduce the urban heat island effect. Though these solutions tackle adaptation at the city-level, workers within buildings will benefit from them greatly through a more comfortable safer workplace.

Sponge cities

Sponge cities are a concept aimed at addressing urban flooding issues, particularly in cities that are prone to heavy rainfall and floods. The idea behind sponge cities is to create urban environments that are designed to absorb, store, and reuse rainwater. By doing so, cities can reduce the amount of runoff water and alleviate the risk of flooding, while also providing other benefits such as improving air quality and reducing the urban heat island effect.

Singapore: The ABC Waters program

The ABC Waters program is a comprehensive initiative implemented by Singapore to create sponge cities. The program aims to enhance the quality of Singapore's water bodies and transform them into beautiful, clean, and lively spaces that can also help mitigate flood risks. The program is designed to promote environmental sustainability while enhancing the quality of life for Singaporeans.

The ABC Waters program is based on the principles of low-impact development and green infrastructure. The program includes a range of measures such as rain gardens, bioretention swales, and green roofs that work together to capture, store, and treat rainwater. The ABC Waters program has been highly successful in transforming Singapore's waterways. According to the National Water Agency of Singapore, over 300 projects have been completed under the ABC Waters program, covering over 4,200 hectares of land. These projects have helped to improve water quality, enhance biodiversity, and reduce flood risks.

CHINA

Wuhan is a city located in central China, which has experienced severe flooding due to its location at the intersection of the Yangtze and Han rivers. In response to this, the Chinese government launched the Sponge City Program in 2015, which aimed to transform 16 cities, including Wuhan, into sponge cities by 2020.

The program aims to use natural methods to absorb, retain, and purify rainwater and reduce the risk of flooding. To achieve this, Wuhan has implemented various measures, including the use of permeable pavements, rain gardens, and green roofs to capture and store rainwater. For example, the city has built more than 300 hectares of permeable pavements and 150 hectares of green roofs to help absorb and retain rainwater. Another key feature of Wuhan's approach is the use of underground storage tanks to collect rainwater. The collected rainwater is then treated and reused for irrigation, firefighting, or other purposes, reducing the strain on the city's freshwater resources.

According to a report by the World Bank, the Sponge City Program has shown promising results in Wuhan, reducing the runoff coefficient by 67 percent and increasing the proportion of infiltrated water by 78 percent. The program has also contributed to the creation of new public spaces and improved water quality. However, the report also notes that challenges remain, including the need for better integration between different departments and agencies involved in the program and ensuring long-term funding and maintenance of the infrastructure.

Wuhan's sponge city program demonstrates the potential for nature-based solutions to mitigate the impact of natural disasters such as flooding while also promoting sustainable water management practices in one of the most climate-affected regions globally.

Sponge city projects can have significant economic and environmental benefits, often leading to increased property values, reduced flood damage, and improved water quality. Additionally, sponge city projects can also create jobs and stimulate economic growth. However, implementing sponge cities can also be a complex and challenging process, requiring significant investment in infrastructure and changes in urban planning and development. Successful implementation requires engagement and participation from a range of stakeholders, including residents, businesses, and local governments, as well as sustained monitoring and evaluation to ensure that desired outcomes are achieved.

THE UNITED STATES: ENHANCING URBAN RESILIENCE THROUGH SUSTAINABLE DEVELOPMENT

Building on a project funded by the US Department of Energy through the Inclusive Energy Innovation Prize, One Montgomery Green, Bethesda Green, Montgomery County Economic Development Corporation (MCEDC), and Global Urban Development (GUD) are working to strengthen economic development efforts and climate resilience through promoting sustainable innovation and inclusive prosperity in a diverse urban community by organizing the Wheaton Sustainable Innovation Zone (WSIZ), based on GUD's Porto Alegre Sustainable Innovation Zone–ZISPOA–in Brazil. Through extensive community engagement, project partners will apply the international lessons of sustainable innovation to the curated needs of an underserved community, creating a model for similar communities in the County, State, and Nation. The team also is working to enable new and enhance existing improvements such as retrofitting buildings for energy efficiency, electrification, and solar power; installing electric vehicle charging stations; recycling and composting food waste; building sustainable parklets; expanding community gardens, urban reforestation, stormwater mitigation, and rain gardens; and much more for resilience and safety,

Financial tools to strengthen occupational safety

Small businesses often struggle with financial constraints and limited resources, making it difficult for them to invest in worker and occupational safety measures. However, disasters such as natural disasters, pandemics, or accidents can have a significant impact on the financial stability of small businesses. In this context, financial tools such as disaster insurance can play a crucial role in strengthening worker and occupational safety.

Disaster insurance is a type of insurance that provides financial protection to businesses in the event of a disaster, including natural disasters such as earthquakes, hurricanes, and floods, as well as pandemics and other unexpected events. Disaster insurance can provide small businesses with financial assistance to cover expenses such as repairs, replacements, and business interruption losses.

Japan's disaster insurance program provides a useful example of how insurance can play a role in improving worker and occupational safety in the face of natural disasters. The program was created in response to the frequent natural disasters that occur in Japan, including earthquakes, typhoons, and floods, which cause significant damage to businesses and disrupt economic activity.

The disaster insurance program provides financial support to SMEs affected by natural disasters. This includes coverage for losses related to property damage, inventory loss, and business interruption losses. The program aims to help SMEs recover quickly after a disaster and to encourage them to invest in disaster preparedness measures that can improve worker and occupational safety.

The disaster insurance program has been successful in encouraging SMEs to invest in disaster preparedness measures such as earthquake-resistant buildings and backup power supplies. In the aftermath of the 2011 earthquake and tsunami in Japan, SMEs that had disaster insurance were able to resume business operations more quickly than those that did not have insurance, ultimately providing stability in employment and income for workers even in times of crisis (Sohn et al., 2004).

Overall, disaster insurance can play a crucial role in improving worker and occupational safety in the face of natural disasters. By providing financial support to SMEs affected by natural disasters and encouraging investment in disaster preparedness measures, such programs can improve the resilience of businesses and protect the livelihoods of workers.

What more can be done?

Continuing to invest in adaptation

The world is experiencing an increase in temperature due to climate change, which has led to a greater demand for air conditioning systems in many parts of the world. As temperatures rise, more people are turning to air conditioning as a way to stay cool and comfortable, particularly in urban areas where the urban heat island effect can exacerbate heat waves. However, air conditioning systems require a lot of energy to operate, which contributes to greenhouse gas emissions and further exacerbates climate change. In fact, according to the International Energy Agency, air conditioning accounts for almost 20 percent of global electricity consumption, and its energy use is expected to triple by 2050.

Energy consumption is a major driver of climate change, as the majority of the world's energy is still derived from fossil fuels like coal, oil, and natural gas, which release greenhouse gases into the atmosphere when burned. This results in a buildup of greenhouse gases, such as carbon dioxide, methane, and nitrous oxide, in the atmosphere, which trap heat and cause the planet to warm up. In addition to air conditioning, energy consumption in other sectors, such as transportation, industry, and buildings, also contributes to climate change.

The link between energy consumption and climate change has prompted a global shift towards renewable energy sources like solar, wind, and hydropower, which generate electricity with minimal greenhouse gas emissions. In addition, energy efficiency measures, such as improving insulation in buildings and using more efficient appliances, can help to reduce energy consumption and lower greenhouse gas emissions. Governments, businesses, and individuals all have a role to play in reducing energy consumption and transitioning to cleaner, more sustainable sources of energy, in order to mitigate the impacts of climate change.

Champion collaborative approaches to preparedness and adaptation

Protecting workers and ensuring disaster preparedness are critical to building resilient communities and businesses. While many initiatives and programs have been implemented to improve disaster preparedness and protect workers, there is still more that can be done. This includes expanding access to training and resources for small businesses, promoting the retrofitting of buildings to improve safety, and facilitating private sector stakeholder consultations to promote knowledge-sharing and coordination activities. It is also important to recognize the unique vulnerabilities of certain groups, such as home-based workers and those in developing countries, and tailor programs and initiatives to meet their specific needs. By taking a comprehensive approach to disaster preparedness and worker safety, we can reduce the impact of disasters on individuals, communities, and economies.

Recovering from a disaster alone can be an intimidating, overwhelming, and financially devastating experience for businesses. However, businesses do not need to act alone. Collaborating and working together can enhance their ability to prepare, respond and recover effectively from disasters.

Resource sharing during a disaster situation is critical as resources like food, water, medical supplies, and power may be in short supply. Businesses that work together can pool their resources to help those in need. Coordination of efforts is another advantage of collaboration, as it can help avoid duplication of efforts or competition for resources. This can ensure that all those affected by the disaster receive the assistance they require.

Resilience building is essential in disaster response and recovery. Collaboration between businesses can help build resilience and reduce the risks of future disasters. They can work together to plan and take long-term actions to prepare for future disasters. Furthermore, collaboration provides businesses with access to expertise that they may not have on their own. For example, working with disaster management experts can help businesses to develop effective disaster response plans.

In Bali, Indonesia, a coalition of hotels has been formed to protect against the threat of tsunamis. The initiative, called the Bali Hotels Association (BHA) Tsunami Awareness Program, was launched in 2014 and includes over 100 member hotels. The program aims to provide education and training to hotel staff, guests, and the local community on how to prepare for and respond to tsunamis.

One of the key features of the program is the installation of early warning systems in hotels, which can alert guests and staff to the threat of a tsunami and provide instructions on what to do in the event of an emergency. The BHA also provides training manuals and conducts regular drills to ensure that hotel staff are prepared to respond in a crisis.

The program has been praised for its proactive approach to disaster preparedness and for its efforts to involve the private sector in DRR. According to a report by the United Nations Office for Disaster Risk Reduction (UNDRR), the BHA Tsunami Awareness Program is a "model for the private sector's involvement in disaster risk reduction in tourism."

The BHA Tsunami Awareness Program in Bali demonstrates the importance of collaboration and proactive measures in DRR. By involving the private sector, the program has been able to leverage resources and expertise to better prepare for and respond to the threat of tsunamis.

Conclusion

If we are not prepared to address the disasters of today, we will be ill-prepared to address those of tomorrow. Indeed, climate change will strengthen weather events and require the adoption of stronger and stronger adaptation measures across all industries. Our global workforce must be protected both at work and at home if we wish to maintain strong economies and protect lives. Fortunately, the world abounds with numerous solutions to strengthen our places of work and enable them to survive and even continue to operate through tough disasters. These solutions must now be adopted more broadly, and policy makers need to put incentives and regulations in place to ensure that big businesses are not the only ones which can afford adaptive measures. SMEs make up the backbone of the global economy and their workforce must be protected against calamities to come. Preparedness, policy, finance, and engineering will all be key in protecting the workforce of tomorrow.

Notes

1 Flash Flood Guidance System with Global Coverage (FFGS) | World Meteorological Organization (wmo.int)
2 Agriculture, forestry, and fishing, value added (% of GDP) – Pakistan | Data (worldbank.org)
3 Untitled 1 (unwomen.org)

References

2022 Monsoon Floods – Situation Report No. 15 (As of 9 March) – Pakistan. (2023, March 9). ReliefWeb. Retrieved from https://reliefweb.int/report/pakistan/pakistan-2022-monsoon-floods-situation-report-no-15-9-march-2023

Bangladesh Center for Workers Solidarity. (2012). Death Trap: The Story of the Triangle Shirtwaist Fire and the Bangladesh Garment Industry. Mimeo.

Bick, A., Blandin, A., and Mertens, K. (2020). Working from home: Estimating worldwide potential. *VoxEU*. Retrieved from https://voxeu.org/article/working-home-estimating-worldwide-potential

CEPR. (2020). Working from home: Estimating the worldwide potential. Retrieved from https://cepr.org/sites/default/files/news/CovidEconomics11.pdf

Fung, J., Butry, D., Sattar, S. , and McCabe, S. (2021). [online], https://doi.org/10.1177/87552930211009522, https://tsapps.nist.gov/publication/get_pdf.cfm?pub_id=928704, (Accessed August 31, 2023). The Total Costs of Seismic Retrofits: State of the Art, Earthquake Spectra floods: What you need to know. (2019, September 4). CNN. Retrieved from https://www.cnn.com/2019/08/08/asia/pakistan-floods-intl-hnk/index.html

Holzer, T. (1995). The 1995 Hanshin-Awaji (Kobe), Japan, Earthquake. *Geological Society of America Today*, 5(8). Retrieved from https://rock.geosociety.org/net/gsatoday/archive/5/8/pdf/i1052-5173-5-8-sci.pdf

HUD. (2021). Stronger Homes, Safer Communities. Strategies for Seismic and Flood Risks. Retrieved from https://www.hudexchange.info/resource/5096/stronger-housing-safer-communities-strategies-for-seismic-and-flood-risks/

ILO. (2013). Improving working conditions in the ready-made garment sector. *International Labour Organization*. Retrieved from https://www.ilo.org/global/topics/geip/WCMS_211237/lang–en/index.htm

ILO. (2018). Safety and health at the heart of the future of work: Building on 100 years of experience. *International Labour Organization*. Retrieved from https://www.ilo.org/wcmsp5/groups/public/—dgreports/—dcomm/documents/publication/wcms_686645.pdf

ILO. (2019a). Women and men in the informal economy: A statistical picture. Retrieved from https://www.ilo.org/global/publications/books/WCMS_626831/lang–en/index.htm

ILO. (2019b). Working on a WARMER planet – The impact of heat stress on labour productivity and decent work. Geneva.

ILO and WHO. (2018). Global estimates of the burden of injury and illness at work, 2016. Retrieved from https://www.ilo.org/wcmsp5/groups/public/—dgreports/—dcomm/documents/publication/wcms_647365.pdf

International Finance Corporation. (2017). SME insurance gaps in developing countries. Retrieved from https://www.ifc.org/wps/wcm/connect/e26c5e63-ade7-4758-b027-12c3820d924c/SME+Insurance+Gaps+in+Developing+Countries.pdf?MOD=AJPERES&CVID=mzRjkLv

Klein, N. (2021, June 10). Extreme Heat Will Change Us. *The New York Times*. https://www.nytimes.com/2021/06/10/opinion/extreme-heat-climate-change.html

Kuhlicke, C., Scolobig, A., and Tapsell, S. et al. (2011). Contextualizing social vulnerability: Findings from case studies across Europe. *Nat Hazards* 58, 789–810. 10.1007/s11069-011-9751-6

Kumar, S. (2014). Cost of Bangladesh factory collapse. *Risk Management*, 61(9), 12.

Mead, L. (2022, May 23). Disaster risk reduction in an unstable world. Retrieved from https://www.iisd.org/articles/disaster-risk-reduction

MLIT. (2019). White Paper on Land, Infrastructure, Transport, and Tourism in Japan, 2019. Ministry of Land, Infrastructure, Transport, and Tourism. Retrieved from 001325161.pdf (mlit.go.jp)

OCHA. (10, March 2023). Pakistan: 2022 Monsoon Floods - Situation Report No. 15 (As of 9 March 2023). Retrieved from Reliefweb: https://reliefweb.int/report/pakistan/pakistan-2022-monsoon-floods-situation-report-no-15-9-march-2023

Ready Business. (2020). Hurricane Toolkit. Retrieved from Ready Business HURRICANE TOOLKIT

SFDBI. (2022). DBI.AnnualReport.2021-2022.pdf (sf.gov) Annual Report. San Francisco Department of Building Inspection. Retrieved from

Sohn, J., Hewings, G. J. D., Kim, T. J., Lee, J. S., and Jang, S. G. (2004). In Okuyama, Y. and Chang, S.E. (eds). Advances in Spatial Science: https://doi.org/10.1007/978-3-540-24787-6_12 Analysis of Economic Impacts of an Earthquake on Transportation Network Modeling Spatial and Economic Impacts of Disasters Springer, Berlin Heidelberg

UNDRR. (2022). Heatwaves: Addressing a sweltering risk in Asia-Pacific (p. 22). Retrieved from https://www.undrr.org/publication/heatwaves-addressing-sweltering-risk-asia-pacific

United Nations. (2023, March 7). UN continues to support Pakistan flood response. Retrieved from United Nations News: https://news.un.org/en/story/2023/03/1134302#:~:text=More%20than%2033%20million%20people%20were%20affected%20overall%2C,acres%20of%20agricultural%20land%20and%202.2%20million%20houses.

WHO and ILO. (2018). *Joint estimates of the work-related burden of disease and injury, 2000-2016: Global monitoring report*. Geneva: World Health Organization.

World Bank. (2016). Building regulation for resilience: Managing risks for safer cities. Retrieved from documents1.worldbank.org/curated/en/326581468337788007/pdf/ACS15966-WP-PUBLIC-BRR-report-002.pdf

Yashinsky, M. and Kwok, K. O. (2017). Earthquake resilience of soft-story woodframe buildings: A review. *Earthquake Spectra*, 33(1_suppl), S67–S84.

17

IMPACT OF COMMUNITY ON CONSTRUCTION PROJECTS

Lessons from South Africa

C.O. Aigbavboa and I.O. Akinradewo

University of Johannesburg, South Africa

Introduction

Construction projects involve complex processes and different internal stakeholders, such as the owner, designer, contractor, and suppliers. However, the involvement of the community, which includes external stakeholders such as neighbours, residents, and other interested parties, is often overlooked in the construction process (Atkin and Skitmore, 2008; Molwus, 2014). The community's impact on construction project site activities cannot be overstated as it can have a significant influence on the project's success (Hussain, Zhu, Ali and Xu, 2017). In recent years, the community has become increasingly involved in construction projects, and their involvement can significantly impact the activities on the construction site (Rathenam and Dabup, 2017). In their study, Rathenam and Dabup (2017) considered the influence and impact of community engagement on the Hammanskraal Pedestrian Bridge project in the City of Tshwane, South Africa. They explored how community engagement, involvement, and collaboration can affect the success or failure of a construction project, particularly a public infrastructure project. Findings from the study revealed that the hawker committee had the highest influence on the project followed closely by the community while local labour was ranked fifth. The study also revealed the factors that influenced delays in the project. Socio-political factors, in the form of strikes and civil unrest by the community, were ranked as major influence, followed by local traders' union interference, lack of support from the local ward councillor, bad public relations practices in dealing with communities, and conflict with local labour on site. All these factors had relative importance index values above 0.574 which indicates that the local community play a significant role in the success or failure of a construction project.

Community involvement can help in identifying potential safety hazards and environmental concerns that can have an impact on the project site activities (Vajjalla et al., 2005). Vajjalla et al. (2005) provided insights into the various effects of construction projects on communities, including social, economic, and environmental impacts. These are: the production of environmental wastes, air pollution, noise pollution, water pollution, traffic disruption, land contamination, and destruction of natural resources. The community's opposition to the construction project can cause delays, threats to workers' lives, delay in

DOI: 10.1201/9781003262671-17

the delivery of materials to site, and, ultimately, budget overruns, and even termination of the project (Rathenam and Dabup, 2017). Construction projects can sometimes be located in densely populated areas, and as a result, the activities on the construction site may have significant impacts on the surrounding community. The level of community involvement and impact can vary depending on the project's site location, size, complexity, and duration (Dartey-Baah, 2022). Therefore, understanding the impact of the community on the construction project site activities and how to manage this impact is essential for project success and the overall performance of the construction industry.

The community's impact on construction project site activities has been a topic of increasing interest globally in recent years (Vajjalla et al., 2005; Celik and Budayan, 2016; Almahmoud and Doloi, 2020; Danku et al., 2020; Budayan and Celik 2021). The construction industry has recognized the importance of the community's involvement in the project, and its impact on the site activities (Barami, Thiruchelvam and Ibrehem, 2019). The community's clamour to be involved in construction projects can be attributed to several factors, including a desire for transparency and accountability in the construction process, concerns about the project's impact on the environment, and the desire to protect the community's interests (Zikargae, Woldearegay and Skjerdal, 2022). Additionally, the community's involvement in construction projects is often a result of regulatory requirements and legal obligations that mandate public participation in certain types of construction projects (Masango, 2001). For instance, in the United States, the National Environmental Policy Act (NEPA) requires federal agencies to involve the public in certain types of construction projects that may have an impact on the environment (Luther, 2008). The project's impact on the community can be either positive or negative, depending on the community's level of involvement and communication with the construction team (Watson et al., 2018). According to Vajjalla et al. (2005), the positive impacts may include increased job opportunities and economic development resulting from the construction project. On the other hand, negative community impacts may include noise, traffic congestion, and dust pollution, which can cause disruptions in the daily activities of the community (Mansour and Aljamil, 2022).

Given the potential positive and negative impacts, it is crucial for the construction team to engage the community in the project activities and manage any potential negative impacts (Agenda, 2016). To achieve this, it is essential to identify the key factors that contribute to the community's impact on the construction project site activities (Dartey-Baah, 2022). These factors may include the level of community involvement in the planning and design stages, the frequency and quality of communication between the construction team and the community, and the mitigation measures put in place to minimize any negative impacts on the community (Zikargae et al., 2022). By identifying these key factors, the construction team can develop effective strategies for managing the community's impact on the construction project site activities, resulting in a successful project outcome. Therefore, it is necessary to identify and analyse the impact of local labour and businesses on construction project site activities to develop effective and efficient methods of engaging the community and mitigating their risks. However, current research on local labour and business management in construction projects is limited, and the existing literature is mainly focused on stakeholder theory and management in construction projects, with minimal attention given to the impact of local labour and businesses (Uribe et al., 2018).

Consequently, there is a gap in the literature concerning the impact of local labour and businesses on construction project site activities, as well as the methods of mitigating their impact. To address this knowledge gap, the current study seeks to examine the impact of local labour and businesses on construction project site activities and provide recommendations for effective community engagement and risk mitigation. Specifically, the chapter explores the positive and negative impacts of the community on construction project site activities and strategies that can be employed to manage the community's impact on construction project site activities. The study adopts a desktop literature review approach.

Overview of stakeholder theory

Stakeholder theory has been widely discussed in the business literature, and it has been adopted by the construction industry in the effort to understand the impact of various stakeholders on construction projects (Collinge, 2012). The theory focuses on the identification and management of stakeholders, which include individuals or groups who have a direct or indirect interest in the project (Benn et al., 2016). The construction industry has used the stakeholder theory to better understand the roles and responsibilities of construction organizations towards various stakeholders involved in projects (Rowlinson, 2004). The theory recognizes that the construction organization must consider the interests of all stakeholders involved in the project, including clients, contractors, subcontractors, employees, suppliers, regulatory agencies, local communities, and the environment (Molwus, 2014), in addition to the regular aim of maximizing profits for the organization and ensuring value for money for the building owners (Baines et al., 2007). The goal of the application of stakeholder theory in construction is to ensure that the needs and perspectives of all stakeholders are considered in decizion making, resulting in more sustainable, equitable, and successful construction projects (Loosemore and Andonakis, 2007).

According to Freeman (1984), stakeholders are individuals or groups who have a stake in an organization and are affected by its actions. Freeman (1984) further proposed the stakeholder theory which asserts that businesses should not only aim to maximize profits for shareholders but also should consider the interests of all stakeholders, including the local community. The construction industry is a stakeholder-intensive sector, and it requires the support of many stakeholders to ensure successful project completion (Benn et al., 2016). In construction projects, external stakeholders, including the local community, can have a significant impact on the project's success (Di Maddaloni and Davis, 2017). External stakeholders are individuals or groups who do not have a direct interest in the project but are affected by its outcome (Akinradewo et al., 2022). The local community is a crucial external stakeholder in construction projects, and their impact can significantly affect the project, depending on how the project is executed (Rathenam and Dabup, 2017).

Over the years, the construction industry has experienced several community-related issues, such as protests, legal action, and forced work disruptions and stoppages, which have resulted in significant project delays and financial losses (Davis and Franks, 2014) and on some occasions, destruction of materials and site equipment. In recent years, stakeholder theory has gained significant attention in the construction industry, and researchers have endeavoured to identify and manage stakeholders to minimize negative impacts on construction projects (Yang, Shen and Ho, 2009). Studies have shown that managing stakeholder relationships can lead to better project outcomes, and effective stakeholder management can help to avoid conflicts, delays, and other issues that can negatively impact project success (Dwivedi and Dwivedi, 2021).

In essence, stakeholder theory is a management philosophy that focuses on the interests of all stakeholders involved in a business, rather than just the interests of shareholders (Valentinov and Chia, 2022). One of the key benefits of stakeholder theory in the construction industry is that it can help to improve project outcomes by reducing conflicts and increasing collaboration (Maier and Aschilean, 2020) between the project owners, contractors and the community most especially. By involving stakeholders in the planning and decision-making process, construction companies and clients can gain a better understanding of their needs and priorities, and hence, work to address these in a way that is mutually beneficial (Davis and Franks, 2014). This can lead to better outcomes for everyone involved, including the construction professionals and project team, the range of stakeholders, and the broader community. Another benefit of stakeholder theory is that it can help to mitigate risk in construction projects. By engaging with the various stakeholders and understanding their concerns, construction companies and clients can identify potential issues early on and work to address them before they become major problems. This can help to prevent delays, cost overruns, and other issues that can impact the project's success (Rathenam and Dabup, 2017).

There are several different approaches to stakeholder theory, each with its own strengths and weaknesses. One of the most well-known approaches is the stakeholder salience model, which suggests that stakeholders are prioritized based on three criteria: power, legitimacy, and urgency (Yacobucci and Jonsson, 2019). This model can be particularly useful in the construction industry, where there are often stakeholders with varying levels of influence and importance. Another approach is the stakeholder value model, which suggests that companies should strive to create value for all stakeholders, rather than just shareholders (Freeman and McVea, 2005). This approach can be particularly useful in the construction industry, where there is often a wide range of stakeholders with differing interests and priorities (Maier and Aschilean, 2020).

In practice, implementing stakeholder theory in the construction industry can be challenging. This is due in part to the complexity of construction projects, which often involve multiple stakeholders with conflicting interests and priorities (Lehtinen et al., 2023). Additionally, there is often a lack of trust between stakeholders, which can make it difficult to engage in a meaningful dialogue and collaboration (Karlsen et al., 2008). However, there are several strategies that can help to overcome these challenges and enable stakeholder theory to be effectively implemented in the construction industry (Collinge, 2012; Yacobucci and Jonsson, 2019; Maier and Aschilean, 2020). These include:

- developing a clear understanding of all stakeholders and their interests and priorities.
- creating a transparent and open communication process that allows for meaningful dialogue between stakeholders.
- establishing clear guidelines and expectations for stakeholder engagement, including how decisions will be made and how conflicts will be resolved.
- providing training and education to all stakeholders to help them better understand the project and their roles and responsibilities.

In summary, stakeholder theory is a management philosophy that has become increasingly important in the construction industry. By focusing on the interests of all stakeholders, rather than just shareholders, construction professionals can improve project outcomes, mitigate risk, and create value for everyone involved.

The role of external stakeholders in construction projects

The construction industry, like other industries, has been faced with the challenges posed by external stakeholders such as the community in which the project is being executed (Molwus, 2014). The role of external stakeholders in construction projects is paramount in ensuring the success of a construction project (San Cristóbal et al., 2018). The project's stakeholders can be classified into primary and secondary stakeholders based on their level of involvement, interest, and influence in a project (Freeman, 1984). Primary stakeholders, such as the community, customers, suppliers, and employees, have a direct impact on the project, while secondary stakeholders, such as government agencies, non-governmental organizations (NGOs), and regulatory organizations, have an indirect impact on the project. Local workers and businesses, who are not directly involved in the project planning but the execution, can affect the project delivery time, quality and cost, and can pose significant risks to the success of the project (Riahi, 2017). Furthermore, as projects have become more complex, the interaction between construction activities and the community has also become more complex, leading to more significant community influence and control (Makhdumi and Taha El Baba, 2017). The impact of the local community and businesses on construction projects is further complicated by the lack of standardization in dealing with these stakeholders, which can result in project failure (Molwus, 2014).

The community is a significant primary external stakeholder that can have impacts on construction project site activities (Frooman, 1999). The community's impact can be both positive and negative, depending on the nature of the project and the level of involvement (Agenda, 2016). Positive community impact may include community support, access to local resources, and employment opportunities. While negative community impact may include protests, litigation, and damage to reputation (Ngwenya, 2015). The construction industry is known for its negative impact on the environment, which has led to several instances of community protests and litigation (Celik and Budayan, 2016). To mitigate this, construction companies are taking initiatives to be more environmentally responsible and are engaging more in environmental and sustainability efforts (Kinnunen et al., 2022). The community's impact on construction projects can also extend to the project schedule and cost (San Cristóbal et al., 2018). Furthermore, the community's interest may not always align with that of the construction professionals, leading to conflicts and project disruptions (Celik and Budayan, 2016). Hence, it is essential for the construction professionals and project team to understand the community's needs and interests and work towards aligning them with their goals for the success of the project, and thus, at the same time, meeting the needs and interests of the community. It is pertinent to note that primary external stakeholders can become internal stakeholders during the construction phase of the project if they are employed to deliver services towards the actualization of the construction project. Hence, it is common to have division among the local community members regarding a construction project (Molwus, 2014).

Community participation in construction projects

The participation of the community is defined as the active involvement of citizens in local or grassroots planning, governance, and development initiatives (Mafukidze, 2009). Public participation has emerged as a crucial element of South African planning and is a recurring topic in various legislative and theoretical papers. For example, the South African Constitution provides a basis for a democratic system that is both representative and participatory (De Villiers, 2010). Also, the Department of Human Settlements (DHA) claims that the South African government aims to create a favourable environment that prioritizes the needs of individuals during the housing settlement provision process (Blaauw et al., 2016). Hence, the establishment of human settlements that provide better housing opportunities and achieve a sustainable housing goal was encouraged (Lategan, 2012). Since the adoption of democracy in South Africa, it has become clear that input from the communities impacted by planning and housing provision is of immense value and the Reconstruction and Development Programme (RDP) underscores the essential role community plays in guaranteeing a democratic process at the grass-roots level (Ludick et al., 2021).

Imparato and Ruster (2003) argue that participation is a crucial element for sustainable development. Therefore, community stakeholders must be involved in decision-making at every phase of a project cycle to ensure a better understanding of development problems and needs (Blaauw et al., 2016). Community organizations, such as civic groups and labour unions, should also be the focus of participation, as they can strengthen democratic participation (Lategan, 2012). The DHA advocates for the mobilization of communities and community-based organizations to enhance engagement with housing programmes as it is essential to raise community awareness about these programmes and provide education (Ludick et al., 2021). According to a gazette of the Department of Public Works (DPW) (CIDB, 2017), community participation should also extend to persons with disabilities to ensure adequate planning and housing solutions that cater for the needs of entire communities. This is because the lack of sufficient community participation may hinder the attainment of self-determination, potentially harming the sustainability of human settlements (CSIR, 2002). The DPW gazette identifies local participation in decision-making as a key indicator of sustainable human settlements (CIDB, 2017).

Skidmore et al. (2005) contend that community participation policies prioritize social capital. This is achieved by allowing communities to engage in governance and service delivery issues to form relationships with public officials and institutions. These relationships can provide access to external financial, political, and supportive resources (Lategan, 2012). According to the DHA, a decentralized approach to participation prioritizes local opportunities and enables broader contributions of skills, labour, creativity, and financial and other resources (Ludick et al., 2021). For successful local economic development to occur, the DPW gazette suggests the participation of the local political sphere, community, and business sector (CIDB, 2017). Hence, Thwala (2008) identifies several objectives of community participation, including empowerment, building beneficiary capacity, improving project effectiveness and efficiency, and sharing project costs. These objectives identify four levels of participation intensity: information sharing, consultation, decision-making, and initiating action (Thwala, 2010). According to Lategan (2012), community participation is generally more effective when the community assumes most of the responsibility, rather than higher-level public agencies attempting to assess consumer preferences through surveys or meetings.

Community participation: Impact on construction projects and the community

Lizarralde (2008) claims that the term 'participation' has been used so frequently that it has lost its meaning in the construction sector as developers use it as a buzzword to gain government approval for their projects. Political and administrative barriers can hinder community-based projects when project goals do not align with authorities, leaving the needs of the community on the back burner (Blaauw et al., 2016). Additionally, public-private partnerships promoted by the government can reduce community participation by transferring large stakes in projects to non-governmental organizations (Lizarralde, 2008). Balancing the interests of all stakeholders will lead to higher rates of success. Both the community and development agencies should receive equal focus. Senyal (2008) argues that focusing solely on community participation and a bottom-up approach could mean that top-down planning, which is a critical institutional mechanism for initiating change, is disregarded. This is because political parties, labour unions, and community enterprises, can facilitate or obstruct development trajectories (Imparato and Ruster, 2003). Participation processes can also be expensive and dependent on political will, time investments, and resources (Morris, 2005). Given the significant demands on governmental resources and the strained nature of delivery, the justification for a commitment to participatory programmes is often a contentious issue. Some argue that funds could be better allocated to more pressing needs such as physical housing and service delivery (De Villiers, 2010).

Also, there are challenges of acceptance of construction material and method alternatives by communities when proposed by project financiers (Ludick et al., 2021). According to the report by the National Department of Human Settlements (2010), in cases where beneficiaries are hesitant to accept alternatives, construction is halted, and cost increases. This requires housing departments to invest in consumer education and intervention strategies to encourage acceptance (Ludick et al., 2021). When new construction materials are to be introduced in low-cost housing, it is crucial to inform and educate the community and its leaders about the advantages and disadvantages of these alternatives (Lategan, 2012). One effective method is to construct prototype or show homes using the new materials and layouts. This allows community members to experience and form opinions about the suggested alternatives. However, the size and finishes of these prototype homes must comply with the final product's quality and size. If misguidance occurs, the participation process will be unsuccessful as it would be based on the previously experienced product (Blaauw et al., 2016). The importance of community participation is further highlighted by empirical research conducted in Rose Valley, Oudtshoorn, which found that community members would only consider alternatives such as higher density living when they understood the benefits (Lategan, 2012). Therefore, it is crucial to ensure that community members understand and agree to the materials and configurations of their homes to promote sustainable human settlements.

The National Department of Human Settlements (2010) highlights the importance of investing in consumer education as the key measure to ensure the successful implementation of alternative building technologies in low-cost housing. For beneficiaries to accept these alternatives, they need to understand the materials, their strengths and weaknesses, and maintenance requirements (Lategan, 2012). Community buy-in should be secured prior to the construction of houses using alternative building technologies to ensure successful implementation. Participation should not only deliver better quality homes but also provide long-term benefits such as economic growth and development. Participation should provide opportunities for skills transfer to take place, empowering communities and individual beneficiaries, as stated in the Housing Code (CIDB, 2017). Low-cost housing projects should, as far as possible, offer the community employment opportunities and a chance to practice the skills learned as a trade. Effective participation and communication between stakeholders can facilitate the transfer of these skills. A change in the perception of the right to housing is also necessary. Due to limited land resources and subsidies, it is impractical to provide every South African in need with a freestanding sizeable home on a separate stand. Convincing communities of this fact should begin with public participation processes and personal interaction with planners, authorities, and political leaders (Lategan, 2012).

CSIR (2002, p. 73) states that the success of community participation varies in different projects, as it can lead to both positive and negative outcomes. While community participation has worked well in the People's Housing Process (PHP) program, it has caused the failure of some water projects and urban service-delivery programs. In South Africa, the PHP program is the primary example of community participation in action, aiming to support community efforts, provide access to subsidies, foster partnerships between government and civil society, facilitate skills transfer and build capacity (Lategan, 2012). Therefore, policy and programme planning should clearly define the level and type of community participation required. However, Mafukidze (2009) warns that community participation can lead to negative outcomes such as mistrust, disillusionment, conflict and fragmentation. To avoid these negative outcomes, community participation should be considered a central component in any development approach and empowering communities should become a key outcome of the process (Morris, 2005).

Local workers and businesses and their influence on construction project site activities

Local labour and businesses, as primary external stakeholders before and after becoming primary internal stakeholders, are critical to the success of a construction project. This section of the chapter discusses the influence of local labour and businesses on construction project site activities. Local labour and businesses are individuals, entities and groups that have a direct interest in the project and are affected by the project's outcomes (Riahi, 2017). They include local residents, businesses, community organizations, and public officials (Renaut et al., 2021). One of the significant positive impacts of local labour and businesses on construction projects is the provision of local resources. Local residents can provide local materials, manpower, and knowledge, which can significantly reduce the project's cost and duration. In addition, local labour and businesses can create a positive project environment through their support and collaboration, which can increase the project's success rate (Molwus, 2014).

Another significant positive impact of local labour and businesses on construction projects is their contribution to the effect on the local economy. When a construction project employs local labour and uses local businesses, it generates economic activity in the local community (Celik and Budayan, 2016). This economic activity includes increased spending on goods and services, increased employment opportunities, and increased tax revenue for the local government. The economic benefits of using local labour and businesses can be significant and can help to build support for the project within the community. Local labour and businesses can also provide the construction project with knowledge and expertise. Local workers often have specialized knowledge and skills that are specific to the local area (Akinradewo et al., 2022). They may be familiar with particular local building regulations, codes and bye-laws, and they may have experience working on similar construction projects in the area. Similarly, local businesses may have a unique understanding of the local market and can provide the project with valuable insights into business practices and access to resources (Zikargae et al., 2022).

On the other hand, local labour and businesses can have a negative impact on construction projects if their interests are not aligned with the project's goals (Lehtinen et al., 2023). Furthermore, the interest of the local labour and businesses in preserving their community can lead to conflicts and project disruptions (Davis and Franks, 2014). For example, local labour may demand higher wages or better working conditions than the construction company is willing to provide. Local businesses may have different expectations about the project timeline or may have concerns about the impact of the project on their operations. Conflicts can lead to delays and increased costs, which can impact the project's success. To mitigate the negative impact of local labour and businesses on construction projects, the construction team should engage with the local labour and businesses (Makhdumi and Taha El Baba, 2017). The engagement should be initiated early in the project life cycle to allow the local labour and businesses to provide their inputs into relevant aspects of the project. This approach can ensure that the project's goals align with the interests of the local labour and businesses (Renaut et al., 2021). Additionally, construction companies can take steps to provide the local labour and businesses with incentives for their participation in the project. Such incentives can include employment opportunities, provision of public amenities, or sponsorship of community projects (Riahi, 2017), thus promoting initiatives of corporate social responsibility to support the local community.

In South Africa, local labourers and businesses have a huge impact on construction project activities. Their participation can have both positive and negative effects on the progress and success of construction projects. For instance, South Africa's building industry, being a major source of employment for workers, can benefit from their familiarity with the local environment, cultural norms, and language, if it hires local labour for construction projects (Oyewobi et al., 2016). When this relationship between local labour and the construction entities is well managed, it results in improved communication, enhanced problem-solving, and good performance on the project (Oyewobi et al., 2016). Also, construction projects can provide local labourers the chance to gain new skills and enhance their old ones (Agumba et al., 2015). This benefits the individuals, and also contributes to the growth of the South African economy by expanding the pool of qualified workers (Oyewobi et al., 2016), considering the limited availability of skilled construction workers in the country. Chileshe and Haupt (2005) noted that when building projects acquire supplies from local firms, they can decrease shipping costs and boost the local economy. Moreover, local materials may be better adapted to the particular environmental conditions of the region, resulting in enhanced physical performance and durability of the items built on the projects (Chileshe and Haupt, 2005). Local labour and businesses can also promote community support and engagement (Ofori 2012). The support of local labour and businesses, when properly managed, can have a favourable impact on the overall progress of a construction project, since community members may be more willing to collaborate and less eager to cause disruptions (Ofori, 2012). Ofori (2012), further informed that local labourers and businesses are more likely to be aware of and concerned about local environmental issues. Recognizing this can result in the use of more ecologically friendly construction processes and the use of sustainable materials, hence decreasing the project's overall environmental effect (Ofori, 2012).

However, there are also potential negative effects linked with the use of local workers and firms in construction projects. Local workers and businesses may have limited resources and capacities, which may result in construction project delays and greater expenses (Chileshe and Haupt, 2005). This can be especially problematic if the project requires specialized equipment or materials that are unavailable locally. Moreover, most times, local workers and businesses lack the essential skills or expertise to meet the building project's demands (Agumba et al., 2015). This may necessitate the use of external labour, which may be more expensive and may hinder the project's performance (Agumba et al., 2015). Also, local workers and businesses may not necessarily adhere to the same quality standards as larger companies or multinational corporations (Chileshe and Haupt, 2005). This might result in substandard construction work, which may subsequently necessitate costly repairs or maintenance to the owners of the built items. Nonetheless, the participation of local labourers and businesses in South African construction projects can have both beneficial and bad effects on the overall performance of the projects (Oyewobi et al., 2016). Good planning, communication, and management can prevent possible problems and ensure that local resources are efficiently utilized, resulting in successful projects that benefit both the local community and the broader economy (Ofori, 2012).

Another primary external stakeholder, mostly from the community, who play a major role in construction projects in South Africa are the "construction mafias" who are also known as "local business forums". Construction mafias are organized groups that engage in illegal activities in the construction industry. They go around construction sites to demand for a stake in the construction project or sums of money, in a form of systemic extortion. These groups can have negative impacts on a construction project in various ways (Flysjö, 2020; van der Heever, 2022). According to the South African Forum of Civil Engineering Contractors (SAFCEC), the country lost R40.7 billion to these construction mafias in 2020. According to Parker (2022), the construction mafias may demand extortionate payments, known as "protection money", from contractors, subcontractors, or suppliers, leading to increased project costs and potential financial losses to the firms. Additionally, the construction mafias can cause project delays by disrupting construction activities, sabotaging work, or interfering with the supply chain of the project, such as deliveries of materials (Jaspers, 2019). Quality and safety concerns may arise as well, as the construction mafias may pressure contractors to use substandard materials or bypass safety regulations to meet their demands (Parker, 2022; van der Heever, 2022). Also, contractual and legal issues may arise, as the construction mafias may engage in corrupt practices to secure contracts or influence decision-making processes related to the project, leading to contractual disputes, legal challenges, and reputational damage (Flysjö, 2020). Moreover, ethical concerns may also arise, as the construction mafias often engage in the exploitation of workers, human rights abuses and environmental violations (Jaspers, 2019). To mitigate these risks, project stakeholders need to implement robust governance mechanisms, promote transparency and accountability, foster a culture of ethical behaviour, and collaborate with law enforcement agencies to combat illegal activities in the construction industry (Parker, 2022; van der Heever, 2022).

Strategies for managing local workers and businesses on construction project sites

Effective management of the members of the local community who are stakeholders in a construction project is critical to the success of the project (Kinnunen et al., 2022). Hence, this section discusses strategies for managing local labours and businesses in construction projects. The first strategy is to identify and engage the local workers and businesses early in the project life cycle. Incorporating the community's needs and expectations into project design is made possible by early engagement, enabling construction professionals and project teams to gather relevant information. Early engagement allows the construction professionals and project team to gather information on the community's needs and expectations and incorporate them into the project design (Maier and Aschilean, 2020). This approach can ensure that the project goals align with the community's interest. Early engagement also creates a positive relationship with the local workers and businesses, which can increase their support for the project (Lehtinen et al., 2023). A practical example of this strategy would be a construction project where a new hospital is being built in a local community. The project team would identify and engage local labourers and businesses early in the project life cycle before construction begins. They would actively involve the local community in the planning process, seeking their input on the hospital's design and gathering information on their needs and expectations. For example, the project team may hold community meetings or workshops to discuss the project and gather feedback from local labourers and businesses. They may also collaborate with local business associations or organizations to ensure representation from local businesses. By engaging with the local labourers and businesses early on, the project team can incorporate their input into the project design, ensuring that the hospital meets the community's needs and aligns with their interests. This early engagement approach can help create a positive relationship between the project team and the local labourers and businesses. It establishes a sense of ownership and involvement among the local community, which can increase their support for the project. For instance, local labourers and businesses may be more willing to provide labour or materials at competitive prices or support the project through local partnerships or sponsorships.

The second strategy is to create a communications plan that outlines the construction professionals' and project team's engagement approach with the local labour and businesses (Riahi, 2017). The communications plan should include the project's goals, timelines, and any potential impacts on the local workers and businesses. The plan should also define the channels and frequency of communication, the stakeholders to be engaged, and the methods of engagement (San Cristóbal et al., 2018). Effective communication can create trust between the construction professionals and project team, and the local workers and businesses and increase their support for the project. For instance, the communications plan may include regular community meetings, newsletters, updates on the project website or via social media, and direct outreach to the local labourers and businesses. The construction professionals and project team would use the communication plan to effectively engage with the local labourers and businesses, providing timely and relevant information about the project. They would actively address any concerns or questions raised by the local community and provide updates on the progress of the project. By maintaining open and transparent communication, the project team can build trust with the local labourers and businesses and demonstrate their commitment to keeping the community informed and involved.

The third strategy is to provide the local workers and businesses with incentives for their participation in the project. Incentives can include employment opportunities, provision of public amenities, or sponsorship of community projects (Makhdumi and Taha El Baba, 2017). Providing incentives can create a sense of ownership and involvement in the project, which can increase the support of the local workers and businesses for the project. Incentives can also create a positive relationship between the construction professionals and project team and the local labour and businesses, which can increase the chances of a project's success (Celik and Budayan, 2016). Another incentive could be to invest in the training and skill development of local labourers which can increase their productivity and enhance the project's overall performance (Agumba et al., 2015). Offering workshops, on-site training, and mentorship programmes can assist in bridging the skills gap and empowering local workers (Oyewobi et al., 2016).

The fourth strategy is to manage the expectations of the local workers and businesses by providing them with accurate information about the project (Frooman, 1999). Managing

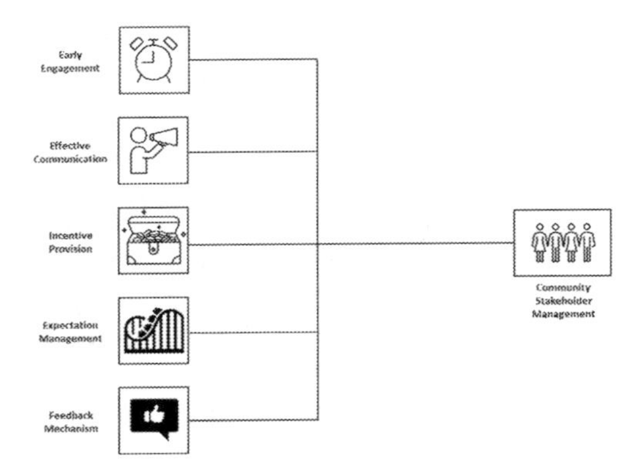

Figure 17.1 Local labour and businesses management framework.

Source: Authors.

expectations involves informing the local workers and businesses of the project's goals, timelines, and potential impacts. Providing accurate information can reduce conflicts and protests by the local workers and businesses, which can increase the project's chances of success (Yacobucci and Jonsson, 2019). Also, expectations can be managed through the implementation of local procurement policies. Implementing local procurement policies that prioritize local suppliers can help create a more sustainable supply chain and stimulate the local economy (Chileshe and Haupt, 2005). Setting targets for local procurement and monitoring compliance can contribute to achieving this goal.

The fifth strategy is to establish a community feedback mechanism that allows the local workers and businesses to provide feedback on the project's progress (Uribe, Ortiz-Marcos, and Uruburu, 2018). The feedback mechanism should define the stakeholders' roles, the methods of feedback, and the frequency of feedback. The mechanism can provide an opportunity for the local workers and businesses to provide input into the project, which can increase their support for the project (Rathenam and Dabup, 2017). It can also create a sense of ownership and involvement in the project, which can increase the likelihood of the project's success (Valentinov and Chia, 2022).

In summary, early engagement, effective communication, provision of incentives, managing of expectations, and establishing of feedback mechanisms are strategies that can be used to manage local labours and businesses in construction projects. These strategies can create a positive relationship between the construction professionals and project team and the local workers and businesses, increase their support for the project, and ultimately increase the project's success rate. Therefore, this is depicted in the framework for local labour and business management developed in Figure 17.1. From the framework, to achieve an effective community stakeholder management, the identified strategies must be implemented from the inception of the project.

Lessons learnt from local labour and businesses on construction project sites

Based on the discussion in this chapter, the following lessons can be inferred. The construction industry is facing challenges from external stakeholders who can greatly impact the chances of success of a project. External stakeholders include individuals or groups outside of the construction industry who can affect, or are affected by, the construction project and can be classified into primary and secondary stakeholders based on their level of involvement, interest, and influence on a project.

One of the most important primary external stakeholders is the community, which can have positive or negative impacts on construction projects. In today's settings, construction professionals are taking the initiative to be more environmentally responsible, and stakeholders are engaging more in environmental and sustainability efforts. However, it is important for construction professionals to understand the community's needs and interests and work towards aligning them with their goals on their projects.

Overall, the construction industry professionals need to recognize the critical role that the external stakeholder, especially the community, plays in the success of a project and be proactive in engaging and collaborating with them to increase the project's success rate.

Conclusion and recommendation

External stakeholders play a crucial role in construction projects, particularly the local labour and businesses who can affect the project's delivery time, quality, cost, and health and safety. Therefore, it is essential to understand their needs and interests, and align them with the project goals through early engagement, and collaboration throughout the project. Moreover, construction professionals can offer incentives to the local labour and businesses, such as employment opportunities or public amenities. This will give the local labour and businesses a sense of engagement in the construction project which will help to reduce negative impact of the community on the construction project. Managing local labour and businesses requires a well-designed and executed strategy that involves identifying and prioritising stakeholders, assessing their level of influence, interest, and power, communicating effectively with them, and engaging in social responsibility and sustainability efforts. The construction industry should establish standards for managing local labour and businesses to minimize their negative impact on construction projects.

Based on the reviews conducted in this study, it is suggested that construction professionals should develop and implement a stakeholder management plan, including a community engagement plan, to identify and prioritize stakeholders, assess their level of influence, interest, and power, and communicate effectively with them. The engagement plan should include a stakeholder analysis, identifying stakeholders' concerns and expectations, providing them with timely and relevant information, and engaging them in meaningful and collaborative activities. Construction companies should also integrate social responsibility and sustainability efforts into their stakeholder management plans, such as reducing the environmental impact of their projects, and supporting community projects. Furthermore, the construction industry should establish standards for managing local labour and businesses to reduce project risks and minimize negative impacts on communities. This can be achieved through collaboration with regulatory bodies and industry associations to develop best practices and guidelines. In summary, effective management of local labour and businesses is critical to the success of construction projects, and construction companies should adopt proactive and collaborative approaches to engage and collaborate with local labour and businesses.

References

Agenda, I. (2016, May). Shaping the future of construction a breakthrough in mindset and technology. In *World Economic Forum*.

Agumba, J. N., Odhiambo, G., & Musonda, I. (2015). Enhancing local participation in construction projects in South Africa. *Journal of Construction Project Management and Innovation*, 5(2), 1191–1205.

Akinradewo, O., Mushatu, W., Mashwama, N., Aigbavboa, C., & Thwala, D. (2022). Learning from Existing Errors: External Stakeholders' Impact on Road Infrastructure Projects. *Human Factors in Management and Leadership*, *55*, 24.

Al Barami, M. A., Thiruchelvam, S., & Ibrehem, A. S. (2019). Factors affecting client's involvement in construction projects. *Test Engineering and Management*, *81*, 3802–3810.

Almahmoud, E., & Doloi, H. K. (2020). Identifying the key factors in construction projects that affect neighbourhood social sustainability. *Facilities*, *38*(11/12), 765–782.

Atkin, B., & Skitmore, M. (2008). Stakeholder management in construction. *Construction Management and Economics*, *26*(6), 549–552.

Baines, T. S., Lightfoot, H. W., Evans, S., Neely, A., Greenough, R., Peppard, J., Roy, R., Shehab, E., Braganza, A., Tiwari, A., Alcock, J. R., J Angus, Bastl, M., Cousens, A., Irving, P., Johnson, M., Kingston, J., Lockett, H., Martinez, V., Michele, P., Tranfield, D., Walton, I., & Wilson, H. (2007). State-of-the-art in product-service systems. *Proceedings of the Institution of Mechanical Engineers, Part B: Journal of Engineering Manufacture*, *221*(10), 1543–1552.

Benn, S., Abratt, R., & O'Leary, B. (2016). Defining and identifying stakeholders: Views from management and stakeholders. *South African journal of business management*, *47*(2), 1–11.

Blaauw, D., De Beer, J., Viljoen, K., & Jarbandhan, D. B. (2016). Low-cost housing finance and delivery challenges in South Africa. *Loyola Journal of Social Sciences*, *30*(2), 169–190.

Budayan, C., & Çelik, T. (2021). Determination of important building construction nuisances in residential areas on neighbouring community. *Teknik Dergi*, *32*(2), 10611–10628.

Celik, T., & Budayan, C. (2016). How the residents are affected from construction operations conducted in residential areas. *Procedia Engineering*, *161*, 394–398.

Chileshe, N., & Haupt, T. C. (2005). Best practices for the implementation of the CIDB"s objectives for local economic development through the construction industry. *Journal of the South African Institution of Civil Engineering*, 47(1), 2–10.

CIDB. (2017). *cidb Best Practice Project Assessment Scheme: Standard for Contract Participation Goals for Targeting Enterprises and labour through Construction Contract Works*. Department of Public Works Gazette.

Collinge, W. H. (2012). Re-thinking stakeholder management in construction: Theory & research. *Project Perspectives*, *34*(1), 16–23.

CSIR. (2002). Sustainability Analysis of Human Settlements in South Africa. http://researchspace.csir.co.za/dspace/bitstream/10204/3522/1/Du%20Plessis_2002.pdf

Danku, J. C., Adjei-Kumi, T., Baiden, B. K., & Agyekum, K. (2020). An exploratory study into social cost considerations in Ghanaian Construction Industry. *Journal of Building Construction and Planning Research*, *8*, 14–29

Dartey-Baah, S. K. (2022). *The relationship between project complexity and project success and the moderating effect of project leadership and roles in the construction industry of an emerging economy* (Doctoral dissertation, Stellenbosch University).

Davis, R., & Franks, D. (2014). Costs of company-community conflict in the extractive sector. *Corporate Social Responsibility Initiative Report*, *66*(1), 6–34.

De Villiers, D. (2010). Stick in the mud. http://www.designindaba.com/article/stickmud

Di Maddaloni, F., & Davis, K. (2017). The influence of local local labour and businesses in mega-projects: Rethinking their inclusiveness to improve project performance. *International Journal of Project Management*, *35*(8), 1537–1556.

Dwivedi, R., & Dwivedi, P. (2021). Role of stakeholders in project success: theoretical background and approach. *International Journal of Finance, Insurance and Risk Management*, *XI* (1), 38–49.

Flysjö, L. (2020). *A review of corruption and organized crime in the construction industry*. (Master's dissertation, Malmo University).

Freeman, R. E., & McVea, J. (2005). A stakeholder approach to strategic management. *The Blackwell handbook of strategic management*, 183–201.

Freeman, R. E. (1984). *Strategic management: A stakeholder approach*. Boston: Pitman.

Frooman, J. (1999). Stakeholder influence strategies. *Academy of Management Review*, *24*(2), 191–205.

Hussain, S., Zhu, F., Ali, Z., & Xu, X. (2017). Rural residents' perception of construction project delays in Pakistan. *Sustainability*, *9*(11), 2108.

Imparato, I. & Ruster, J. (2003). Slum Upgrading and Participation Lessons from Latin America. http://www.schwimmer.ca/francesca/slum%20upgrading.pdf

Jaspers, J. D. (2019). Business cartels and organised crime: Exclusive and inclusive systems of collusion. *Trends in Organized Crime*, *22*, 414–432.

Karlsen, J. T., Græe, K., & Massaoud, M. J. (2008). Building trust in project-stakeholder relationships. *Baltic Journal of Management*, *3*(1), 7–22.

Kinnunen, J., Saunila, M., Ukko, J., & Rantanen, H. (2022). Strategic sustainability in the construction industry: Impacts on sustainability performance and brand. *Journal of Cleaner Production*, *368*, 133063.

Lategan, L. G. (2012). *A study of the current South African housing environment with specific reference to possible alternative approaches to improve living conditions* (Doctoral dissertation, North-West University, Potchefstroom Campus).

Lehtinen, J., Kier, C., Aaltonen, K., & Huemann, M. (2023). A complexity perspective on project stakeholder management. *Research Handbook on Complex Project Organizing*, *243*.

Lizarralde, G., & Massyn, M. (2008). Unexpected negative outcomes of community participation in low-cost housing projects in South Africa. *Habitat international*, *32*(1), 1–14.

Loosemore, M., & Andonakis, N. (2007). Barriers to implementing OHS reforms–The experiences of small subcontractors in the Australian Construction Industry. *International Journal of Project Management*, *25*(6), 579–588.

Ludick, A., Dyason, D., & Fourie, A. (2021). A new affordable housing development and the adjacent housing-market response. *South African Journal of Economic and Management Sciences*, *24*(1), 1–10.

Luther, L. G. (2008). *The national environmental policy act: Background and implementation.* Washington, DC: Congressional Research Service.

Mafukidze, J. K., & Hoosen, F. (2009). Housing shortages in South Africa: A discussion of the after-effects of community participation in housing provision in Diepkloof. In *Urban forum.*, *20*, 379–396. Springer Netherlands.

Maier, A., & Aschilean, I. (2020). Importance of stakeholder management for sustainable construction industry. In *Proc., 36th IBIMA Conference.*

Makhdumi, Z. A. F., & Taha El Baba, A. (2017). Project management approaches in mega construction projects in developing countries: cases from Pakistan.

Mansour, A. I., & Aljamil, H. A. (2022). Investigating the effect of traffic flow on pollution, noise for urban road network. In *IOP Conference Series: Earth and Environmental Science* (Vol. 961, No. 1, p. 012067). IOP Publishing.

Masango, R. (2001). *Public participation in policy-making and implementation with specific reference to the Port Elizabeth Municipality* (Doctoral dissertation, University of South Africa).

Molwus, J. J. (2014). *Stakeholder management in construction projects: a life cycle based framework* (Doctoral dissertation, Heriot-Watt University).

Morris, J. (2005). Removing the barriers to community participation. http://www.feantsa.org/files/freshstart/Working_Groups/Participation/2009/NonFEANTSA_docs/compendium/values_participation/05_Removing_the_barriers_to_commu nity_participation.pdf

National Department of Human Settlements. (2010). The Use of Alternative Technologies in Low Cost Housing Construction: Why The Slow Pace Of Delivery? http://www.dhs.gov.za/Content/Publications/Human%20Settlements%20Review%20Chapters/Human%20Settlements%20Review%20The%20use%20of%20alternative.pdf

Ngwenya, N. (Ed.). (2015). *Community Protest: Local Government Perceptions*. SALGA Publications.

Ofori, G. (2012). Developing the construction industry in Ghana: The case for a central agency. *Journal of Construction in Developing Countries*, *17*(1), 15–28.

Oyewobi, L. O., Windapo, A. O., & Jimoh, J. A. (2016). Evaluating construction project performance: A case of construction SMEs in South Africa. *International Journal of Construction Management*, *16*(2), 177–187.

Parker, D. (2022). *Failure to tackle the construction mafia will have dire consequences, report warns.* Accessed on 06 April 2023 https://www.engineeringnews.co.za/article/failure-to-tackle-the-construction-mafia-will-have-dire-consequences-report-warns-2022-06-20

Rathenam, B. D. C., & Dabup, N. L. (2017). Impact of Community Engagement on Public Construction Projects--Case Study of Hammanskraal Pedestrian Bridge, City of Tshwane, South Africa. *Universal journal of management*, *5*(9), 418–428.

Renaut, F., Comín, F., Armas, I., De Maeyer, P. De Villiers, J. , & Thijsen, G. (2021). Nature-based solutions for natural hazards and climate change. *Frontiers in Environmental Science*, 9, 678367.

Riahi, Y. (2017). Project stakeholders: Analysis and management processes. *SSRG International Journal of Economics and Management Studies*, *4*(3), 37–42.

Rowlinson, S. (2004). *Construction safety management systems.* Routledge.

San Cristóbal, J. R., Carral, L., Diaz, E., Fraguela, J. A., & Iglesias, G. (2018). Complexity and project management: A general overview. *Complexity, 2018.*

Senyal, M. (2008). Community participation in housing delivery in South Africa: A critical analysis. Urban Forum, *19*(3), 223–242.

Skidmore, P., Bound, K. & Lownsbrough, H. (2005). Community participation, Who benefits? http://www.jrf.org.uk/system/files/1802-community-network-governance.pdf

Thwala, W. D. (2008). Employment creation through public works programmes and projects in South Africa: Experiences and potentials. *Acta Commercii*, *8*(1), 103–112.

Thwala, W. D. (2010). Community participation is a necessity for project success: A case study of rural water supply project in Jeppes Reefs, South Africa. *African Journal of Agricultural Research*, *5*(10), 970–979.

Uribe, D. F., Ortiz-Marcos, I., & Uruburu, Á. (2018). What is going on with stakeholder theory in project management literature? A symbiotic relationship for sustainability. *Sustainability*, *10*(4), 1300.

Vajjalla, M., Koehn, E., Kumar, K., & Ravikanth (2005). Effects of construction on communities. *Proceedings of the 2005 ASEE Gulf- Southwest Annual Conference Texas A &M University-Corpus Christi*, 1–12.

Valentinov, V., & Chia, R. (2022). Stakeholder theory: A process-ontological perspective. *Business Ethics, the Environment & Responsibility*, *31*(3), 762–776.

Van der Heever, B. (2022). *A thorough report on the construction mafia in South Africa.* Accessed on 06 April 2023 https://www.asaqs.co.za/news/609452/A-thorough-report-on-the-construction-mafia-in-South-Africa.htm

Watson, R., Wilson, H. N., Smart, P., & Macdonald, E. K. (2018). Harnessing difference: a capability-based framework for stakeholder engagement in environmental innovation. *Journal of Product Innovation Management*, *35*(2), 254–279.

Yacobucci, I., & Jonsson, N. (2019). A more sustainable society through stakeholder salience: Furthering stakeholder theory by exploring identification and prioritization processes with a focus on intraorganizational perceptions in an SME (Masters dissertation, Malmo University).

Yang, J., Shen, Q., & Ho, M. (2009). An overview of previous studies in stakeholder management and its implications for the construction industry. *Journal of facilities management.*

Zikargae, M. H., Woldearegay, A. G., & Skjerdal, T. (2022). Assessing the roles of stakeholders in community projects on environmental security and livelihood of impoverished rural society: A nongovernmental organization implementation strategy in focus. *Heliyon*, *8*(10), e10987.

18

LABOUR, THE BUILT ENVIRONMENT AND HUMAN SETTLEMENTS

Lessons from the book

George Ofori and Edmundo Werna
LONDON SOUTH BANK UNIVERSITY

Introduction

The chapter

This chapter ends the book by providing a summary of what has been discussed, drawing linkages among the contents of the earlier chapters in the book, and considering the future development of the field of knowledge of the potential subject: labour, the built environment and human settlements. First, the intentions behind the production of the book, in the form of the aims and objectives, are revisited. Second, after a brief analysis of the contents of each of the chapters, focusing on the conclusions, its links with the other chapters in the book most closely related to it are discussed. The contents of each of the chapters are placed in the context of the book, and how it contributes to the body of knowledge of the subject is outlined. Finally, the extent to which the book succeeds in the building of that body of knowledge is discussed.

Revisiting the intentions

Considering the aim and objectives of the book

The aim of the book was to present a state-of-the-art review of the complex and dynamic relationship between labour and the built environment and indicate possible new directions for the subject, also including human settlements which the built environment creates. The objectives of the book were to:

1 present a comprehensive review of the literature on the main topics of the relationship between labour and the built environment, focussing on construction, the largest segment of the sector which produces the built environment;
2 explore the fundamental issues and underpinnings of the subject such as the critical nature of the built environment, prevailing and recurrent deficits of labour, changes in technology, employment relations, and conditions of work and of living; and

DOI: 10.1201/9781003262671-18

3 propose new topics and sub-topics to be explored with the view to enabling the subject to grow.

It is suggested here that, from the discussions in the chapters, it is evident that the aim and objectives of the book were met. This view is tested in the next section, where key points of the chapters are presented and analysed, with emphasis on the conclusions, supplemented with consideration of some emerging issues to contribute to developing the field. Particular attention is paid to drawing out the links among the chapters in order to show that the field is a cohesive whole.

Analysis of the chapters of the book

In Chapter 1, there is a section which weaves the chapters together, going from one chapter to the next. My objective was to introduce the chapters and show where each one is going in relation to the book. The present chapter also attempts to weave the chapters together, but there is a great difference between what was done in Chapter 1 and what is being done in this chapter. Here the intention is to show where each chapter arrived and to analyse the conclusions and recommendations.

After the introduction of the book (Chapter 1), which set out the whole scene, Chapter 2 is also broad in scope, as it links the subject of the book to overall national socioeconomic development. It argues that, considering the strategic importance of construction labour, it should be continually developed and its performance improved in order to enable construction activity and the built environment it creates to contribute to the optimal extent to national development and have the beneficial impacts that it has the potential to deliver, to the economy in general. The chapter notes the need to make construction more attractive to workers and, at the same time, to keep pace with the increasingly fast changes in technology.

Following Chapter 2, which gives a broad view of development, the subjects of these two parts of the book, which are noted in Chapter 1, are covered: (i) the workers who produce the built environment – the situation of, and problems, and possible solutions related to employment and working conditions of construction workers; and (ii) how the built environment impacts workers in all sectors of the economy. These two parts of the book are now considered.

Part I: Workers of the built environment

Part I of the book is on the workers of the built environment. It starts with Chapter 3, which provides a broad analysis of the construction industry and its labour and, as such, serves as an umbrella to Part I. It argues that the nature of construction and how the industry responds to it have major implications for workers, both in the construction industry and other sectors of the economy. The chapter suggests that actions are required in developing construction labour to enable labour in other sectors of the economy and society as a whole to derive optimum benefit from the economic and social potential of construction. The construction industry, its processes and products have specific characteristics which, together, make its products and processes unique. They have to be taken into account in developing policies which will be successful in leading to change.

Both Chapters 2 and 3 mention the employment generation potential of construction – without which there is no work, and which can be considered to enhance the possible number of jobs from each project. This is taken up in Chapter 4, which concentrates on this topic. The chapter analyses several challenges to create full employment and argues that government intervention is necessary, giving examples of public employment guarantee programmes. This is an important conclusion because it challenges the common view that the fully fledged free market economy by itself has led to full employment. Hence, public intervention is necessary, although the market also has its role to play. Chapter 4 also brings to the fore the urban dimension (human settlements), which is another key aspect of the book and is particularly considered in Part II.

Workers need training, which is dealt with in Chapters 5 and 6. Chapter 5 analyses training in the context of low-energy construction (LEC). It divides training into two broad systems applied in the countries whose programmes are analysed. It concludes by pointing out the pros and cons of each system, showing that, in any set of circumstances, the inclusion of climate literacy in professional training curricula represents an important means to empower construction workers, providing both motivation for building a zero-carbon economy and a threat to the status quo. Above all, achieving equity remains a critical issue in construction and should be given due consideration by construction unions if they are to be a positive force in the transition to a green economy. Finally, the chapter notes that valuing labour is key to valuing the environment and combating climate change. Chapter 5 considers new issues, including labour agency, equity and training specifically focused on LEC. The latter resonates the argument about technological changes, a subject noted in Chapter 2 (although that chapter did not necessarily cover green technology). The points about equity and the status quo interface with the argument of Chapter 3 about labour problems being related to the present nature of the construction industry (which is the status quo). Therefore, from this point in the book, the different chapters start to interface with each other – while at the same time keeping their respective individuality.

Chapter 6 follows up the discussion in Chapter 5 by also considering an approach to green training in occupations related to the built environment. A difference is that, while Chapter 5 focuses on unionised workers and top vocational education and training, often at the national level, Chapter 6 approaches training and jobs provided by local actors for vulnerable/low-income populations. Like in Chapter 5, the issue of equity surfaces (here, with a focus on climate justice). From a review of the literature, Chapter 6 draws the conclusion that environmental conservation may take place without social development – or worse, it can be used as a smokescreen to cover the lack of social policies. This underlines the importance of climate justice. Programmes of green job training for vulnerable groups constitute an important set of actions to increase equity.

Equity also depends on how decently workers are paid. Chapter 7 unveils the existence of major problems in this respect – not only delays in payment and underpayment but also non-payment of workers. It demonstrates the magnitude of this set of problems with examples from several countries in different parts of the world. These specific problems, such as others noted in some of the previous chapters, are here also linked to the status quo (in terms of the business models used in the industry). Governments in all of the countries examined (with the exception of India) see wage abuse as a challenge they must address. The chapter suggests a set of actions: institute wage protection systems that monitor and detect non-payment in real time and provide a record of wages paid; develop separate, ring-fenced project accounts for wages; improve the payment of project funds down the subcontracting chain with the right to withdraw labour until paid; and shift the liability for wages up the subcontracting chain to the users of the labour and make the main contractor or client liable. If these actions are taken in combination and accompanied by an active labour inspectorate, there may be a chance of deterring and possibly even preventing wage abuse before it happens.

Chapter 7 notes that migrant workers are a specific target group (although not the only one) for wage theft, due to the fragile circumstances of many migrants. The discussion in that chapter interfaces with that in Chapters 8 and 10, the latter being specifically on that group of workers.

Wage theft has an impact on the quality of life of the workers. Many times, it also affects their working conditions. Chapter 8 deals with working conditions, considering many topics, including inappropriate conditions on site, occupational safety and health, child labour, plus different groups of workers, in a discussion which interfaces with that in many chapters. It also deals with implications of technological change, interfacing with Chapters 2, 5 and 6. It uses Guy Standing's concept of the precariat (precarious proletariat) as an umbrella concept. The chapter proposes policy options to governments, employers and workers. These include labour-market reforms, the promotion of equal access to decent employment for women and youth, inclusion of social clauses in public procurement, development of employment programmes and passing due diligence regulations.

Many of the chapters mentioned earlier highlight the need for engagement of employers (construction companies) and changes to be made. Chapters 9 and 10 deal with organisational issues from different approaches: human resource management (HRM) and culture in construction, respectively. Chapter 9 demonstrates how construction has wrongly used HRM methods applied in other sectors of the economy and that this has not worked well, due to the many specificities of the construction industry. It examined how contemporary HRM has changed in response to the above. HRM is moving away from antiquated and traditional practices and towards more innovative, technologically advanced and effective methods of operation. There is still space for improvement, especially in emerging nations, where it is important to promote a culture and environment that encourages technical advancement. To assist in building practical and useful HRM solutions for construction businesses, it is necessary to increase the capacity and competencies of construction firms. Construction companies in developing nations might also be able to grow through several technological phases and catch up with the developed nations as a result of technological breakthroughs.

Chapter 10, in turn, focuses on the impact of differences in business cultures, making the management of construction processes challenging. It highlights the organisation-related aspects from a labour viewpoint and within an international scope. Based on lessons from the past, the chapter concludes that this is an increasingly important and challenging outlook for today's construction industry and that of the future, and the processes therein, with their increasingly globalising characteristics, leading to a growing mix of multi-cultural processes and projects, creating an increasingly challenging business environment and labour contexts. Thus, there is a need for people with increased awareness, learning the competencies and skills needed to handle the challenges. In sum, the multi-cultural dimension of construction – with workers of different origins – needs to be taken into consideration for the benefit of all.

The discussions on work quality throughout Part I frequently draw attention to women and migrants as groups of workers with specific challenges in the construction industry. Chapter 11 addresses the first group. It has strong interfaces with Chapters 2, 3 and 8 because of the challenges of decent work for women, and an interface with Chapter 10 because the challenges faced by women in construction have a strong cultural dimension. There are also other linkages with other chapters, including some in Part II.

Chapter 11 argues that the pace of change concerning gender distribution in the construction industry is slow. This particularly applies to skilled trades. The chapter provides evidence of barriers that women tend to face in construction, including biased recruitment and promotion processes, sexual harassment and inequalities in the allocation of work tasks based on perceptions of traditional gender roles and differences in geographical locations. Also, it demonstrates that construction workplaces present a masculine space where women are described as not belonging. The theme emphasises that construction organisations are continuously shaped by gendered processes and practices. Masculine identities are described as being closely linked to organisational processes and a macho culture. The chapter discusses how masculinities are shaped. The bulk of the literature on gender in construction tends to focus mainly (or only) on specific problems faced by women. Thus, the analysis of masculinity in Chapter 11 is different from the norm. Finally, the chapter shows that comparing gender analyses concerning the construction industry with those of other male-dominated industries reveals that there are several similarities. However, other industries, such as mining and forestry, seem to have moved from physically demanding work tasks towards new forms of work, but these tendencies are not reported in construction. Moreover, while the problems faced by women have been exhaustively studied, it should be noted that working conditions in construction are described to be harsh for men as well. This challenges the argument that construction is not easy for female workers as opposed to male workers. There is a great need for change if female and male workers are to flourish in gender-equal construction workplaces.

Chapter 12 provides a historical account of migration and argues that abuse has taken place throughout the history of humankind. The chapter argues that mistreatment of construction migrants is likely to persist. Structural changes in the modern-day construction industry, reinforced by discriminatory legislation and judicial systems of host countries, have further entrenched the undesirable mistreatment of migrants. The chapter suggests that construction labour discrimination can only be overcome through a combination of short-, medium- and long-term policies. Moreover, all relevant international organisations and non-government organisations should collaborate in the search for effective solutions, along the same lines as their combined investigation of labour abuse in Qatar which culminated in the first inter-governmental agreement to address all aspects of international migration comprehensively under the Global Compact for Safe, Orderly and Regular Migration which was endorsed by the United Nations General Assembly in 2018.

Part II: Impact of the built environment on workers

The book then goes into its Part II and considers the impact of the built environment on workers. Chapters 13 and 14 are about the development of human settlements, taking labour into account. Chapter 13 suggests that research and policy making regarding the linkages between territorial development, livelihoods and labour must be revisited considering the increasing influence of what is called "platform capitalism". In particular, technological advances, as reflected in the rise of cloud computing, the increased capacity to collect, process and use big data, and the changing roles of corporate providers of information and communication technology (ICT) have combined to significantly change the contemporary production of urban space that transcends the usual considerations of smart or intelligent cities.

Cities and metropolitan areas form the backbone of the emerging platform economy. They concentrate the economies of agglomeration, driven by the variety of labour, presence of specialised supplier firms and the large-scale, rapid circulation of data and information, which represents the prime raw material for corporate platforms. The rise of platform urbanism is a great transformation in the role of information technology in the planning, management and finance of cities. The chapter describes the more optimistic aspects of platform urbanism structured around the sharing economy, where a common infrastructure is being allocated efficiently over an increasing number of networker users. It stressed dimensions of the supply and demand side of platform capitalism that will not necessarily benefit cities and urban workers and require further empirical research.

The first among the possible research topics is the consideration of alternative explanations for the existence and profitable utilisation of rent gaps as the driver behind the transformation of the built environment through rental housing and platforms. Second, the chapter challenges the argument that finance capital and real estate platforms are penetrating the market for informal housing in low-income settlements, explaining that there are barriers and the process is not so straightforward. A third theme is that platforms have become attractive options for transnational investment elites. In countries of the Global South, there should be an investigation regarding the role of national investment elites, considering the drastic inequalities in income and wealth. Fourth, empirically based work on how real estate platforms and finance emerge as new builders and operators of cities in the Global South will be a useful research topic. It should be studied with new theories on how knowledge regarding geographical and historical contexts can inform research on how the penetration of digital technologies and big data, in combination with finance, is able to transform the built environment on a global scale. Finally, the illustration of the impacts of real estate platforms on urban space and the affordability of housing represents one of the possible threats that is posed by the platform economy to the right to the city. Platforms have emerged as the privileged device to extract rent from all. In sum, Chapter 13 raises red flags about the sophisticated development of the aforementioned platforms in cities, with consequences for the urban workers. It may perpetuate inequalities. This should remind one of the discussions in the chapters in Part I about (in)equity and the status quo. There, focus was mainly on the built environment; Chapter 13 considered it at the level of human settlements.

Chapter 14 seems to emerge from the broad discussion of Chapter 13, although it was written separately by different authors. The chapter focuses on the relationships between housing markets and labour markets. The two markets are strongly connected. Therefore, policies on each of these markets need to be linked for mutual benefit. The chapter demonstrates that macroeconomic analyses have presupposed the existence of one housing market and one labour market in any given urban area and sought to examine how they interact. In fact, there are several markets of each of them (housing and labour) in the same city. Formulating policies on the basis of such a narrow view is limited. Chapter 14 in this book argues that, while the respective literature on housing markets and labour markets is extensive, the collection of works which link both of them is, somehow, incipient and, at the same time, oriented to the Global North. The chapter highlights key aspects of such combined literature and suggests areas for further research on topics such as the pluralism of both markets, housing as an asset in the labour market, the economics of housing construction, and the relationship between housing and workers' well-being. The chapter introduces innovation and adds value and by analysing the housing-labour market nexus through the lens of their respective complexities and by bringing developing countries into the forefront.

Human settlements are formed not only of buildings but also of open spaces. Chapter 15 considers the phenomenon where, in many cities in the Global South, over a long period, many workers resort to working on the streets and other public spaces. Due to many factors including lack of formal employment and financial difficulties relating to having a proper workplace, many self-employed workers tend to take two different courses of action. Some took to working on the streets and public open spaces; others took the course of working at home. Thus, while Chapter 15 concentrates on working in public spaces, it also provides a brief analysis of home-based enterprises. The chapter analyses in-depth the problems faced by those who work in public spaces, including frictions with street users and the authorities. It argues that the planning and design of public spaces should be done with the participation of the people who use them to ensure that the design or rearrangement of workplaces sets aside sufficient and appropriate space for other public space users, for example, through the provision and management of facilities such as bathrooms, toilets, waste collection arrangements, lighting, and drainage to support the use of public spaces in the variety of ways that is currently observable and the diversity of which is likely to increase. The chapter states that work in public spaces is not only traditional, but it is here to stay. Therefore, there is need for a policy shift to enable the formalisation of the use of public spaces as working spaces, rather than fight the tendency to use them in this way. Moreover, taxation mechanisms could be introduced as a way of regulating work in public space as well as generating revenue for city governments.

Chapter 16 focuses on the resilience and security of workplaces. Workers are subject to the risks of the built environment in which they operate, and these risks are particularly evident in developing countries. Resilient workplaces are crucial for protecting the lives of workers and also for maintaining economic stability in the face of disasters. The lack of building codes and standards to inform urban land consumption, planning and construction and the lack of enforcement of codes and quality control to ensure that the construction of buildings prioritises safety and resilience are factors contributing to the vulnerability of workplaces to disasters. Globalisation can also contribute to the lack of appropriate regulations and enforcement, as governments may prioritise attracting businesses and creating a favourable business environment over protecting workers' rights and safety. As the world continues to grapple with the impacts of climate change, the safety of the built environment in general and workplaces in particular is becoming an increasingly urgent concern. Chapter 16 argues that if the world is not prepared to address the workplace disasters of today, it will be ill-prepared to address those of tomorrow. The workforce must be protected both at work and at home to safeguard lives and incomes and maintain strong economies in their countries. Examples of systemic failures in protection from disasters demonstrate the importance of strong regulatory frameworks and enforcement of safety regulations to safeguard worker safety. They also show that disaster insurance, structural retrofits, comprehensive building codes, and the development of business continuity plans can help communities and local economies to withstand and recover from disasters. It suggests that a coordinated and collaborative approach involving local communities, businesses and government agencies is necessary to develop and invest in comprehensive disaster management plans that prioritise public safety and resilience while reducing collateral economic losses. The chapter calls for good practices and recommends that such solutions must now be adopted more broadly. Policy makers should put incentives and regulations in place to ensure that all businesses can afford adaptive measures, including the small- and medium-sized enterprises whose workforce must also be protected.

Protecting workers and ensuring disaster preparedness are critical to building resilient communities and businesses. Additional measures which can be implemented include expanding access to training and resources for small businesses, promoting the retrofitting of buildings to improve safety and facilitating consultations among private-sector stakeholders to promote knowledge sharing and collaboration, and co-ordinating such cooperative activities. It is also important to recognise the vulnerabilities of certain groups, such as home-based workers and those in developing countries, and tailor programmes and initiatives to meet their specific needs. On their part, businesses can collaborate to enhance their ability to prepare for, respond to and recover effectively from disasters. Co-ordination of their efforts and sharing of resources can help to avoid duplication of efforts or competition for resources and ensure that all those affected by the disaster receive the assistance they require. Collaboration among businesses to plan and take long-term actions to prepare for future disasters can help build resilience and reduce the risks of such disasters. There are many examples of good practice in disaster risk reduction, preparation and response by either governments, public-private collaboration or the private sector. The solutions required to strengthen places of work and enable workers to survive and even continue to operate through disasters should be adopted more broadly, with regulations, policies and incentives to ensure that all businesses, especially the more numerous SMEs, can afford the adaptive measures. Preparedness, policy, finance and engineering will all be key in protecting the workforce of tomorrow.

It is worth recapping that Chapters 5 and 6 analyse the training of construction workers, with a particular focus on environmental protection. Good training is fundamental for the delivery of good-quality products by the workforce (in this case, buildings). Poor practices of construction companies have also been discussed under Part I (for example, in Chapters 2, 3 and 7). Many buildings turn out to be substandard with respect to quality due to the low level of competence of the builders. Chapter 15 includes a short analysis of home-based workers; thus, it interfaces with Chapter 16 in these regards.

Chapter 17 explores the impact of the actions of external stakeholders (the local labour, businesses and community) on work on construction project sites. These actions can affect the project's delivery time, cost, quality and health and safety performance. Therefore, it is essential to understand their needs and interests and align them with the project goals through early engagement and collaboration with them throughout the project. Construction professionals can offer incentives to the local labour and businesses, such as by providing them with employment and business opportunities. Notably, managing local labour and businesses requires a well-designed and executed strategy that involves identifying and prioritising all the stakeholders of the project; assessing their level of influence, interest, and power; communicating effectively with them; and engaging in social responsibility and sustainability efforts. Therefore, the construction industry should establish standards for managing local labour and businesses to minimise their negative impact on construction projects. Construction companies should also integrate social responsibility and sustainability efforts into their stakeholder management plans, such as taking measures to reduce the project's environmental impact, and supporting projects for the community. Furthermore, the chapter recommended that the construction industry should establish standards for managing local labour and businesses to reduce project risks and minimise negative impacts of the works on the site on communities. This can be achieved through collaboration with regulatory bodies and industry associations to develop best practices and guidelines. Therefore, Chapter 17 considers the construction site and nearby communities together. This way, it links construction with human settlements – with a labour perspective – although the coverage is limited to the nearby milieu.

Summary of analysis of chapters

The above analysis shows the contribution of each chapter and the several linkages among the chapters. It can be stated that a start has been made in this book to study the subject of labour, the built environment and human settlements together, and a contribution to making it a distinct area of research has been made. Many suggestions of topics for new research to broaden the latter are embedded in the analysis of the chapters.

By having separate authors for each chapter, it was expected that the book would bring value as a set of independent contributions, each one dealing with an important issue of the labour-built environment-human settlement nexus. This has been accomplished. In addition, the analysis in this section shows many interfaces among chapters, which demonstrate that they are more than a fragmented set of contributions.

There are some broad conclusions, such as the point that the book brings to the fore issues which have not always been considered together. The main example is the 'two sides' of the relationship between labour and the built environment: (i) the workers who produce the built environment; and (ii) how the built environment, particularly the products of construction, impact on workers. As the built environment is the founding element of human settlements, the book also includes inferences to such settlements. One should also note that Part II is about all urban workers, and this also includes construction workers, i.e., the very workers who produce the built environment are also affected by it. Another example of bringing together separate issues is decent work (in terms of work quality) and company organisation.

Concluding remarks

As the agents of production, workers should have pride of place in the discussion on the contribution of the built environment and the construction industry which creates it to national socio-economic development. Every nation should act to ensure that it has a workforce which is adequate in quality and quantity to meet its physical needs. This calls for the institution of an appropriate regime of training, effective deployment, continuous development, and appropriate working conditions, employment relations and overall welfare. This will also mean ensuring decent jobs, rewarding careers and the possibility of construction workers being effective players in the market for products of the built environment sector.

Considering the significant proportions of the total workforce of every country constituted by construction workers and the forward and backward linkages, plus indirect and induced jobs and the families of workers, in effect, a sizeable number of people in every country depend on jobs in, or related to, construction. However, considering the project-based nature of construction, its jobs are inherently unstable. The industry typically goes through periods of boom, possible overbuilding and declines in activity. Moreover, changes in employment relations in many countries have weakened the bargaining position of workers and led to the worsening of the conditions in which they work. The embarking of the construction industry on technological change as it seeks to shed its image as a laggard in this regard and boost productivity, adopting the principles and advances being made available by the Fourth Industrial Revolution, is likely to lead to a 'migration' of jobs from construction to other sectors, such as manufacturing and ICT and also among countries, and an increase in mechanisation (including automation) might mean that many construction workers lose their jobs. The book sought to analyse the present status of, and the future prospects for, employment in construction, and to make appropriate recommendations for necessary interventions.

It was shown in most of the chapters in the book (Chapters 2, 3, 4, 7, 8, 9 and 12) that many construction workers do not have decent jobs. There are significant shortfalls in these regards in many countries, and occurrences such as low wages, delays in payment of wages or non-payment; poor conditions of work including unsatisfactory safety and health, long and often unsocial working hours; discrimination on various grounds such as gender and nationality, and especially in countries where construction has a poor social image, effective exclusion and mistreatment and abuse of rights, especially of foreign workers (Chapters 7, 8 and 12). A new term coined to explain an emerging class of people facing insecurity, moving in and out of precarious work that gives little meaning to their lives, the 'precariat' aptly describes the construction workers around the world (and was applied to construction labour in Chapter 8). The construction workers are better- or worse-off depending on the characteristics of the construction process, as discussed in Chapter 3.

At the same time, workers in other sectors of the economy have strong needs related to the elements of the built environment produced by construction (Chapters 13–16). The quality of such facilities influences their work and all other aspects of their lives and can influence their health and security in the premises. The cost, speed and quality with which the built items are completed influence activities in all sectors of the economy.

It was noted in Chapter 1 that the prospects for the reduction of poverty which is the main task under the current global development agenda are linked to the creation of good-quality employment. The reduction of poverty needs elements of the built environment, such as workplace facilities, other items of infrastructure and other associated elements, to be in place. These points were confirmed in the discussion in the book. Thus, it is necessary to consider labour in the built environment and human settlements in a strategic manner because synergies would result from looking at the inter-related issues together. This would enable effective policies and initiatives to be developed. The book made such a consideration possible, as it presents chapters which, together, constitute a critical examination of the complex issues in the three fields and the relationships among them and showcase the exploration of some new issues and under-explored ones.

The discussion in this book has shown that there is a bona fide field of knowledge which can be called Labour, the Built Environment and Human Settlements. The extension of the title: "The Built Environment at Work", which has a loaded and deep meaning, is an appropriate expression. It is necessary to build the body of knowledge for this field. This book has set a foundation on which further development of the field can be built. The development of this field should be of interest to the authorities in cities, districts and neighbourhoods. It will be of direct benefit to them in their work in planning, controlling the provision of infrastructure and buildings, and managing the built environment to optimise both economic and social benefits.

INDEX

Abuse: of labour rights 116, 130; of substance 162, 170; of alcohol 162, 170
Accessibility 45, 53, 56
Accident 41, 138, 140–2, 151, 153, 154, 285, 293
Accountability 298, 306
Adaptation 284, 291, 293, 294, 295
Affordability 56, 318, 319, 320
Agriculture 125, 126, 192, 285
Apprenticeship 12, 27, 77, 81, 84, 85, 86, 87, 88, 90, 91, 93, 152

Balance wheel of the economy 51
Beneficiary 48
Biodiversity 56, 100
Bond 124, 126
Build-Up skillls 79, 90
Buildability 53
Building 101, 103–6, 142, 144, 146–8, 150, 151, 230, 232, 233, 236, 239; code 275, 276, 277, 278, 279, 281, 282, 287; control 53, 240; envelope 79, 83, 92; information modelling (BIM) 29, 150; materials 105, 106, 139, 149, 153; regulations 52, 101, 105, 237, 282; standards 89
Building it Green 77
Built environment 1, 2, 3, 4, 5, 6, 7, 12, 13, 14, 15, 17, 26, 56, 97–100, 103–109, 137, 138, 141, 143, 149, 151, 154, 177, 224, 225, 235, 238–40, 263, 264, 275, 276, 277, 282, 287, 313, 314, 314, 315, 317, 318, 319, 321, 322; cluster 15
Business continuity plan 286, 288, 289, 290

Capacity 47, 48, 53, 153, 224–6, 237
Capital 78; access to 289; intensive 3
Capitalism 78
Capitalist 267
Carbon: embodied 76; emissions 76, 77, 79, 84, 91, 92
Career 102, 108, 146, 195, 196, 197, 200, 201, 202, 204; progression 196, 200, 201, 204
Casual, Casualisation of, labour 3, 24, 46, 140, 144, 154
Central Business District 269, 272
Child labour 148, 149, 152
City 275, 281, 282, 285, 286, 287, 291, 292, 317, 318, 319, 322; authority 263, 265, 270, 271, 272, 319; earthquake-prone 281; growth 281; Sponge 291, 292
Civil unrest 297
Client 46, 48, 49, 119, 131, 133, 139, 140, 144, 226, 228
Climate: affected 292; change 5, 7, 79, 80, 81, 82, 83, 86, 88, 97–107, 150, 275, 276, 277, 285, 293, 294, 295, 315, 319; crisis 276, 284; disaster 285; emergency 92; justice 5, 92, 97–103, 106–7, 109, 315; literacy 76, 81, 82, 86, 87, 88, 91, 92, 93, 315; migrant 286; refugee 286; resilience 292
Collective: agreement 88, 164; bargaining 164, 165, 170
Communication 51, 82, 92, 226
Community 6, 13, 21, 23, 32, 41, 42, 44, 47, 56, 99–100, 106,109, 225, 226, 236, 237, 275, 276, 277, 279, 286, 287, 288, 290, 292, 297, 300, 301, 302, 306, 329, 320; awareness 302; benefit 87, 89; buy-in 303; coastal 283; collaboration 297; employment opportunities 303; engagement 8, 56, 292, 297, 298, 299, 309; feedback 308; impact

297, 298, 299, 301, 309; influence 301; infrastructure levy 43; interest 298, 301, 306; involvement 297, 298; led movement 279; marginalised 278, 286; needs 301, 306, 307, 309; opposition 297; organisation 304; participation 41, 43, 56, 301, 302, 303, 304; project 305, 307, 309; protest 301; resilience 276, 282, 294; services 304; stakeholders 302, 308; support 301, 305; training 98; underserved 292, 293, 295
Compensation 160, 161, 162, 167, 169, 170
Competences 79, 81, 83, 90, 92
Competition 46, 143, 199
Competitive strategies 114
Competitiveness 18, 152, 237
Conflict 162, 177, 179, 181, 183, 184, 185, 189, 297, 299, 300, 301, 304, 305, 308; management 164, 165, 170
Considerate Constructors Scheme 44
Constructability 53
Construction: 4.0, 29, 149; cluster 15; code 84; company 39, 40, 45, 47, 144; development 52–53; industry 1, 4, 12, 15, 16, 17, 19, 20, 25, 26, 27, 39, 114, 115, 116, 117, 121, 123, 124, 125, 127, 131, 139–154, 158, 159, 160, 161, 162, 163, 164, 165, 166, 167, 168, 169, 170, 171, 177, 178, 179, 180, 183, 185, 189, 190, 192, 193, 194, 195, 196, 197, 198, 199, 200, 201, 202, 203, 204, 207, 210, 213, 214, 215, 216, 217, 218, 298, 299, 300, 301, 306, 308, 309, 314, 315, 316, 317, 320, 321; equipment-based 40; labour 314, 317; mafia 42, 306; market 159, 163; process 39, 105, 144, 148; sector 15, 78, 177, 180, 192, 210, 212, 213, 214, 215, 218, 299, 302
Contact, contract and conflict (3C) 181
Continuing education 81, 88
Contractor 79, 89, 90, 91, 115, 128, 129, 132, 139, 143, 145, 151, 154, 299, 300, 306; development 40; general 115, 121, 123, 124, 128, 129; lower tier 114, 117, 130; main 48, 119, 123, 126, 129, 131, 133, 151, 161, 315; petty labour 122, 123; principal 216
Corporate social responsibility 43
Corruption 56, 279
Cost overrun 54
COVID-19 14, 21, 25, 27, 29, 46, 119, 140, 141, 150, 167, 177, 209, 214, 216, 217, 276, 283
Credit 280; access to 283, 284
Culture 6, 40, 51, 56, 103, 105, 140, 147, 150, 151,158, 159, 163, 171, 177, 178, 179, 180, 181, 183, 184, 185, 186, 187, 188, 189, 190, 192, 195, 197, 201, 202, 203, 305, 306, 316, 317; workplace 196, 197, 200, 201
Curriculum 81, 82, 83, 84, 86, 87, 88, 92, 93
Cyber-physical system 29

Decent work 30, 56, 97, 102, 137, 138
Demand 103, 108, 138, 140, 150, 153, 225, 228, 231, 232, 233, 235, 238; derived 47; for construction 164; for labour 101, 104, 144, 145, 151, 165, 225, 228
Design 14, 16, 48, 55, 226, 234, 235
Developer 43, 182, 180, 231, 235
Developing country 30, 115, 137, 139, 141–4, 146, 147, 149, 150, 159, 160, 161, 163, 165, 170, 171, 192, 203, 271, 272, 275, 276, 278, 279, 280, 281, 282, 283, 285, 286, 288, 289, 294, 316, 318, 319, 320
Development 301, 303, 304, 314, 318, 319; assistance 57, 153; economic 298, 302, 303; national 15; project 207, 208, 317; socio-economic 1, 56, 212, 314, 315, 321, 322; sustainable 302
Dimension of human behaviour 178, 179
Disaster 276, 278, 280, 283, 284, 286, 289, 294; climate 285; impact 277; insurance 276, 280, 290, 293, 319; management 282, 286, 294, 319; planning 289; preparedness 275, 276, 284, 287, 290, 291, 293, 294, 295, 320; protection 277; recovery 294; resilience disaster 277, 286, 287, 288, 290; response 275, 278, 282, 290, 291, 294; risk 275, 277, 278, 283, 290, 292; risk management 286, 290, 291; risk reduction 275, 277, 295, 320
Discontinuity 46, 47, 48, 50
Discrimination 6, 146, 147, 164, 195, 197, 200, 204, 317, 322
Diseases 138, 140, 148
Diversity 26, 88, 103, 144, 153, 195, 200, 202
Downtime 47
Durability 55

Earthquake 275, 281, 282, 283, 287, 288, 290; design standard 287, 288; preparedness 282; related fatality 281, 287; resilience 287; resistance 287, 293; response 281; risk 287, 290; scenario planning 290
Economic: activity 138; cost 153; development 1, 4, 12, 16, 103, 137, 232, 236, 239, 276, 277, 278, 285, 292; effect, push and pull 19, 150; expansion 158; impact 40, 57, 226, 227, 235, 237; loss 276, 278, 280, 284, 286; regeneration 44, 225, 230; resilience 289, 291
Economy 18, 45, 97, 99–104, 106–109, 147, 148, 150, 151, 225–8, 231, 234, 237, 238, 240, 275, 277, 284, 285, 286, 288, 289, 294, 295; dual 266; equitable 280; gig 276; global 277, 278, 295; green 315; informal 276,

283, 286; local 276; of agglomeration 318; sharing 318
Ecosystem 40, 57, 100
Education system 76
Emergency 279, 295; evacuation plan 291; funds 284; response 277, 285, 290
Emerging nation 160, 170, 171
Employability 78
Employee: protection 159; relations 164, 165, 170; welfare 164, 170
Employer: -based system 80, 84, 93; -driven initiative 76, 78, 91; association 81, 82, 89, 90, 91
Employment 39, 212, 213, 215, 101, 103–6, 108, 137–154, 234, 239; benefits 114, 115, 116, 119, 126, 128, 129, 130, 132; conditions 138, 139, 146, 154; contract 115, 116, 120, 121, 124, 125, 126; decent 207, 216, 316, 322; direct, indirect, induced 20; elasticity 40; fair 207, 216, 218; full 315; generation 20, 24, 53, 138, 158, 224, 238, 239, 314; insecure 207, 214; irregular 210; opportunities 320; relations 213, 313, 321; rights 115, 116, 120, 121, 122, 125, 126, 128, 130, 132, 196, 200, 203
Empowerment 41, 302, 303, 304, 307, 315
Energy 56, 149, 150, 230; clean, renewable 97–99, 101–102, 105, 107, 150, 151; concern 158; consumption 77; efficiency 55, 70, 81, 83, 86, 100, 101; efficient building 80, 90; embodied 55; literacy 76; production 56; transition 79
Environment 56, 86, 297, 298, 299, 301, 302, 305, 306, 309, 315, 320
Environmental: attribute 44; concerns 158; footprint 55, 56; impact 40, 41, 100, 102, 150, 237; justice 89, 98, 99, 108; performance 53; protection 83; sustainability 13, 79, 81; technology 91
Equality 6, 77, 99, 102
Equity 77, 315

Facility Management 55
Factory: collapse 280; fire 278, 279
Female participation 192, 193, 194, 195, 196, 199, 203, 204
Feminine 199
Femininity 196, 197, 199, 203
Finance 227, 229–39
Financial: difficulty 46; management 46
Flexibility: in employment 49; in labour practice 114, 116
Fluctuating workload 49
Foreign worker 47, 50, 143
Formal sector 16, 144
Fourth Industrial Revolution 2, 3, 29, 30, 321

Gearing 54; potential 54
Gender 6, 26, 28, 82, 88, 97, 99, 146, 147, 152, 163, 164, 192, 193, 195, 196, 200, 203, 204, 317, 322; bias 197; diversity 194, 200; equality 195, 197, 199, 200, 201, 202, 203, 204; identity 198; segregated 192, 200; stereotype 195
Gendered 198, 203; culture 198, 199; organisation 196; workplace 196
General Agreement on Trade in Services 213
Geography-bound mobilisation 40
Gestation 47; period 47, 54
Ghana Beyond Aid 57
Glass: ceiling 195; wall 195
Global: Climate Risk Index 284; Compact 218, 317; construction labour migration 212; Contact for Safe, Orderly and Regular Migration 6; media 207; North 77, 265, 318; South 2, 225, 228, 236, 239, 263, 272, 318, 319; warming 93, 215, 285, 286
Globalisation 319
Government 47, 54; initiative 41, 153; legislation, policies 39, 100, 105, 145, 151,152, 154, 233; local 98, 101, 104, 106, 107, 225, 226, 228, 237, 238; role 51–52, 150, 151
Green: building 13, 80, 101; construction 92; job 5, 89, 97–110, 153; skills 88, 91; technology 86; transition 79
Greenhouse Gas Emissions 98, 103, 105, 150
Gross development product (GDP) 17, 19, 100, 158, 285; per capita 4

Harassment 6, 195, 196, 200, 203, 204, 267, 317
Hawker 264, 268, 271, 272, 297
Health 13, 17, 99, 100, 103, 140, 141, 145, 148
Health and safety 3, 26, 41, 48, 52, 53, 56, 115, 127, 139–141, 149–152, 162, 163, 164, 168, 169, 170, 198, 199, 278, 280, 289, 309, 316, 320, 322
Heat: exposure 285; extreme 286; stress 215, 285, 286; wave 284, 285, 291, 293
Heating, ventilation and air conditioning 80
Historically disadvantaged 42
HIV Aids 41
Hollowing out 114, 115
Home-based enterprise 268, 269, 271
Housing 7, 13, 14, 16, 21, 22, 27, 29, 31, 55, 275, 283, 284, 287, 288; affordable 225, 228, 229, 231, 240, 275; development 138; formal private rental 225, 229–39; green 101, 105; informal subdivision 236; public rental 226; market 229–31, 237; slum,

squatter settlement 236; unconventional delivery 229
HS2 44
Human: behaviour 6, 177, 178, 179, 180, 181, 183, 185; made 12; resource 146, 159, 162, 163, 165, 167, 169, 170; resource development 47, 53, 159, 160, 167, 170; resource management 5, 158, 159, 162, 163, 165, 167, 168, 169, 170, 171, 316; resource optimisation 51; rights 99, 102, 145, 163, 207, 216, 306; settlement 1, 3, 4, 8, 177, 302, 303, 313, 315, 317, 318, 319, 320, 321, 322
Hurt 189, 90
Hygiene 41

Image 196, 197, 199, 200
Indigenous workers 85, 86
Industrial relations 83
Informal: economy 263, 266, 272, 283, 286; employer 115, 122; employment 276, 278, 283, 284; livelihood 265, 272; network 195, 196, 201; sector 144, 151, 153, 263, 266, 267, 276, 283, 284; settlement 283; work 276; worker 263, 266
Immigrant: undocumented 210
Incentive 280, 282, 284, 286, 295
Inclusive Vocational Education and Training for Low Energy Construction 82
Indoor air: pollution 14; quality 14
Inequality 277, 285, 318
Inflation 54
Informal: employment 276; labour 276; sector 16, 21, 22, 24; settlement 275
Information and Communication Technology (ICT) 52, 224–6, 232, 317, 321
Infrastructure 3, 4, 16, 17, 18, 20, 21, 28, 29, 30, 41, 52, 55, 101, 103–109, 137, 145, 165, 177, 185, 207, 208, 209, 224, 225, 227, 234, 236–40, 275, 276, 277, 284, 287, 288, 289, 291, 292, 297, 318, 322; plan 47
Innovation 159
Instability (political, social, employment) 54, 139
Insurance 280, 284; disaster 226, 280, 293
Interest rate 54, 233
Interface conflict 41
International Labour Office/Organisation 98, 138
Investment 7, 98, 100, 101, 104–6, 108, 152, 226–237, 239, 264, 267, 272

Job 137, 138, 137–42, 145, 147, 152–4, 232; creation 4, 101, 103, 104, 107, 150, 153, 228; green 315; insecurity 164; opportunity 266, 272; satisfaction 28, 195, 202; security 286

Just transition 89

Knowledge, skills and competences 77

Labour 39, 44, 45, 54, 97, 101–4, 106–9, 137–9, 142, 143, 145–54, 161, 165, 224–226, 228, 231, 237, 238, 240, 314, 316, 321, 322; abundant 207, 213; abuse 209, 213, 216, 217, 218, 317; agency 315; agreement 164, 213; bonded, forced 142–4, 148, 152, 217; contingent 120, 128; contract 122, 123; contracting 122; cost 161; discrimination 207, 218, 317; environment 189, 190; equity 315; exploitation 215; exporting 213; flow 207, 208, 209, 210, 211, 212, 213; force 163, 208; forced 208, 214, 215; -intensive 30, 40, 45, 130, 151, 154, 159; local 297, 298, 299, 304, 305, 306, 307, 308, 309, 320; maltreatment of 217; management of 56; market 83, 164, 316, 318, 106, 109, 116, 137, 139, 145, 146, 151, 152, 154, 164, 228, 237, 238, 208, 213, 316, 318; market intermediary 114, 115, 116, 117, 118, 121, 130, 132, 179, 189; migrant 207, 209; migration 208, 209, 210, 212, 217; mobility 212; movement 211; in other sectors of the economy 314; receiving 212, 217; reform 217; saving 50, 207, 209; sending 212; shortage 209, 210, 212; situation 177, 178, 180, 189, 190; starved 209; supply of 116; supplier 115, 117, 118, 119, 125; turnover 162, 170; unskilled 217
Language barrier 50
Leadership 47, 98; in Energy and Environmental Design 84, 86
Levy-funded 82, 83, 91
Liability 119, 129, 130, 131, 133
Linkage effect (backward, forward) 19, 20, 47, 56
Livelihood 97, 224, 225, 229, 238, 275, 277, 285, 293, 317
Living conditions 165, 207, 215, 216
Local industry 40
Location 99, 103, 108, 148, 229–32, 234, 235, 237, 239, 264, 265, 266, 268, 269, 270, 271; specificity 40, 44, 45, 49, 56; spontaneous 268
Locational sustainability 55
Logistics 40, 235, 240
Loss: economic 277, 280, 285; of life 275, 276, 277, 278, 281, 282, 283, 284, 287, 288, 297
Low: carbon 80, 88; -cost housing 303; economy 89; energy construction 5, 76, 77, 79, 81, 82, 83, 86, 88, 89, 90, 91, 92, 93, 315; energy design 92; -income 88, 267, 268, 269, 272, 280, 283, 285, 315; -income

settlements, communities 99, 101, 103, 106, 107, 229, 236, 237, 239; -income workers 103, 109, 232, 233

Macho 199, 202, 204; culture 202, 204, 317
Maintenance 15, 20, 21, 22, 103, 138, 142, 147, 150, 152, 225, 226, 230, 235, 239, 240
Male-dominated 163, 164, 170, 194, 195, 198, 199, 204
Man-Year Entitlement 51
Management (of labour) 147, 236, 228, 229, 231, 234, 236, 238, 239, 240
Manufacturing 40, 149, 226, 233
Market 40, 78, 101, 102, 106, 145, 149–52, 154,163, 170, 183, 225, 228–36, 238–40, 264, 265, 266, 268, 269, 272; authorised 265; behaviour 263; building 180, 182; built-up 265; concept 263, 264; Economy, free 263, 267, 315; Economy, Coordinated 77, 78, 80, 83, 84, 85, 89, 92, 93; Economy, Liberal 77, 78, 80, 84, 89, 92, 93; formal 267, 269; housing 318; institution 264; online 265, 272; open-air 265, 269; perceived 264; physical 264, 265; real 264
Manliness 198, 203
Masculine 198, 203, 317; culture 201, 202; identity 203, 204; space 196, 203
Masculinity 6, 192, 197, 198, 199, 203, 204
Matrix 16, 56, 182
Mental: health 27; illness 162, 170
Micro-business 92
Micro, small and medium enterprises 20, 42, 139, 141, 144, 145
Migrant 137, 139, 141, 143, 145, 146, 228, 238; construction 213, 214, 215, 216, 217, 317; illegal 216; internal circular 121; mistreatment of 212, 316, 317; trail 207, 210; unauthorised 210; worker 78, 79, 116, 117, 119, 120, 121, 122, 123, 125, 126, 127, 128, 129, 132, 160, 161, 162, 163, 165, 170, 196, 207, 208, 209, 210, 211, 212, 213, 214, 215, 216, 218; rights 218
Migration 5, 6, 114, 119, 120, 124, 125, 127, 128, 131, 145, 161, 317, 321; chain 120; industry 120
transient contract 124, 125
Minority 195, 203
Misclassification (of workers) 115, 128, 132
Motivation 159
Modern Methods of Construction 29
Mortality 46, 139
Mortgage 53, 232–4, 238
Movement: illegal 210
Multi-skills 47
Multiplier 20

Nearly zero energy building 82, 89
Neoliberalism 267
Net zero 27, 55, 80, 84, 85
Non-government organisation (NGO) 6, 99

Obsolescence 55; economic 55; functional 55; physical 55; sustainable 55
Occupant wellbeing 13
Occupation 76, 78, 79, 81, 82, 83, 84, 85, 90, 91, 92, 93, 159, 192, 193, 194, 198, 200
Occupational 159, 162, 163, 168, 170, 199; safety and health 5, 141, 153, 154; Safety and Health Administration 278, 280
Offsite 209
On-the-job learning 78
Ownership 307, 308

Paris Agreement 77, 80
Passport confiscation 215
Pay 195, 196, 215, 216; equal 195; when paid 117, 119, 121, 214
Payment 145–7, 160, 161, 170, 234, 235, 227; delay 46, 50, 54, 138–40; non- 50, 140, 214
Performance 162, 315; evaluation 167; indicators 166, 170; management 165, 166, 167, 169, 170; measurement 169; objective 165
Planning 14, 16, 43, 48, 55, 99–101, 107, 153, 224, 225, 226, 228, 238
Platform: capitalism 7, 225, 227, 228, 238, 317, 318; economy 318; urbanism 225, 228, 237, 238, 240, 318
Pollution 41, 97, 100, 103–105, 109
Poverty 41, 103, 108, 125, 126, 128, 148, 153
Precariat 3, 137, 138, 143, 149, 153, 154, 316, 322
Precarious (workers) 137–40, 142–44, 147–49, 151–4, 195, 200
Preferential Procurement Regulations 42
Procurement 52, 79, 152, 154, 308
Production approach 77, 78
Productivity 3, 14, 17, 18, 20, 26, 27, 28, 30, 40, 46, 47, 49, 53, 55, 56, 145, 147, 159, 162, 165, 166, 197, 198, 285, 321
Profit: margin 46, 143, 161, 166; sharing 169
Project 103, 106, 107, 137, 139, 140, 146, 151, 153, 154, 226, 229, 239; -based industry 45, 47, 56; cost 302, 304, 309, 320; cost increase 303, 305, 306; cost overrun 300, 301; delay 48, 55, 139, 297, 299, 300, 305, 306; disruption 298, 299, 301, 305; environment 304; goal 301, 302, 303, 304, 305, 306, 307, 308, 309, 320; management 48, 79, 82, 92, 140, 197; performance 298, 305, 306, 307; quality 320; success 297, 298, 299, 300, 301, 303, 304, 305, 306, 307, 308, 309

Projectification 56
Protectionism 50
Protest group 42
Public: health 13, 103; Private Partnership 51; space 6, 7, 13, 230, 263, 264, 265, 266, 267, 268, 269, 270, 271, 272, 273, 319; workers 267, 270, 271, 272

QualiBuild 90
Quality 275, 284, 287; air 291; control 276; of employment 4, 138, 151–3; of life 1, 100, 291; of water 292

Real estate 7, 44, 229–36, 238, 239, 318; platform 225, 228, 229, 231, 234, 237–40
Recruitment 4, 6, 147, 163, 164, 165, 168, 170, 195, 196, 203, 210, 216, 317
Refurbishment 55, 137, 150
Regulation 276, 277, 278, 282, 289, 295; labour 279; labour market 276; safety 277, 278, 279, 280, 286
Remittance 207, 212, 213
Rent 227–40
Research and development 30, 150
Resident 41, 44, 98, 103, 105, 108, 109, 229, 230, 232, 237
Resilience 290, 294; building 56, 275, 276, 277, 282, 286, 288; built environment 275, 276, 277, 282, 287; business 290, 293; community 276, 282, 288, 293, 294; economic 280, 281, 289, 291; earthquake 287, 288; home 284, 287, 288; infrastructure 56; public 286; urban 292; workplace 7, 275, 293
Retrofitting 89, 91, 92, 151, 239, 276, 281, 282, 283, 286, 287, 288, 289, 292, 294, 320
Reward 160, 169, 170; system 160, 169, 170
Risk 4, 46, 53, 56, 99, 103, 105, 114, 115, 116, 121, 122, 123, 127, 128, 139–44, 150–2, 159, 160, 163, 164, 166, 169, 170, 196, 198, 199, 202, 203, 231, 233, 235–7, 276, 277, 279, 286, 301, 320; from heatstroke 285; mitigation 275, 298, 299, 300, 306; of workers 279, 281, 289; posed by building 281, 282, 288; reduction 275, 287, 295, 309
Robot 164, 169, 170
Role model 195, 197, 202

Safeguarding 44, 53, 150
Safety: building 275, 277, 281, 283, 286, 287, 288; concern 278, 306; equipment 278; hazard 279, 297; inspection 279; investment 279, 280; law 278, 279; management 163, 168, 169, 280; measure 278, 279, 280; net 284; norm 279; occupant 281; occupational 289, 293; protocol 168, 279; public 275, 282, 286; regulation 278, 279, 280, 286, 319; standards 207, 215, 278, 279, 280, 286, 289; training 278, 279, 280; workplace 275, 278, 279, 280
San Marino Declaration 56
School-based system 80
Security of Payment 46
Self: -determination 302; -employment 28, 49, 56, 92, 115, 132, 140, 143, 144, 150, 200, 213, 215, 319
Skill 47, 56, 57, 77, 79, 86, 89, 101, 102, 104, 105, 145, 146, 149–50, 152, 153, 197, 200, 224, 226, 237, 238, 302, 303, 305; development 307; transfer 41, 146, 303, 304; -based approach 78, 84
Slum 269; dweller 269
Small business 276, 280, 289, 290, 293, 294
Small-: and medium-sized enterprise 40, 280, 289, 290, 293, 295; micro-, and medium-sized enterprise 144, 264, 272, 320
Smart city 7, 224,226, 228, 237, 240, 270, 317
Social 229, 231, 232, 235, 238; acceptance 271; attribute 44; capital 268; cohesion 56; concern 158; dialogue 102, 138, 141, 150, 228; hierarchy 270; image 322; impact 267; innovation 263, 271; justice, inclusion, equity 77, 92, 99, 100, 101, 107, 108, 232; legislation, policies 42, 109, 152, 154; media 227; network 42, 102, 107, 195, 201; partner 77, 78, 80, 81, 82, 83; protection 102, 138, 141, 153; relations 270; sector 18; security 2, 141, 144; space 273; ties 42; value 42, 43; Value Act 42; welfare 267; well-being 265
Socially responsible procurement 42
Soft story building 281, 282, 286, 287
Sponsorship system 116, 118
Stakeholder 8, 42, 48, 76, 77, 78, 87, 90, 93, 159, 199, 226, 278, 290, 292, 320; analysis 309; consultation 290, 294; engagement 300; influence 320; management 8, 298, 299, 308, 309, 320; network 56; prioritisation 300; legitimacy 300; power 300, 309, 320; urgency 300; relationship 299; salience model 300; value model 300; theory 298, 299, 300
Strategy 158, 163, 166, 167, 170
Street 263, 264, 266, 268, 269, 270, 271, 272; manager 270; social 273; space 265, 270; trading 270, 272, 319; user 270, 272, 319; vendor 263, 264, 265, 266, 270, 271, 272
Stereotype 6, 143, 147
Stress 26, 148
Subcontract Act (Japan) 46
Subcontracting 28, 46, 56, 114, 116, 117, 119, 121,

122, 123, 129, 130, 131, 132, 133, 214, 215, 216, 217, 228, 315; chain 114, 116, 117, 118, 119, 120, 122, 123, 124, 129, 130, 131, 132, 133; labour 130; multi-level 50, 139, 144, 151
Subcontractor 46, 48, 49, 50, 91, 117, 122, 123, 130, 131, 139, 151, 161, 166, 214, 215, 216; labour 214, 215
Suicide 26, 27
Supply chain 57, 91, 123, 139, 149, 150, 152, 235, 279
Sustainable: construction 52, 55, 57; development 81; Development Goals 17, 55, 56; resources 81
Sustainability 43, 103, 105, 108, 109, 301, 302, 309, 320; social 79
Synergy 3

Teamwork 56
Technical, vocational education and training 5, 24, 25, 28, 53, 76, 77, 78, 79, 80, 81, 82, 83, 84, 86, 87, 89, 90, 91, 93, 151
Technology 167, 168,169, 171
Temporary organisation 56
Territorial development 6, 7, 228, 317
Training 4, 5, 26, 27, 28, 31, 32, 45, 47, 49, 50, 53, 54, 56, 77, 82, 85, 86, 87, 88, 89, 90, 91, 92, 98, 99, 101–3, 105, 106, 108–10, 116, 140, 150, 151, 160, 162, 163, 164, 167, 168, 169, 170, 200, 213, 315, 320, 321; approach 77, 83; infrastructure 90; on-the-job 167
Transparency 298, 300, 306, 307

Underemployment 267
Unemployment 18, 21, 24, 27, 29, 41, 141, 267, 272
Union 76, 77, 79, 81, 82, 83, 85, 86, 87, 88, 89, 90, 92, 93, 97, 102, 103, 107, 108, 115, 116, 122, 127, 128, 129, 131, 164, 165, 238, 279, 315; anti- 217; led training 84; non- 164
United Nations 6, 97, 138, 151, 153; Office for Disaster Risk Reduction 295
Uniqueness 45, 47
United States Federal Emergency Management Agency 289
Urban 98–108, 147, 149, 224–6, 228–31, 237–40; centre 265; congestion 14, 140; design 267, 270, 272; development 276, 277, 287, 292; economy 6, 102, 108; governance 267; green areas 56, 102–104; heat island effect 291, 293; management 267, 270; planning 267, 270, 272, 292; policy 267, 272; population 272; redress 41; resilience 292; space 267, 317, 318; sprawl 275; sustainability 13; transparency 56; worker 275, 318, 321
Urbanisation 17, 145, 224, 236
Urbanism: neo-liberal 267
User 41, 224–8, 237, 238
Utility 55

Value 54, 55, 178, 183, 185, 225, 226, 227, 229, 232, 234, 240, 300, 302; added 41, 226, 239; chain 266, 271; for money 177, 299
Vulnerable (groups, people) 97–9, 102, 103, 105, 108, 109, 228–30

Wage 100, 102, 114, 115, 116, 117, 118, 119, 120, 121, 122, 123, 124, 125, 126, 126, 127, 129, 130, 131, 132, 133, 138–40, 143, 145, 214, 216, 217, 236, 237; abuse 131, 132, 133, 315; exploitative 207, 213; late payment of 115, 116, 117, 118, 120, 123, 127, 132, 160; low-wage jobs 278, 286; minimum 118, 125, 128, 129, 131; non-payment of 116, 119, 123, 127, 132; protection 315; Protection System (Bahrain) 217; secure 114, 127; theft 114, 116, 120, 127, 128, 129; unpaid 215; violation 5, 115, 116, 127, 129, 132, 139, 140, 143, 145, 147, 150, 152; withholding 214, 215
Waste 41, 49, 81, 150, 151
Welfare 49, 53, 100, 200
Well-being 162, 163, 199
Work: -family conflict 162, 195; -life balance 161, 162, 170
Worker 53, 97, 102–5, 107–9, 137–54, 224, 225, 228–33, 238; dispatch 115; home-based 276, 283, 284, 294; illegal 216; in other sectors of the economy 56, 314, 316, 321, 322; Minimum Standards of Housing and Amenities Act, 2019 (Malaysia) 217; quality of life 316; rights 277, 279, 285; Specific Skilled (Japan) 217
Workforce 43, 44, 77, 82, 83, 86, 92, 97, 98, 101, 107–10, 139, 143–7, 149, 159, 160, 161, 162, 166, 168, 169, 210, 319, 320, 321; construction 207, 209; core 214; diversity 82, 162, 164, 166; low-skilled 115, 132, 211; permanent 114, 115, 122, 126; semi-skilled 120, 122, 193, 209; shifting 167; skilled 115, 122, 126, 200, 203, 209; temporary 116, 126, 130, 132; undocumented 127, 128, 132, 163; unskilled 115, 116, 120, 122, 125, 126, 129, 132, 203, 209, 211, 217
Working: at home 319; clothes 196; conditions 79, 102, 137–40, 143, 147–9, 151, 154, 160,

163, 164, 170, 195, 196, 201, 204, 213, 215, 217, 238, 269, 278, 279, 285, 314, 316, 317, 319, 321; environment 47, 50, 56, 195, 201, 202, 276, 286; from home 276, 283; hour 195, 196, 197, 201, 279, 285; space 264, 266, 267, 268, 269, 270, 272, 319
Workplace 6, 7, 44, 78, 139, 141, 142, 151, 153, 154, 235, 263, 264, 265, 268, 270, 271, 272, 275, 276, 279, 283, 284, 285, 291; fissured 114; online 272; resilience 319, 320; safety 278, 279, 280, 291; security 319, 322; vulnerability 281
Works council 84
World Trade Organisation 213

Zero carbon 79, 88, 89, 93

Printed in the United States
by Baker & Taylor Publisher Services